VOLUME SEVENTY SEVEN

ADVANCES IN BOTANICAL RESEARCH

Nitric Oxide and Signaling in Plants

ADVANCES IN BOTANICAL RESEARCH

Series Editors

VOLUME SEVENTY SEVEN

ADVANCES IN BOTANICAL RESEARCH

Nitric Oxide and Signaling in Plants

Volume Editor

DAVID WENDEHENNE

University of Bourgogne Franche-Comté, UMR 1347 Agroécologie, Dijon, France

ELSEVIER

AMSTERDAM • BOSTON • HEIDELBERG • LONDON
NEW YORK • OXFORD • PARIS • SAN DIEGO
SAN FRANCISCO • SINGAPORE • SYDNEY • TOKYO

Academic Press is an imprint of Elsevier

Academic Press is an imprint of Elsevier
125 London Wall, London EC2Y 5AS, UK
The Boulevard, Langford Lane, Kidlington, Oxford OX5 1GB, UK
50 Hampshire Street, 5th Floor, Cambridge, MA 02139, USA
525 B Street, Suite 1800, San Diego, CA 92101-4495, USA

First edition 2016

Notices
Knowledge and best practice in this field are constantly changing. As new research and experience broaden our understanding, changes in research methods, professional practices, or medical treatment may become necessary.

Practitioners and researchers must always rely on their own experience and knowledge in evaluating and using any information, methods, compounds, or experiments described herein. In using such information or methods they should be mindful of their own safety and the safety of others, including parties for whom they have a professional responsibility.

To the fullest extent of the law, neither the Publisher nor the authors, contributors, or editors, assume any liability for any injury and/or damage to persons or property as a matter of products liability, negligence or otherwise, or from any use or operation of any methods, products, instructions, or ideas contained in the material herein.

ISBN: 978-0-12-801074-7
ISSN: 0065-2296

CONTENTS

CONTRIBUTORS

Alexandra Ageeva
Helmholtz Zentrum München – German Research Center for Environmental Health, Institute of Biochemical Plant Pathology, München-Neuherberg, Germany

Magdalena Arasimowicz-Jelonek
Department of Plant Ecophysiology, Faculty of Biology, Adam Mickiewicz University, Poznań, Poland

Juan B. Barroso
Group of Biochemistry and Cell Signaling in Nitric Oxide, Department of Biochemistry and Molecular Biology, University of Jaén, Jaén, Spain

Juan C. Begara-Morales
Group of Biochemistry and Cell Signaling in Nitric Oxide, Department of Biochemistry and Molecular Biology, University of Jaén, Jaén, Spain

Diana Bellin
Dipartimento di Biotecnologie, Università degli Studi di Verona, Verona, Italy

Gerald A. Berkowitz
Department of Plant Science and Landscape Architecture, University of Connecticut, Storrs, CT, USA

Alexandre Boscari
Institut Sophia Agrobiotech (ISA), INRA, CNRS, Université Nice Sophia Antipolis, Sophia Antipolis, France

Renaud Brouquisse
Institut Sophia Agrobiotech (ISA), INRA, CNRS, Université Nice Sophia Antipolis, Sophia Antipolis, France

Claude Bruand
Laboratoire des Interactions Plantes-Microorganismes (LIPM), INRA, CNRS, Castanet-Tolosan, France

Mounira Chaki
Group of Biochemistry and Cell Signaling in Nitric Oxide, Department of Biochemistry and Molecular Biology, University of Jaén, Jaén, Spain

Francisco J. Corpas
Group of Antioxidants, Free Radicals and Nitric Oxide in Biotechnology, Food and Agriculture, Department of Biochemistry, Cell and Molecular Biology of Plants, Estación Experimental del Zaidín, CSIC, Granada, Spain

Natalia Correa-Aragunde
Instituto de Investigaciones Biológicas, Consejo Nacional de Investigaciones Científicas y Técnicas (CONICET), Universidad Nacional de Mar del Plata (UNMdP), Mar del Plata, Argentina

Laura De Gara
Centro Integrato di Ricerca, Università Campus Bio-Medico di Roma, Rome, Italy

Maria Concetta de Pinto
Dipartimento di Biologia, Università degli Studi di Bari "Aldo Moro", Bari, Italy

Massimo Delledonne
Dipartimento di Biotecnologie, Università degli Studi di Verona, Verona, Italy

Zuzana Drábková Trojanová
Department of Botany, Faculty of Science, Palacký University in Olomouc, Olomouc, Czech Republic

Jolanta Floryszak-Wieczorek
Department of Plant Physiology, Poznań University of Life Sciences, Poznań, Poland

Noelia Foresi
Instituto de Investigaciones Biológicas, Consejo Nacional de Investigaciones Científicas y Técnicas (CONICET), Universidad Nacional de Mar del Plata (UNMdP), Mar del Plata, Argentina

Pierre Frendo
Institut Sophia Agrobiotech (ISA), INRA, CNRS, Université Nice Sophia Antipolis, Sophia Antipolis, France

Lucas Frungillo
Department of Plant Biology, Institute of Biology, University of Campinas, Campinas/SP, Brazil

Alok Kumar Gupta
National Institute of Plant Genome Research, New Delhi, India

Kapuganti J. Gupta
National Institute of Plant Genome Research, New Delhi, India

John T. Hancock
Faculty of Health and Applied Sciences, University of the West of England, Bristol, UK

Imène Hichri
Institut Sophia Agrobiotech (ISA), INRA, CNRS, Université Nice Sophia Antipolis, Sophia Antipolis, France

Zahra Imanifard
Dipartimento di Biotecnologie, Università degli Studi di Verona, Verona, Italy

Karolina Izbiańska
Department of Plant Ecophysiology, Faculty of Biology, Adam Mickiewicz University, Poznań, Poland

Aardra Kachroo
Department of Plant Pathology, University of Kentucky, Lexington, KY, USA

Pradeep Kachroo
Department of Plant Pathology, University of Kentucky, Lexington, KY, USA

Eva-Esther König
Helmholtz Zentrum München – German Research Center for Environmental Health, Institute of Biochemical Plant Pathology, München-Neuherberg, Germany

Izabella Kovacs
Helmholtz Zentrum München – German Research Center for Environmental Health, Institute of Biochemical Plant Pathology, München-Neuherberg, Germany

Lucie Kubienová
Department of Biochemistry, Faculty of Science, Palacký University in Olomouc, Olomouc, Czech Republic

Aprajita Kumari
National Institute of Plant Genome Research, New Delhi, India

Lorenzo Lamattina
Instituto de Investigaciones Biológicas, Consejo Nacional de Investigaciones Científicas y Técnicas (CONICET), Universidad Nacional de Mar del Plata (UNMdP), Mar del Plata, Argentina

Aleš Lebeda
Department of Botany, Faculty of Science, Palacký University in Olomouc, Olomouc, Czech Republic

Gah-Hyun Lim
Department of Plant Pathology, University of Kentucky, Lexington, KY, USA

Christian Lindermayr
Helmholtz Zentrum München – German Research Center for Environmental Health, Institute of Biochemical Plant Pathology, München-Neuherberg, Germany

Tengfang Ling
Dipartimento di Biotecnologie, Università degli Studi di Verona, Verona, Italy

Ruitao Liu
Dipartimento di Biotecnologie, Università degli Studi di Verona, Verona, Italy

Vittoria Locato
Centro Integrato di Ricerca, Università Campus Bio-Medico di Roma, Rome, Italy

Lenka Luhová
Department of Biochemistry, Faculty of Science, Palacký University in Olomouc, Olomouc, Czech Republic

Yi Ma
Department of Plant Science and Landscape Architecture, University of Connecticut, Storrs, CT, USA

Eliane Meilhoc
Laboratoire des Interactions Plantes-Microorganismes (LIPM), INRA, CNRS, Castanet-Tolosan, France

Sonal Mishra
National Institute of Plant Genome Research, New Delhi, India

Annalisa Paradiso
Dipartimento di Biologia, Università degli Studi di Bari "Aldo Moro", Bari, Italy

Marek Petřivalský
Department of Biochemistry, Faculty of Science, Palacký University in Olomouc, Olomouc, Czech Republic

María C. Romero-Puertas
Department of Biochemistry and Cellular and Molecular Biology of Plants, Estación Experimental del Zaidín (CSIC), Granada, Spain

Wilma Sabetta
Spin-off SINAGRI srl, Università degli Studi di Bari "Aldo Moro", Bari, Italy

Ione Salgado
Department of Plant Biology, Institute of Biology, University of Campinas, Campinas/SP, Brazil

Beatriz Sánchez-Calvo
Group of Biochemistry and Cell Signaling in Nitric Oxide, Department of Biochemistry and Molecular Biology, University of Jaén, Jaén, Spain

Luisa M. Sandalio
Department of Biochemistry and Cellular and Molecular Biology of Plants, Estación Experimental del Zaidín (CSIC), Granada, Spain

Michaela Sedlářová
Department of Botany, Faculty of Science, Palacký University in Olomouc, Olomouc, Czech Republic

Steven H. Spoel
Institute of Molecular Plant Sciences, School of Biological Sciences, University of Edinburgh, Edinburgh, UK

Raquel Valderrama
Group of Biochemistry and Cell Signaling in Nitric Oxide, Department of Biochemistry and Molecular Biology, University of Jaén, Jaén, Spain

Elodie Vandelle
Dipartimento di Biotecnologie, Università degli Studi di Verona, Verona, Italy

Aakanksha Wany
National Institute of Plant Genome Research, New Delhi, India

Matthew Whiteman
University of Exeter Medical School, University of Exeter, Exeter, UK

PREFACE

Living cells possess intricate signaling systems which play crucial roles for sensing and interacting with their environment. The complexity of these systems is exemplified in higher eukaryotes in which physiological functions require coordinated communications between organs at the whole organism level. Understanding these systems is a formidable task and of fundamental importance. Deciphering cell signaling processes requires identification of the actors and resolution of their spatial and temporal dynamics of mutual and coordinated interactions. Although the elucidation of cell signaling mechanisms seems an unresolvable problem as ever more data come to hand, the discovery that nitric oxide (NO) displays signaling function in many, if not all, living organisms is considered as a major breakthrough.

Nitric oxide is a reactive diatomic gas. According to some authors, it may have played an early role during evolution in primitive organisms by neutralizing ozone and by serving as a defence mechanism against reactive oxygen species. Furthermore, the possibility that the release and sensing of NO could represent an intra- and extracellular mechanism before specialized receptors evolved has been proposed. Paradoxically, its involvement as a physiological mediator in mammals was not recognized until the mid-1980s. Nevertheless, over the past 30 years, our understanding of the role of NO in animals has been enriched by fascinating discoveries. First identified as an antimicrobial effector molecule, numerous studies have highlighted that NO is also a unique biological messenger involved in major processes including neuronal signal transmission, blood vessel dilatation and both innate and adaptive immunity. The discovery of the enzyme nitric oxide synthase (NOS) catalysing its synthesis, as well as those of key mechanisms controlling the expression of NOS and mediating NO effects has greatly expanded our understanding of its role. The search for NO functions remains an active area of investigation in mammals with applications in medicine.

From the end of the 1990s, NO was also revealed to be a surprisingly ubiquitous signaling molecule in plants. Indeed, NO possesses an unusually repertoire of physiological functions in plants. It is involved in all major physiological processes such as seed dormancy and germination, root growth, stomatal closure, flowering, iron homoeostasis, hormonal signaling, senescence and contributes to immunity as well as to the adaptive response to abiotic stresses. Therefore, it is not surprising that this gas has become a dominant part of physiological, cellular and molecular biology research. Accordingly, five Plant Nitric Oxide International Meetings have been held since 2006, and the sixth will be organized in 2016 in Granada in Spain. Interestingly, an essential part of the data generated to date indicates that NO is a major component of plant cell signaling and, compared to animals, there is a remarkable degree of similarity in the molecular mechanisms underlying its effects. The field of NO research in plants is still in its infancy and numerous questions remain as yet unanswered or poorly answered, notably those relating to its synthesis and its basis for specificity.

This focus section of Advances in Botanical Research presents a series of 14 chapters highlighting recent insights into NO in plants. We learn from these contributions that NO cooperates with other signaling compounds including hydrogen sulphide and reactive oxygen species, notably the superoxide anion and hydrogen peroxide. As mentioned by J. Hancock and M. Whiteman (Chapter 1), these interactions might be viewed as an interactive and complex web of reactive species determining cellular responses. Other authors illustrate the importance of such interplay in root growth and developmental processes (N. Correa-Aragunde et al., Chapter 3), in programmed cell death (V. Locato et al., Chapter 9), in the establishment of symbiosis (I. Hichri et al. Chapter 10) and in immunity (E. Vandelle et al., Chapter 11; P. Kachroo et al., Chapter 12). In addition to the reactive species, crosstalk also operates between NO and hormones, lipids and the second messengers Ca^{2+} and cyclic GMP. Examples of the occurrence of such crosstalk in auxin signaling and plant immunity are provided by N. Correa-Aragunde et al. (Chapter 3), P. Kachroo et al. (Chapter 12) and Y. Ma and G. Berkowitz (Chapter 14).

Another main issue discussed in these chapters concerns the specificity of NO action. It appears that the specificity of NO signaling is partly governed by its ability to modulate the activity of target proteins through posttranslational modification (PTM) including S-nitrosylation and tyrosine nitration. These NO-dependent PTMs play a fundamental role in the regulation of gene expression (I. Kovacs et al., Chapter 2), auxin signaling (N. Correa-Aragunde et al., Chapter 3), nitrogen assimilation (L. Frungillo et al., Chapter 4), responses to abiotic stresses (F. Corpas et al., Chapter 5) including cadmium exposure (M. Arasimowicz-Jelonek et al., Chapter 6), metabolism (M. C. Romero-Puertas and L. Sandalio, Chapter 7; A. K. Gupta et al., Chapter 8), programmed cell death (V. Locato et al., Chapter 9), symbiosis and immunity (I. Hichri et al. Chapter 10; E. Vandelle et al., Chapter 11; Y. Ma and G. Berkowitz, Chapter 14). Importantly, NO-dependent

PTMs and, more generally, NO functions are dictated by the chemistry of NO in biological contexts and the production of derived species such as nitrosoglutathione resulting from the interaction of NO and reduced glutathione as summarized by F. Corpas et al. (Chapter 5).

Most chapters provide an informative overview of NO functions in a particular physiological process. These contributions underline the importance of combining all the 'omic' sciences together with physiological and genetic approaches. Deciphering the functions and mode of action of NO will also require the application of metabolomic approaches as suggested by M. C. Romero-Puertas and L. Sandalio (Chapter 7), the search of proteins regulated by NO-dependent PTMs in specific cellular compartments (see, for instance, I. Kovacs et al., Chapter 2; A. K. Gupta et al., Chapter 8) as well as a better understanding of the crosstalk between distinct PTMs as discussed by several authors (see, for instance, J. Hancock and M. Whiteman, Chapter 1). Finally, NO being a diffusing gas and a ubiquitous messenger, there is a need to consider how NO is perceived and/or produced by pathogenic and symbiotic microorganisms infecting plants. These poorly investigated but fascinating aspects of NO biology are nicely discussed by I. Hichri et al. (Chapter 10) and M. Sedlářová et al. (Chapter 13).

It has been my pleasure to act as Guest Editor for this focus issue about NO signaling in plants. Also, I am grateful to all the authors for their excellent manuscripts and to Pr. Jean-Pierre Jacquot for inviting me in editing this issue.

David Wendehenne

CHAPTER ONE

Alone NO Longer: Interactions of Nitric Oxide with Reactive Oxygen Species and Hydrogen Sulfide

John T. Hancock*[,1], Matthew Whiteman[§]
*Faculty of Health and Applied Sciences, University of the West of England, Bristol, UK
[§]University of Exeter Medical School, University of Exeter, Exeter, UK
[1]Corresponding author: E-mail: john.hancock@uwe.ac.uk

Contents

Abstract

Nitric oxide (NO) is a hugely important signaling molecule in both animals and plants. In plants, it has been implicated in the control of a host of cellular and physiological events, from roots to leaves, from germination to senescence. It is known to be made by enzymes found in plants and to have downstream targets in cell signaling pathways. However, events leading to the initiation of the involvement of NO in signaling are often common to those which cause increases in reactive oxygen species and other reactive compounds such as hydrogen sulfide (H_2S). The interaction of all these reactive compounds will be at several levels. They may react directly together given favourable conditions, with the potential to produce further signaling molecules. These mediators may interfere with each other's accumulation. They may control the enzymes that produce each other. Finally, they may compete for downstream targets, such as thiols on proteins. Therefore, NO signaling should be considered as part of an interactive web of reactive species, working together and competing with each other to lead to the final desirable outcome for the cell. Understanding this better will enable the development of chemicals which can manipulate such signaling, perhaps leading to control of plant diseases, better crops or better postharvest storage of plant materials.

Advances in Botanical Research, Volume 77
ISSN 0065-2296
http://dx.doi.org/10.1016/bs.abr.2015.10.012

1. INTRODUCTION

Cell signaling is vitally important to all organisms and cells within an organism, with a wide array of components working together to bring about a coordinated response in cells (Hancock, 2010). This is important not only for the continuing functioning of cells, but also it allows them to adapt and survive, perhaps in times of stress. In the 1980s, it was suggested that a rather unusual component was involved in such signaling and that was nitric oxide (NO) (Palmer, Ferrige, & Moncada, 1987). This of course paved the way for others to think of a wider range of small reactive components in signaling pathways (reviewed by Apel & Hirt, 2004; Hancock, 2009). Today the list of such mediators includes those known as the reactive oxygen species (ROS), which includes hydrogen peroxide (H_2O_2) and superoxide anions ($O_2^{\bullet -}$), other reactive nitrogen species (RNS) alongside NO, such as peroxynitrite ($ONOO^-$), along with gases such as hydrogen sulfide (H_2S) and carbon monoxide. Such gases have been dubbed the gaseotransmitters (Wang, 2002, 2003).

What has become very evident through the research since the 1980s is that mediators such as NO are critically important for the control of a host of physiological functions and cellular activities (Baudouin & Hancock, 2014; Mur et al., 2012; Wilson, Neill, & Hancock, 2008). In fact, the range is so large that it begs the question as to how a cell can interpret what an increase in NO accumulation may mean, and what the appropriate response should be. Furthermore, NO has been found to be produced and have an effect in a wide range of organisms, from plants to humans (Wendehenne, Pugin, & Durner, 2001). Also, what is very apparent from looking at the biology of NO across this range of organisms is that the events that initiate NO signaling, the proteins involved and the responses that ensue, have a similarly. Often, for example, NO is involved in stress responses, both in animals and plants. NO is also often involved in either the onset or the reduction in disease in organisms (Pacher, Beckman, & Liaudet, 2007). However, it is also important that NO does not work alone. Cell signaling is not a set of components working in isolation or even as a neat linear line of interactions, but rather a tangled web of interactions, allowing divergence, convergence and crosstalk in pathways (Hancock, 2010). This is, of course, true for NO too, and therefore this chapter will discuss the ways in which NO may interact with other reactive molecules that are generated in the same place and the same time as NO, initiation of the production of such things being stimulated by the same challenge, for example stress.

2. GENERATION AND ACCUMULATION OF NO

In animals, particularly mammals, the main enzymes which are responsible for the generation of NO are well characterized. In humans, there are three genes which encode NO synthase (NOS) enzymes (reviewed by Förstermann & Sessa (2012)). These encode three proteins are referred to as eNOS (endothelial), nNOS (neuronal) and iNOS (inducible, previously referred to as mNOS after macrophage). These dimeric enzymes have binding sites for nicotinamide cofactors, flavins (FAD and FMN) and haem prosthetic groups and have a capacity to carry out redox reactions. In this case the substrate is L-arginine, which is converted to a nonreleased intermediate, hydroxyarginine, with the final products being L-citrulline and NO. By using radioactive L-arginine, the activity of this enzyme may be monitored by measuring the production of radioactive L-citrulline but it should be noted that this does not actually measure the release of NO and care in the interpretation of data from such assays has been called for (Tischner et al., 2007).

Although there are many other enzymes which may generate NO (see below), it has been suggested for over 15 years (Barroso et al., 1999) that an NOS-like enzyme is present in plants. However, despite several papers claiming to find such an enzyme such work was questioned (Zemojtel et al., 2006), and to-date no definitive NOS has been isolated and characterized from a higher plant. Indeed, early proteins thought to be identified as plant NOS have since been found to have GTPase activity (Moreau, Lee, Wang, Crane, & Klessig, 2008). This topic has recently been reviewed (Hancock & Neill, 2015; see also other chapters of this issue).

Having said that finding an NOS enzyme in higher plants has not been successful, there is evidence that some early plant species may have a gene encoding an NOS in their genome. Fairly recent studies of the algae Ostreococcus have shown that sequences of a likely NOS exist, which are homologous to a human enzyme (Correa-Aaragunde, Foresi, & Lamattina, 2013; Foresi et al., 2010). In *Ostreococcus tauri*, the similarity is reported to be 42%. The putative enzyme contains reductase and oxygenase domains, as well as a calmodulin binding site. Cofactor binding sites have also seemed to exist, along with a Cys-X3-Cys domain for zinc binding. The structure is thought to be most similar to iNOS in mammals. Therefore, it appears that lower plants may have enzymes similar to NOS, although higher plants seem to have lost it (reviewed by Hancock & Neill (2015)).

With the failure of finding the true plant NOS, other enzymes have been studied to account for the generation of NO in higher plants. The most prominent of these is nitrate reductase (NR), originally studied because of its role in nitrogen assimilation (Horchani & Aschi-Smiti, 2010). This enzyme was found to have the capacity to produce NO both *in vitro* and *in vivo* (Kaiser et al., 2002; Rockel, Strube, Rockel, Wildt, & Kaiser, 2002) and has been found to be important in physiological responses in Arabidopsis, for example, where it is involved in stomatal closure (Desikan, Griffiths, Hancock, & Neill, 2002) and in flowering (Seligman, Saviani, Oliveira, Pinto-Maglio, & Salgado, 2008). There are two versions of this enzyme in Arabidopsis and it has been investigated as to which is most important in cell signaling events. Studies have used double mutants, *nia1nia2*, for example, in freezing tolerance (Zhao, Chen, Zhang, & Zhang, 2009) but this does not distinguish the relative roles of the different forms. Other reports do try to determine which NR isoform is involved in signaling, with reports on stomatal closure suggesting that NR1 was the major player (Wilson et al., 2009). This was supported by work on *Camptotheca acuminata* treated with a fungal elicitor (PB90 from *Phytophthora bochmeriae*) which also found that NR1 was more important in this system (Lu et al., 2011).

Other sources of NO in plants include those that are nonenzymatic, but it is not likely that this will be controllable in a way that is useful for cell signaling (Zweier, Wang, Samouilov, & Kuppusamy, 1995). However, other enzymes can make NO too, such as xanthine oxidoreductase, particularly under anaerobic conditions (Millar et al., 1998) or PM-bound nitrite: NO reductase (Stohr, Strube, Marx, Ullrich, & Rockel, 2001). NO from organelles has also been reported, both from mitochondria (Igamberdiev, Ratcliffe, & Gupta, 2014; see chapter 8) and chloroplasts (Galatro, Puntarulo, Guimet, & Simontacchi, 2013; Misra et al., 2014; Tewari, Prommer, & Watanabe, 2013).

Regardless of the enzymes which are involved, a question which needs to be asked is how such enzymes may be controlled, and of particular relevance to the discussion here, whether other reactive chemicals are involved in such control. As will be discussed below, ROS, RNS and H_2S are all capable of reacting with proteins, and it is quite conceivable that such reactions could be involved in the modification of proteins, including those involved in NO generation, altering their structures and therefore activities and functions. For most of these potential interactions, the studies have not been carried out, and until they have been done a full understanding of the interactions of all these compounds, and enzymes which may be able to make them, will not be gained.

As well as altering the proteins themselves, the expression of the genes for these enzymes may also be under the control of reactive signaling compounds. It has been known for some time, for example, that the transcription factor NF-kB can be controlled by the presence of ROS (Schreck, Rieber, & Baeuerle, 1991), and it has more recently been shown that H_2S may also modify this protein and its activity (Sen et al., 2012). Large-scale transcriptomic studies have been carried out following ROS treatment in plants, and it was shown that many genes were expressed in response to their presence while many others had their expression depressed (Desikan, Mackerness, Hancock, & Neill, 2001). Similar studies have been carried out with NO (Parani et al., 2004), and in both cases a large proportion of the genes encode proteins involved in signaling, many of which will be involved in NO signaling. Therefore, it is clear that NO signaling, and perhaps more importantly, a cell's capacity in the future to enlist NO signaling through the increased levels of NO-generating enzymes, is under the influence of other reactive compounds such as H_2O_2.

Of course, it is not just the generation of NO which will determine whether it accumulates and therefore has a potential effect in cells. All cell signaling components need to be removed too, else such signaling would persist. NO will convert to nitrite and nitrate, but also react with other reactive compounds such as $O_2^{\bullet -}$, as discussed further below. However, other reactive chemicals such as H_2O_2 and H_2S also need to be removed. The former, and other ROS, are removed by what appears to be an army of enzymes and chemicals (reviewed by Apel & Hirt, 2004), usually lumped together under the umbrella term "antioxidants". This includes small compounds such as ascorbate and α-tocopherol, but also includes a range of enzymes which includes superoxide dismutases (SOD) and catalase. The levels of the activities of such enzymes can, of course, be controlled on two levels, either by the amount of the proteins present or by their specific activities. The former could be modulated through gene expression or by controlling protein breakdown, perhaps by ubiquitination (Moon, Parry, & Estelle, 2004). The latter could be by protein modification, as discussed more below. What is known is that NO impinges on this. NO has been shown to affect antioxidant levels in cells (Groß, Durner, & Gaupels, 2013), and therefore levels of NO will alter the levels of ROS resulting potential signaling. H_2S, in plants, can be removed by *O*-acetylserine (thiol) lyase (Tai & Cook, 2000; Youssefian, Nakamura, & Sano, 1993), and the effects of ROS and NO on such enzymes would be interesting to determine.

3. INTERACTIONS BETWEEN REACTIVE MEDIATORS

NO is widely referred to as a gaseous free radical, denoted as $NO^{\bullet}$. However, it does have the capacity to lose this unpaired electron or to gain another, in both cases losing its free radical status, to become NO^{+} and NO^{-} respectively. Therefore, the chemistry of NO is quite complex (Lancaster, 1997; see also chapter 5), and even more complex when it is considered that it can interact with other reactive chemicals used by cells in signaling, such as ROS. NO can react with $O_2^{\bullet-}$ to produce $ONOO^{-}$ which itself is not only a very reactive molecule but also one which has been shown to be instrumental in cell signaling events (Carbellal et al., 2011). Similarly, NO can react with H_2S to form nitrothiols, which again have the capacity to be compounds used by cells in signaling (Whiteman et al., 2006). As has already been mentioned above, initiation of NO production may be triggered by the same cues that initiated the generation of ROS and H_2S; so it is likely that a gamut of such compounds will be produced at a same time and at the same place in cells, allowing interactions to be likely and common. Therefore, downstream products of NO with compounds such as H_2S should not be dismissed.

As well as the generation of new compounds, it has to be considered what the other consequences of these reactions of NO with other things might be. If NO is needed to have an effect in the cell, but before it is able to carry out this role it reacts with $O_2^{\bullet-}$ or H_2S, then it is no longer available to carry out its original role (discussed in Hancock & Whiteman, 2014). Cell signaling molecules, when measured in cells never seem to be absent altogether, and NO is no exception. Most 'control' levels seen in papers are not zero. DAF-2DA (4,5-diaminofluorescein diacetate) images of controls often show some apparent NO before a trigger or stimulation. Therefore to have an effect, NO has to rise from a low level to a higher level, but one which is recognized as being a signal. It needs to reach a threshold of concentration, beyond which causes an effect, below which signaling would be postponed (Jordan, Landau, & Iyengar, 2000). Reaction of NO with other compounds such as H_2S may well be able to keep NO below such a threshold, as previously argued (Hancock & Whiteman, 2014), modulating signaling by NO (and ROS), to stop it going awry at inappropriate times. Certainly, in support of such an idea, it appears that NO is increased in during some stress responses and diseases, and treatment with H_2S seems to have an ameliorating effect. In this vein, it has been suggested that H_2S treatment may be a good cure of many ailments (Xu, Liu, & Liu, 2014; Zhang et al., 2013), with the action no doubt being through interactions with NO, at least partly. In plants too, H_2S treatment has been suggested, especially for postharvest storage (Hu et al., 2012), where H_2S effects were suggested to be mediated by antioxidants, so again H_2S will almost certainly be impinging on ROS and NO levels, hence the effects seen.

The same can be argued the other way round of course. If H_2S is being made but removed by NO, so its bioavailability goes down, then its signaling capacity is compromised too. Therefore, it is very likely that there is an interplay between all these reactive compounds, with the end result of the signaling being the way local concentrations and kinetics allow the signaling to ensue.

4. NO EFFECTS ON PROTEINS

As has been alluded to above, NO has the ability to interact directly with proteins. Such studies stemmed from the work in ROS. It has long been known that certain ROS, such as H_2O_2, have the capacity to oxidize amino acids in proteins (reviewed by Hancock (2009)). This is particularly true for the thiol groups of cysteine residues. Here, oxidation of the thiol may result in several outcomes (see Figure 1). Firstly, if there are more than one thiol present, and in the correct three-dimensional orientation, a disulfide bond may be formed, so altering, perhaps stabilizing, the protein structure. However, single thiol groups can be oxidized too. With relatively low H_2O_2, the thiol group is converted to the sulfenic acid group. This will have the capacity to alter the structure of the protein and hence alter the

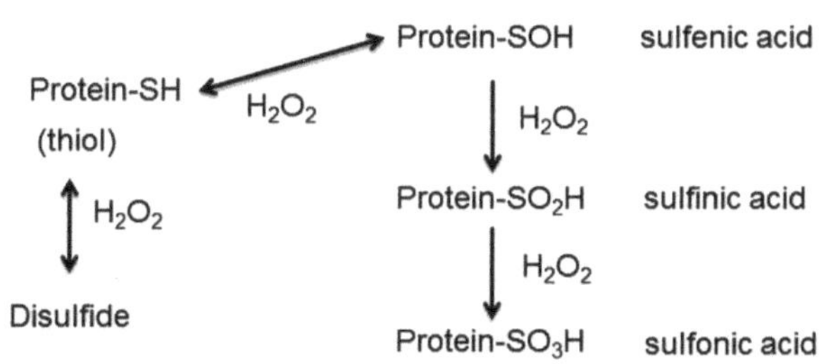

Figure 1 A simplistic scheme to show the oxidation of thiol groups in proteins, which may give a variety of outcomes, some which are reversible and some which are not.

activity or function, and this modification seems to be reversible. Add more H_2O_2 and the sulfinic acid group is formed, again probably reversible and altering protein activity. With relatively high concentrations of H_2O_2 then the sulfonic acid group is formed. This too may alter the activity of the protein but in this case it seems to be irreversible. Of course, the exact concentration needed is determined by the local microenvironment of the thiol group, while the concentration such thiol groups are exposed to will be determined by the local activities of the generation and removal enzymes (such as NADPH oxidases, SOD, and catalase). One of the most well characterized examples of oxidation of a protein by H_2O_2 is the tyrosine phosphatase 1B (Salmeen et al., 2003), where in a circular interaction the sulfenyl-amide intermediate is formed. It is known that increased ROS, such as during oxidative stress increases phosphorylation levels in cells and this can partly account for this observation, as the phosphatase is inhibited, so reducing the dephosphorylation rates of tyrosine residues in cells.

It is not only with ROS that thiols can react with, but similar reactions have been seen with NO, glutathione and H_2S. For NO, the reaction is referred to as nitrosation (also commonly called *S*-nitrosylation; see also other chapters of this issue). Large-scale studies have been carried out to assess the level of this thiol modification in plants (Grennan, 2007; Lindermayr, Sallbach, & Durner, 2005). It was found that proteins identified included those involved in metabolism, signaling and stress responses. One of the key methods employed to determine levels of *S*-nitrosation in plants is known as the biotin switch assay (Jaffrey & Snyder, 2001). However, other methods to determine thiol modifications are also used, such as tagging the thiols with fluorescent markers (Hancock et al., 2005; Williams et al., 2015; Wu, Kwon, & Rhee, 1998). Interestingly, one of the proteins highlighted by such studies in plants is glyceraldehyde 3-phosphate dehydrogenase (GAPDH: Hancock et al., 2005). This enzyme seems to be modified by the presence of H_2O_2, and once reacted it appears to leave the cytoplasm of the cell and translocate to the nucleus, where it can partake in the control of gene expression (Zaffagnini, Fermani, Costa, Lemaire, & Trost, 2013). However, protein thiols and the catalytic activity of this enzyme can be modified by NO too, with nuclear translocation, and in animal cells appears to partake in apoptosis (Hara et al., 2005). This example highlights how reactive compounds such as H_2O_2 and NO may be competing for the same targets. It seems sensible to suggest therefore that other thiol modifying compounds, such as H_2S and glutathione, may also be competing for thiols on proteins, as indicated in Figure 2.

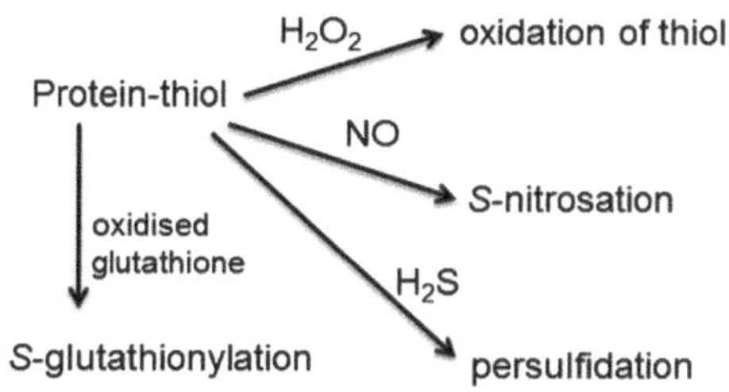

Figure 2 An overview of the competition of different reactive compounds to potentially modify the thiol groups of proteins, each with a potentially different outcome.

If it is known that proteins can be modified by NO, and that cysteine thiols are the target, can the domain structure or sequence around that cysteine be characterized. Recently, working with a focus on iNOS, it was found that a conserved sequence of I/L-X-C-X_2-D/E motif was necessary but also sufficient for *S*-nitrosation in their system (Jia et al., 2014). This opens the way to find other further proteins which are potentially *S*-nitrosylated, and for the search for other similar consensus sequences. It would also be interesting to see if such sequences were needed and sufficient for oxidation or modification by H_2S and glutathione. As this chapter is being written, Prosite database of consensus sequences appears not to have a prorule for nitrosation or sulfide modification of proteins (e.g. persulfidation). The sequence database entry in NCBI for cGAPDH for Arabidopsis (accession number P25858.2) indicates two cysteines (at 156 and 160) which are potentially modified by oxidized glutathione and NO, but neither are within this consensus sequence published by Jia et al. (2014), while similar searches in other sequences known to be nitrosylated (Belenghi et al., 2007; Palmieri, Lindermayr, Bauwe, Steinhauser, & Durner, 2010; Tada et al., 2008; Wang et al., 2009) yield similar results (Table 1), indicating that other factors are clearly important. No doubt differences in such sequences will dictate the relative reactivity of the cysteine residues with NO, and also with glutathione, ROS and H_2S. It will be important to understand if all such sequences in these proteins can be modified under all these treatments and the relative reactivity of each.

NO can also interact with other amino acids too, such as nitration reactions (Lozano-Juste, Colom-Moreno, & Leon, 2011). All such reactions need to be taken into account to understand the full level of the effect of NO on proteins, but as emphasized above, NO will be competing in many cases with other reactive compounds.

Table 1 Sequences in Arabidopsis Proteins Known to Be *S*-Nitrosated, with Relevant Cysteine Residue Underlined

Name of Enzyme	Accession#	Amino Acid Modified	Sequence Surrounding Cysteine (Underlined)
GAPDH C1	P25858.2	156	VSNAS<u>C</u>TTNCLAPLA
		160	SCTTN<u>C</u>LAPLAKVIN
S-adenosylmethionine synthase 1	P23686.2	114	FTK<u>C</u>PEEIGA
Regulatory protein NPR1	P93002.1	156	ADENC<u>C</u>HVAC
Metacaspase-9	Q9FYE1	147	TMISDS<u>C</u>HSG
Beta carbonic anhydrase 1	P27140.2	280	FEDQCGR<u>C</u>EREAVN
Glycine dehydrogenase (decarboxylating) 1 (mitochondrial)	Q94B78.2	98	QTHMAKF<u>C</u>GFDHIDSLIDAT
		402	HIRRDKATSNI<u>C</u>TAQALLAN
		463	QELPFFDTVKIK<u>C</u>SDAHAIA
N.B. sequences here shown to react with glutathione		777	NLHKTF<u>C</u>IPH
		943	TESESKAELDRF<u>C</u>DALISIR
		1022	YGDRKLV<u>C</u>TLLPEEEQ

N.B, Note that.

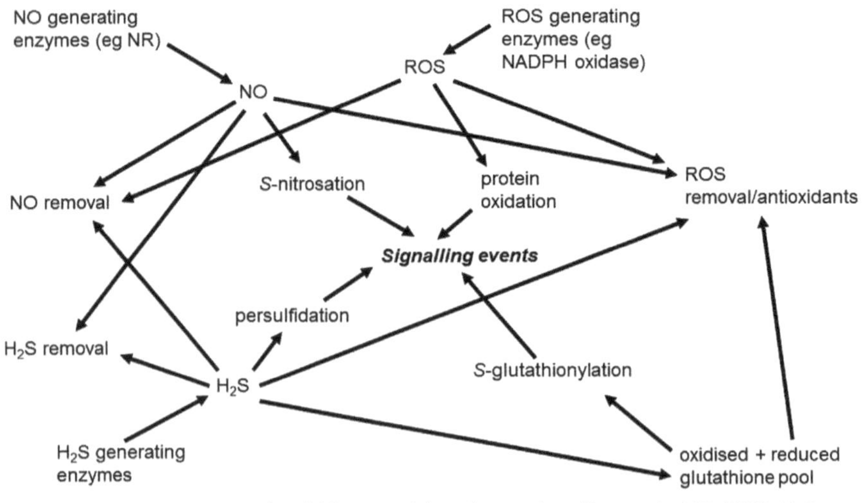

Figure 3 NO interactions should be considered as a signaling web. NO, ROS, H_2S, etc., can have direct signaling through protein modifications, but they interfere with each other too. NO can alter ROS levels as well as alter antioxidant activities, H_2S and NO can interact, reducing each other's levels. H_2S has been shown to alter antioxidant activities in plants as well as contributing to the glutathione pool.

5. CONCLUSIONS

NO is undoubtedly an extremely important molecule which contributes to the cell signaling events and control of cellular activities. There are enzymes which are able to generate it, and there as numerous responses which have been reported when NO is involved. However, NO will not be working in isolation. Triggers which lead to NO production and accumulation in cells will also initiate the generation of other reactive compounds in cells, such as ROS and H_2S. Such reactive molecules can interact at several different levels: they may react directly together; they may alter the enzymes and mechanisms that generate each other; they may compete for the control of downstream targets. Therefore, when considering NO effects in cells a holistic view is required to fully understand how NO may fit into the signaling involving other chemicals, especially ROS, H_2S, and glutathione (Figure 3). Such an understanding should lead to the development of compounds which can manipulate signaling which involves these players, and therefore lead to the control of cellular function during stress and disease, in both animals and plants. The latter may be instrumental in the improvement of both crops and postharvest storage and transport of plant products.

REFERENCES

Apel, K., & Hirt, H. (2004). Reactive oxygen species: metabolism, oxidative stress, and signal transduction. *Annual Review of Plant Biology, 55*, 373–399.

Barroso, J. B., Corpas, F. J., Carreras, A., Sandalio, L. M., Valderrama, R., Palma, J. M., et al. (1999). Localisation of nitric oxide synthase in plant peroxisomes. *Journal of Biological Chemistry, 274*, 36729–36733.

Baudouin, E., & Hancock, J. T. (2014). Nitric oxide signalling in plants. *Frontiers in Plant Sciences, 4*, 553.

Belenghi, B., Romero-Puertas, M. C., Vercammen, D., Brackenier, A., Inze, D., Delledonne, M., et al. (2007). Metacaspase activity of *Arabidopsis thaliana* is regulated by S-nitrosylation of a critical cysteine residue. *Journal of Biological Chemistry, 282*(2), 1352–1358.

Carballal, S., Trujillo, M., Cuevasanta, E., Bartesaghi, S., Möller, M. N., Folkes, L. K., et al. (2011). Reactivity of hydrogen sulfide with peroxynitrite and other oxidants of biological interest. *Free Radical Biology and Medicine, 50*, 196–205.

Correa-Aaragunde, N., Foresi, N., & Lamattina, L. (2013). Structure diversity of nitric oxide syntheses (NOS): the emergence of new forms in photosynthetic organisms. *Frontiers in Plant Science, 4*, 232.

Desikan, R., Griffiths, R., Hancock, J., & Neill, S. (2002). A new role for an old enzyme: nitrate reductase-mediated nitric oxide generation is required for abscisic acid-induced stomatal closure in *Arabidopsis thaliana*. *Proceedings of the National Academy of Sciences of the United States of America, 99*, 16314–16318.

Desikan, R., Mackerness, S. A.-H., Hancock, J. T., & Neill, S. J. (2001). Regulation of the Arabidopsis transcriptome by oxidative stress. *Plant Physiology, 127*, 159–172.
Foresi, N., Correa-Aragunde, N., Parisi, G., Calo, G., Salerno, G., & Lamattina, L. (2010). Characterization of a nitric oxide synthase from the plant kingdom: NO generation from the green alga *Ostreococcus tauri* is light irradiance and growth phase dependent. *The Plant Cell, 22*, 3816–3830.
Förstermann, U., & Sessa, W. C. (2012). Nitric oxide synthases: regulation and function. *European Heart Journal, 33*, 829–837.
Galatro, A., Puntarulo, S., Guimet, J. J., & Simontacchi, M. (2013). Chloroplast functionality has a positive effect on nitric oxide level in soybean cotyledons. *Plant Physiology and Biochemistry, 66*, 26–33.
Grennan, A. K. (2007). Protein S-nitrosylation: potential targets and roles in signal transduction. *Plant Physiology, 44*, 1237–1239.
Groß, F., Durner, J., & Gaupels, F. (2013). Nitric oxide, antioxidants and prooxidants in plant defence responses. *Frontiers in Plant Science, 4*, 419.
Hancock, J. T. (2009). The role of redox mechanisms in cell signaling. *Molecular Biotechnology, 43*, 162–166.
Hancock, J. T. (2010). *Cell signalling* (3rd ed.). Oxford, UK: Oxford University Press.
Hancock, J. T., Henson, D., Nyirenda, M., Desikan, R., Harrison, J., Lewis, M., et al. (2005). Proteomic identification of glyceraldehyde 3-phosphate dehydrogenase as an inhibitory target of hydrogen peroxide in *Arabidopsis*. *Plant Physiology and Biochemistry, 43*, 828–835.
Hancock, J. T., & Neill, S. J. (2015). NO synthase in plants? *CAB Reviews, 033*, 1–9.
Hancock, J. T., & Whiteman, M. (2014). Hydrogen sulfide and cell signaling: team player or referee? *Plant Physiology and Biochemistry, 78*, 37–42.
Hara, M. R., Agrawal, N., Kim, S. F., Cascio, M. B., Fujimuro, M., Ozeki, Y., et al. (2005). *S*-nitrosylated GAPDH initiates apoptotic cell death by nuclear translocation following Siah1 binding. *Nature Cell Biology, 7*, 665–674.
Horchani, F., & Aschi-Smiti, S. (2010). Prolonged root hypoxia effects on enzymes involved in nitrogen assimilation pathway in tomato plants. *Plant Signaling and Behavior, 5*, 1583–1589.
Hu, L.-Y., Hu, S.-L., Wu, J., Li, Y. H., Zheng, J. L., Wei, Z. J., et al. (2012). Hydrogen sulfide prolongs postharvest shelf life of strawberry and plays an antioxidative role in fruits. *Journal of Agricultural and Food Chemistry, 60*, 8684–8693.
Igamberdiev, A. U., Ratcliffe, R. G., & Gupta, K. J. (2014). Plant mitochondria: source and target for nitric oxide. *Mitochondrion, 19*, 329–333.
Jaffrey, S. R., & Snyder, S. H. (2001). The biotin switch method for the detection of *S*-nitrosylated proteins. *Science's STKE, 2001*, l1.
Jia, J., Arif, A., Terenzi, F., Willard, B., Plow, E. F., Hazen, S. L., et al. (2014). Target-selective protein S-nitrosylation by sequence motif recognition. *Cell, 159*, 623–634.
Jordan, J. D., Landau, E. M., & Iyengar, R. (2000). Signaling networks: the origins of cellular multitasking. *Cell, 103*, 193–200.
Kaiser, W. M., Weiner, H., Kandlbinder, A., Tsai, C. B., Rockel, P., Sonoda, M., et al. (2002). Modulation of nitrate reductase: some new insights, an unusual case and a potentially important side reaction. *Journal of Experimental Botany, 53*, 875–882.
Lancaster, J. R., Jr. (1997). A tutorial on the diffusibility and reactivity of free nitric oxide. *Nitric Oxide, 1*, 18–30.
Lindermayr, C., Sallbach, G., & Durner, J. (2005). Proteomic identification of *S*-nitrosylated proteins in Arabidopsis. *Plant Physiology, 137*, 921–930.
Lozano-Juste, J., Colom-Moreno, R., & Leon, J. (2011). *In vivo* protein tyrosine nitration in *Arabidopsis thaliana*. *Journal of Experimental Botany, 62*, 3501–3517.

Lu, D., Dong, J., Jin, H., Sun, L., Xu, X., Zhou, T., et al. (2011). Nitrate reductase mediated nitric oxide generation is essential for fungal elicitor-induced camptothecin accumulation of *Camptotheca acuminata* suspension cell cultures. *Applied Microbiology and Biotechnology, 90*, 1073–1081.

Millar, T. M., Stevens, C. R., Benjamin, N., Eisenthal, R., Harrison, R., & Blake, D. R. (1998). Xanthine oxidoreductase catalyses the reduction of nitrates and nitrite to nitric oxide under hypoxic conditions. *FEBS Letters, 427*, 225–228.

Misra, A. N., Vladkova, R., Singh, R., Misra, M., Dobrikova, A. G., & Apostolova, E. L. (2014). Action and target sites of nitric oxide in chloroplasts. *Nitric Oxide, 39*, 35–45.

Moon, J., Parry, G., & Estelle, M. (2004). The ubiquitin-proteasome pathway and plant development. *The Plant Cell, 16*, 3181–3195.

Moreau, M., Lee, G. I., Wang, Y., Crane, B. R., & Klessig, D. F. (2008). AtNOS/AtNOA1 is a functional *Arabidopsis thaliana* cGTPase and not a nitric oxide synthase. *Journal of Biological Chemistry, 283*, 32957–32967.

Mur, L. A. J., Mandon, J., Persijn, S., Cristescu, S. M., Moshkov, I. E., Novikova, G. V., et al. (2012). Nitric oxide in plants: an assessment of the current state of knowledge. *AoB Plants, 5*, pls052.

Pacher, P., Beckman, J. S., & Liaudet, L. (2007). Nitric oxide and peroxynitrite in health and disease. *Physiological Reviews, 87*, 315–424.

Palmer, R. M. J., Ferrige, A. G., & Moncada, S. (1987). Nitric oxide release accounts for the biological activity of endothelium-derived relaxing factor. *Nature, 327*, 524–526.

Palmieri, M. C., Lindermayr, C., Bauwe, H., Steinhauser, C., & Durner, J. (2010). Regulation of plant glycine decarboxylase by *S*-nitrosylation and glutathionylation. *Plant Physiology, 152*, 1514–1528.

Parani, M., Rudrabhatla, S., Myers, R., Weirich, H., Smith, B., Leaman, D. W., et al. (2004). Microarray analysis of nitric oxide responsive transcripts in *Arabidopsis*. *Plant Biotechnology Journal, 2*, 359–366.

Rockel, P., Strube, F., Rockel, A., Wildt, J., & Kaiser, W. M. (2002). Regulation of nitric oxide (NO) production by plant nitrate reductase *in vivo* and *in vitro*. *Journal of Experimental Botany, 53*, 103–110.

Salmeen, A., Anderson, J. N., Myers, M. P., Meng, T.-C., Hinks, J. A., Tonks, N. K., et al. (2003). Redox regulation of protein tyrosine phosphatase 1B involved a sulphenyl-amide intermediate. *Nature, 423*, 769–773.

Schreck, R., Rieber, P., & Baeuerle, P. A. (1991). Reactive oxygen intermediates as apparently widely used messengers in the activation of the NF-κB transcription factor and HIV-1. *EMBO Journal, 10*, 2247–2258.

Seligman, K., Saviani, E. E., Oliveira, H. C., Pinto-Maglio, C. A., & Salgado, I. (2008). Floral transition and nitric oxide emission during flower development in *Arabidopsis thaliana* is affected in nitrate reductase-deficient plants. *Plant and Cell Physiology, 49*, 1112–1121.

Sen, N., Paul, B. D., Gadalla, M. M., Mustafa, A. K., Sen, T., Xu, R., et al. (2012). Hydrogen sulfide-linked sulfhydration of NF-κB mediates its antiapoptotic actions. *Molecular Cell, 45*, 13–24.

Stohr, C., Strube, F., Marx, G., Ullrich, W. R., & Rockel, P. (2001). A plasma membrane-bound enzyme of tobacco roots catalyses the formation of nitric oxide from nitrite. *Planta, 212*, 835–841.

Tada, Y., Spoel, S. H., Pajerowska-Mukhtar, K., Mou, Z., Song, J., Wang, C., et al. (2008). Plant immunity requires conformational changes [corrected] of NPR1 via S-nitrosylation and thioredoxins. *Science, 321*, 952–956.

Tai, C. H., & Cook, P. F. (2000). O-acetylserine sulfhydrylase. *Advances in Enzymology and Related Areas of Molecular Biology, 74*, 185–234.

Tewari, R. K., Prommer, J., & Watanabe, M. (2013). Endogenous nitric oxide generation in protoplast chloroplasts. *Plant Cell Reports, 32*, 31–44.

Tischner, R., Galli, M., Heimer, Y. M., Bielefeld, S., Oka-Moto, M., Mack, A., et al. (2007). Interference with the citrulline-based nitric oxide synthase assay by argininosuccinate lyase activity in Arabidopsis extracts. *FEBS Journal, 274*, 4238–4245.

Wang, R. (2002). Two's company, three's a crowd: can H_2S be the third endogenous gaseous transmitter? *The FASEB Journal, 16*, 1792–1798.

Wang, R. (2003). The gasotransmitter role of hydrogen sulfide. *Antioxidants and Redox Signaling, 5*, 493–501.

Wang, Y. Q., Feechan, A., Yun, B. W., Shafiei, R., Hofmann, A., Taylor, P., et al. (2009). S-nitrosylation of AtSABP3 antagonizes the expression of plant immunity. *Journal of Biological Chemistry, 284*, 2131–2137.

Wendehenne, D., Pugin, A., & Durner, J. (2001). Nitric oxide: comparative synthesis and signaling in animal and plant cells. *Trends in Plant Science, 6*, 177–183.

Whiteman, M., Li, L., Kostetski, I., Chu, S. H., Siau, J. L., Bhatia, M., et al. (2006). Evidence for the formation of a novel nitrosothiol from the gaseous mediators nitric oxide and hydrogen sulfide. *Biochemical and Biophysical Research Communications, 343*, 303–310.

Williams, E., Pead, S., Whiteman, M., Wood, M. E., Wilson, I. D., Ladomery, M. R., et al. (2015). Detection of thiol modifications by hydrogen sulfide. *Methods in Enzymology, 555*, 233.

Wilson, I., Ribeiro, D. M., Bright, J., Confraria, A., Harrison, J., Barros, R. S., et al. (2009). Role of nitric oxide in regulating stomatal apertures. *Plant Signaling and Behavior, 4*, 467–469.

Wilson, I. D., Neill, S. J., & Hancock, J. T. (2008). Nitric oxide synthesis and signalling in plants. *Plant, Cell and Environment, 31*, 622–631.

Wu, Y., Kwon, K. S., & Rhee, S. G. (1998). Probing cellular protein targets of H_2O_2 with fluorescein-conjugated iodoacetamide and antibodies to fluorescein. *FEBS Letters, 440*, 111–115.

Xu, S., Liu, Z., & Liu, P. (2014). Targeting hydrogen sulfide as a promising therapeutic strategy for atherosclerosis. *International Journal of Cardiology, 172*, 313–317.

Youssefian, S., Nakamura, M., & Sano, H. (1993). Tobacco plants transformed with the O-acetylserine (thiol) lyase gene of wheat are resistant to toxic levels of hydrogen sulphide gas. *Plant Journal, 4*, 759–769.

Zaffagnini, M., Fermani, S., Costa, A., Lemaire, S. D., & Trost, P. (2013). Plant cytoplasmic GAPDH: redox post-translational modifications and moonlighting properties. *Frontiers in Plant Science, 4*, 450.

Zemojtel, T., Frohlich, A., Palmieri, M. C., Kolanczyk, M., Mikula, I., & Wyrwicz, L. S. (2006). Plant nitric oxide synthase: a never-ending story. *Trends in Plant Science, 11*, 524–525.

Zhang, Y., Tang, Z.-H., Ren, Z., Qu, S. L., Liu, M. H., Liu, L. S., et al. (2013). Hydrogen sulfide, the next potent preventive and therapeutic agent in aging and age-related diseases. *Molecular and Cellular Biology, 33*, 1104–1113.

Zhao, M. G., Chen, L., Zhang, L. L., & Zhang, W. H. (2009). Nitric reductase-dependent nitric oxide production is involved in cold acclimation and freezing tolerance in Arabidopsis. *Plant Physiology, 151*, 755–767.

Zweier, J. L., Wang, P., Samouilov, A., & Kuppusamy, P. (1995). Enzyme-independent formation of nitric oxide in biological tissues. *Nature Medicine, 1*, 804–809.

CHAPTER TWO

S-Nitrosylation of Nuclear Proteins: New Pathways in Regulation of Gene Expression

Izabella Kovacs, Alexandra Ageeva, Eva-Esther König, Christian Lindermayr[1]
Helmholtz Zentrum München – German Research Center for Environmental Health, Institute of Biochemical Plant Pathology, München-Neuherberg, Germany
[1]Corresponding author: E-mail: lindermayr@helmholtz-muenchen.de

Contents

Abstract

Nitric oxide (NO) is a reactive free radical with pleiotropic function that is not only involved in regulation of plant growth and development, but also in the response to biotic and abiotic stressors. It mainly acts by posttranslationally modifying proteins. The most important mode of action of NO is protein S-nitrosylation, the covalent attachment of an NO group to the thiol side of protein cysteine residues. Other major types of NO-dependent modifications are metal nitrosylation and tyrosine nitration. NO can regulate gene expression at different levels. On one side, it can initiate signaling cascades or modify proteins involved in signal transduction pathways. On the other side, redox-sensitive transcription factors can be also target for S-nitrosylation, and NO can also affect redox-dependent nuclear transport of some proteins. This suggests that NO plays a pivotal role in regulating transcription and/or general nuclear metabolism in plants.

Advances in Botanical Research, Volume 77
ISSN 0065-2296
http://dx.doi.org/10.1016/bs.abr.2015.10.003

1. INTRODUCTION

Developmental programs and environmental factors affect and/or regulate the gene expression machinery. Because not all genes are active in all cells at all times, the regulation of a proper gene set needs a very precise aligned mechanism. Developmental and environmental signals are perceived by regulatory regions of genes that react by initiating or suppressing their expression. Inside the regulatory region, the promoter contains different sequence elements, which are specific for binding transcription factors and RNA polymerase II to initiate transcription.

Important signaling molecules include reactive oxygen species (ROS), salicylic acid (SA), abscisic acid (ABA), jasmonate (JA) and ethylene – dependent on the stage of development or type of environmental stimuli. Since several years we know that nitric oxide (NO) is also a key player in stress response signaling, but also an important signaling molecule during plant development. Despite the extensive studies on NO function in different processes, the whole picture of NO impact on living cells, including production, activity and metabolism of NO still has to be completed. Different mode of action mechanisms of NO signaling has also been reported in plants. The most studied is protein S-nitrosylation, the covalent attachment of an NO group to the thiol side of protein cysteine (Cys). Protein S-nitrosylation, as a reversible posttranslational modification (PTM), can affect protein activity (activation or inhibition), translocation and protein function. In addition, other major types of modifications of NO have been also reported, such as metal nitrosylation or tyrosine (Tyr) nitration (Astier & Lindermayr, 2012; Martinez-Ruiz, Cadenas, & Lamas, 2011; Toledo & Augusto, 2012). The later one is an irreversible reaction of a nitrating agent with a Tyr residue of a target protein.

NO is able to influence gene expression at multiple levels. Since NO is a diffusible gas, it can be present in all extra- and intracellular spaces, where it easily interacts with the surrounding environment (Figure 1). In this way, NO can initiate signaling cascades and modify proteins involved in signal transduction pathways, which results in altered gene expression. The other possibility of NO to affect gene expression is through a direct regulation of transcription factors or other regulatory elements on gene promoters. Moreover, the NO-dependent modification of the chromatin structure, affecting the accessibility of the DNA, is described. These different possibilities and their effects on gene transcription are discussed in this chapter.

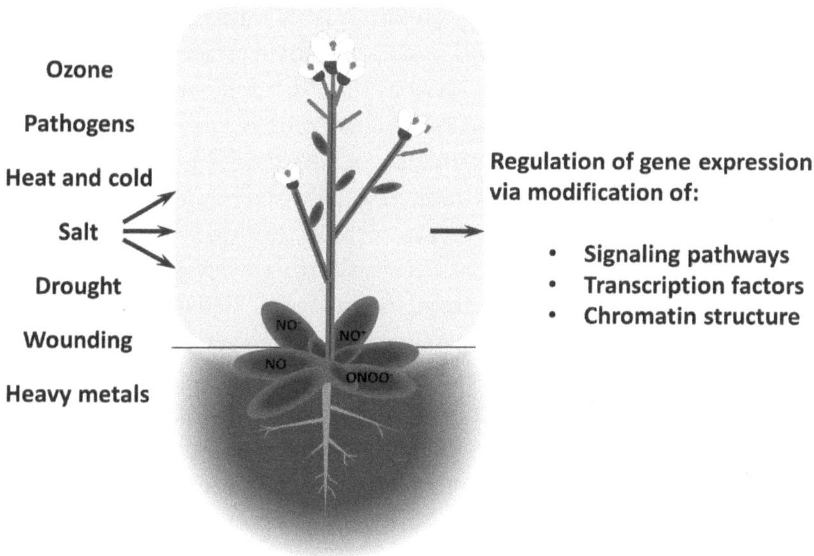

Figure 1 Nitric oxide (NO)-signaling in response to biotic and abiotic environmental stressors. Biotic or abiotic stress-induced NO production affects gene expression and results in stress response.

2. REGULATION OF GENE EXPRESSION VIA MODIFICATION OF SIGNALING PATHWAYS

The first evidence of the signaling function of NO in plants was related to its role in mediating defence responses against pathogens (Delledonne, Xia, Dixon, & Lamb, 1998; Durner, Wendehenne, & Klessig, 1998). The underlying connection between NO signals and immune responses was highlighted by the induction of defence-related genes in several studies (Delledonne et al., 1998; Durner et al., 1998; Feechan et al., 2005; Zeidler et al., 2004). For example, NO donor treatments of tobacco plants or suspension cells induced the expression of pathogenesis-related 1 (PR1) and phenylalanine ammonia lyase (PAL) genes (Durner et al., 1998). This induction of defence genes was a result of NO-mediated signal involving cyclic GMP and cyclic ADP-ribose as second messengers similarly to mammals. In parallel, it was shown that NO promotes the induction of hypersensitive cell death in soya bean cells leading to the induction of genes encoding protein related to secondary metabolism like PAL or chalcone synthase (Delledonne et al., 1998). Similarly, lipopolysaccharide (LPS)-induced rapid NO production has been shown to activate various defence or stress-related genes including glutathione S-transferases, cytochrome P450 and *PR* genes in *Arabidopsis thaliana* (Zeidler et al., 2004). The activated gene expressions were abolished in the NO-deficient *Atnos1* mutant, suggesting a functional link between LPS-induced NO production and gene induction. The loss-of-function mutation of AtGSNOR1 (*gsnor1-3*) encoding nitrosoglutathione (GSNO) reductase provided the first genetic evidence for the crucial role of S-nitrosothiols (SNOs) in plant defence (Feechan et al., 2005). The *gsnor1-3* plants with elevated SNO levels displayed disease susceptibility (in basal, R-mediated defence and non-host resistance), which correlated with reduced and delayed expression of SA-dependent *PR1* gene. However, enhanced basal resistance against the oomycete *Peronospora parasitica* Noco2 accompanied by constitutive activation of *PR1* was detected using antisense GSNOR plants (Rusterucci, Espunya, Diaz, Chabannes, & Martinez, 2007). Systemic immunity has been also shown to be affected by SNO/NO in a concentration-dependent manner (Espunya, De Michele, Gomez-Cadenas, & Martinez, 2012; Wang et al., 2014). SA, as an immune activator, coordinates gene transcription networks to induce defence signaling. The central player in the SA-dependent response is the non-expresser of pathogenesis-related genes 1 (NPR1) protein. The function of NPR1 is regulated by different factors at multiple levels in response to pathogens (Pajerowska-Mukhtar, Emerine, & Mukhtar, 2013). As a redox-regulated protein, NPR1 undergoes conformational changes dependent on the redox environment, S-nitrosylation or on the activity of thioredoxin proteins (Mou, Fan, & Dong, 2003; Tada et al., 2008). S-Nitrosylation of Cys residue-156 facilitates oligomer formation of NPR1 as an inactive form to maintain protein homeostasis in the case of pathogen challenge, while the reduction of oligomer to the active monomer is catalyzed by thioredoxins (TRX-H5 and TRX-H3) (Tada et al., 2008). Then, the monomeric form of NPR1 is translocated to the nucleus where, as a transcriptional co-activator, it interacts with the TGACG motif binding factor (TGA) family of basic domain/Leu zipper (bZIP) transcription factors and regulates their DNA-binding activity to induce a set of defence genes (Despres et al., 2003; Lindermayr, Sell, Muller, Leister, & Durner, 2010). The connection between NO signaling and the gene transcription was demonstrated by the GSNO-mediated S-nitrosylation of TGA1 transcription factor, which enhanced its DNA-binding activity in the presence of NPR1 (Lindermayr et al., 2010). Underlying this positive effect, GSNO was shown to trigger the expression of *PR* genes, resulting in an induced resistance of *A. thaliana* against *Pseudomonas syringae* (Kovacs, Durner,

& Lindermayr, 2015). The NO/GSNO signaling was linked to glutathione (GSH), a major thiol compound, which concentration is crucial to maintain redox balance in the cell. The NO/GSNO-induced induction of GSH level resulted in an increase in free SA and, consequently, in the induction of SA-dependent defence genes (Kovacs et al., 2015).

Similarly, increasing evidences indicate an important role of NO in the plant response to abiotic stressors including drought, salt, heat and cold stress (Corpas et al., 2011; Yu, Lamattina, Spoel, & Loake, 2014). Several candidate proteins for S-nitrosylation have been identified by proteomic studies (Abat & Deswal, 2009; Camejo et al., 2013; Lin et al., 2012; Puyaubert, Fares, Reze, Peltier, & Baudouin, 2014; Tanou et al., 2012); however, the connection to the change of gene expression is less known. Exogenous application of NO resulted in increased cold tolerance in various plants such as wheat, maize and tomato (Neill, Desikan, & Hancock, 2003). Furthermore, NO was shown to be produced rapidly following cold exposure in a nitrate reductase (NR)-dependent manner in *A. thaliana* (Cantrel et al., 2011; Zhao, Chen, Zhang, & Zhang, 2009). Transcriptomic data have revealed that low temperature induces a complex response by reprogramming gene expression to adapt to cold stress, including a high number of transcription factors (Thomashow, 2010). Impaired NO level in the *nia1nia2* NR mutant and in *A. thaliana* plants overexpressing the non-symbiotic haemoglobin 1 (AHb1) inhibited the expression of specific cold-responsive genes, like C-repeat-binding factors *CBF1* and *CBF3* (Cantrel et al., 2011). Members of the AP2/ERF family of transcription factors bind to the CRT/DRE regulatory element in the promoter region of cold-responsive genes. The expression of CBF target genes, like cold regulated 15a, low temperature-induced gene 30 and 78, was also compromised in NO-deficient plants exposed to cold stress, indicating that NO regulates gene expression through a CBF-dependent pathway in response to low temperature (Cantrel et al., 2011).

Salinity composes a major problem in the agriculture worldwide. Pretreatment with the NO donor sodium nitroprusside (SNP) or H_2O_2 showed enhanced tolerance to salt stress in citrus plants and 49 S-nitrosylated proteins were identified in response to salinity (Tanou et al., 2009). Moreover, transcriptional study of NO-related genes in leaves and roots from citrus plants exposed to salt stress disclosed complex tissue- and time-specific mechanisms regulating NO homeostasis (Tanou et al., 2012). Heat sensitive mutant *hot5* (sensitive to hot temperature 5) was identified by a forward genetic screen aimed at identifying genes involving in thermotolerance

(Lee, Wie, Fernandez, Feelisch, & Vierling, 2008). The characterized mutation in this line corresponded to the gene coding for GSNOR and the heat sensitivity of *hot5* alleles was associated with increased NO-related species. The consequence of the higher endogenous NO level on gene expression in the *hot5* mutant was investigated by microarray analysis (Xu, Guerra., Lee, & Vierling, et al., 2013).

2.1 Large-Scale Transcript Profiling Studies of NO-Treated Plants

Based on the various effects of NO in response to multiple environmental and developmental processes, a large number of changes in the gene expression profile were expected in plants following exposure to NO. An early transcriptional analysis of NO-induced genes was performed by cDNA-amplification fragment length polymorphism (AFLP) technique from *A. thaliana* infiltrated with NO donor SNP (Polverari et al., 2003). An altered expression of 120 cDNA corresponding to genes involved in signal transduction, disease resistance, ROS generation and removal, photosynthesis, cellular trafficking and basic metabolism was observed. A cDNA microarray experiment on *A. thaliana* cell cultures treated with NO donor NOR-3 has revealed that several defence-related and antioxidant genes were regulated by NO (Huang, von Rad, & Durner, 2002). Among these genes, that one encoding alternative oxidase 1a (AOX1) showed a highest expression level, which could be suppressed by removal of NO using the NO scavenger cPTIO. The NO-induced activity of the AOX1 protein has been shown to play a role in NO tolerance of *A. thaliana* (Huang et al., 2002). The first large-scale whole genome microarray analysis was performed on *A. thaliana* plants treated with the NO donor SNP at 0.1 mM or 1 mM (Parani et al., 2004). 342 upregulated and 80 downregulated genes were identified following SNP treatment, 162 of them showing a dose-dependent induction to SNP. Interestingly, 10% of the NO-regulated genes encoded transcription factors, like members of the ethylene response factor (ERF) family, WRKY-type transcription factors, zinc finger proteins and Myb-related transcription factors. Additionally, genes involved in cellular detoxification, defence, signal transduction and biosynthesis of ethylene, JA, lignin and alkaloids were differentially expressed, illustrating diverse biological functions of NO in plants. Previous reviews have discussed data from studies in which NO-responsive genes related to stress responses induced by wounding, plant–pathogen interactions and to non-stress-related pathways like flowering, symbiosis and iron homeostasis have been searched

(Besson-Bard et al., 2009; Grun, Lindermayr, Sell, & Durner, 2006). The impact of the crosstalk between NO and H_2O_2 on gene expression was studied by cDNA-AFLP analysis in catalase-deficient tobacco plant (CAT1AS) treated with the NO donor SNP (Zago et al., 2006). The enhanced H_2O_2 level induced by high light exposure strongly potentiated the NO-mediated cell death in CAT1AS but not wild type (WT) plants. However, only 16 transcripts whose induction were dependent on the presence of both H_2O_2 and NO were identified. Among them were ethylene and inositol pathway elements. Thirty-six transcripts specifically induced by NO were involved in signal transduction, defence responses and metabolism processes. Similarly, the role of NO in the O_3-induced cell death was investigated in a study using a whole genome microarray of SNP-treated and O_3-treated *A. thaliana* plants (Ahlfors, Brosche, Kollist, & Kangasjarvi, 2009). Most of the SNP-regulated genes were also regulated by O_3 in the same manner, and the gene enrichment analysis showed an overrepresentation of genes involved in various biotic and abiotic stresses and hormone signaling. The combination of SNP and O_3 treatment resulted in a reduced expression of SA- and defence-related genes compared to the O_3-induced expression values. The enhancement of the O_3-induced cell death phenotype by NO, together with attenuated gene expression levels, indicated that NO could directly affect the cell death program. A cDNA-AFLP analysis-based study, used to monitor gene expression changes under pathogenic and symbiotic conditions, revealed 999 NO-responsive genes from *Medicago truncatula* roots treated with two NO donor SNP and GSNO (Ferrarini et al., 2008). Interestingly, the comparison of the transcriptional response to the different NO donors in leaves and in roots indicated a low correlation between the two NO donor treatments in the same organ. This is probably due to the different nature of the reactive nitrogen molecules released by NO donors. Moreover, the by-products after NO release are different. The authors suggested that these differences need to be considered in the interpretation of the data. Moreover, a combined treatment with NO donors and NO scavengers as a control was suggested.

A robust change in the abundance of NO-related genes has been detected during the hypersensitive response or in the symbiotic nodule formation in *M. truncatula*, underlying the significant role of NO in these processes (Ferrarini et al., 2008). A root-specific microarray experiment was performed in *A. thaliana* plants treated with SNP (Badri et al., 2008). Only 87 regulated genes were identified from root tissues, belonging to similar functional categories as classified in the previous studies; however, the comparison of microarrays from root treated with SA or JA showed a few overlapping genes regulated commonly by NO, SA and JA. GSNO-responsive genes were identified in a large-scale study by RNA sequencing (RNA-seq) from roots and leaves of *A. thaliana* plants treated with 1 mM GSNO (Begara-Morales et al., 2014). 3263 genes were regulated by GSNO including WRKY and MYB family transcription factors and genes participating in disease resistance. Moreover, genes related to abiotic stress like wounding, heat and oxidative stress also responded to GSNO. Interestingly, several non-coding miscellaneous RNAs were identified from root in response to GSNO treatment, suggesting the role of NO in controlling the level of mRNAs through chromatin remodelling and silencing processes. The impact of endogenous NO on the gene transcription was also investigated in a GSNOR null mutant (*hot5-2*) (Xu et al., 2013) which contains higher level of SNOs compared to WT plants (Feechan et al., 2005; Lee et al., 2008). A microarray analysis has identified 99 up- and 170 downregulated genes which were enriched in 'stress response', 'redox' and 'signaling' categories. Six members of ROXY-class glutaredoxins related to redox signaling and three basic helix-loop-helix (bHLH) transcription factors playing a role in iron homeostasis were upregulated. In correlation with pathogen sensitivity of the GSNOR null mutant, 56 of 170 downregulated genes could be linked to pathogen responses (Xu et al., 2013).

A bioinformatics approach was performed to analyze the promoter elements of genes that co-expressed in response to NO (Palmieri et al., 2008). Using the microarray data of NO-treated *A. thaliana* plants and cell cultures, eight families of transcription factor binding sites (TFBSs) were identified. These TFBSs were at least 15% more enriched in the promoter region of NO-regulated genes (Palmieri et al., 2008). WRKY-, GBOX-, OCSE-, L1BX-, MYCL- and OPAQ-elements were overrepresented in upregulated genes, while TBPF- and MIIG-motifs were detected with a higher rate in downregulated genes. The GBOX-, OCSE- and OPAQ-elements contain a core motif for basic region/leucine zipper motif (bZIP) transcription factors, which are involved in biotic and abiotic signaling and in different developmental processes (Schutze, Harter, & Chaban, 2008). Similar to bZIP family members, genes of WRKY family members have been also reported to be regulated by different NO donors, suggesting that NO might directly affect transcription factors or transcriptional regulators probably by S-nitrosylation. However, an indirect regulation of gene expression via modification of signaling pathway cannot be excluded.

2.2 Protein S-Nitrosylation-Mediated Nuclear Translocation

NO can influence protein activity, localization and thereby induce or inhibit downstream signaling pathways. S-Nitrosylation of the cytosolic glycolytic enzyme glyceraldehyde-3-phosphate dehydrogenase (GAPDH) resulted in inhibition of protein activity in vitro (Holtgrefe et al., 2008; Lindermayr, Saalbach, & Durner, 2005). Two Cys residues (Cys-155 and Cys-159) were identified by mass spectrometry to be S-nitrosylated, which were located in the active centre of *A. thaliana* cytosolic GAPDH (Holtgrefe et al., 2008). Interestingly, the cytosolic isoforms of GAPDH have been found also in the nucleus and recently has been shown to localize to the plasma membrane and the endomembrane system (Henry, Fung, Liu, Drakakaki, & Coaker, 2015; Vescovi et al., 2013). In vivo evidence for S-nitrosylation was presented in tobacco BY-2 cells exposed to salt stress, which transiently increased the S-nitrosylation level of GAPDH (Wawer et al., 2010). Two tobacco isoforms of GAPDH (NtGAPCa and NtGAPCb) interact with the *Nicotiana tabacum* osmotic stress-activated protein kinase (NtOSAK) either in the cytosol (NtGAPCa) or in the cytosol and the nucleus (NtGAPCb). Although S-nitrosylation of GAPDHs did not influence their interactions, cysteine mutations in the site for S-nitrosylation abolished the nuclear localization of the NtOSAK/NtGAPCb complex (Wawer et al., 2010). Similar mechanisms for nuclear translocation of GAPDH in response to different stressors have been also reported in the animal field (Tristan, Shahani, Sedlak, & Sawa, 2011). Stress-induced S-nitrosylation of GAPDH interacts with the E3 ubiquitin ligase Siah1, leading to nuclear translocation of the SNO-GAPDH-Siah1 complex. Then, the stabilized SNO-GAPDH-Siah1 complex facilitates the degradation of the nuclear co-repressor N-CoR, resulting cell death (Hara et al., 2005). Further study showed that SNO-GAPDH in the nucleus is able to transnitrosylate nuclear proteins, like the deacetylating enzyme sirtuin-1, the histone deacetylase-2 and the DNA-activated protein kinase (DNA-PK) (Kornberg et al., 2010). By these unique mechanisms, SNO-GAPDH can regulate downstream gene expression changes under different stress conditions. Another glycolytic enzyme, cytosolic aldolase 2 (cALD2), was identified in a proteomic study aiming at identifying S-nitrosylated proteins from *A. thaliana* (Lindermayr et al., 2005). S-Nitrosylation by GSNO or SNP and S-glutathionylation by GSSG inactivate cALD2 enzyme in vitro (van der Linde et al., 2011). Similarly to GAPDH, cytosolic ALD2 was also detected in the nucleus in plants and in various animal tissues (van der Linde et al.,

2011; Saez & Slebe, 2000). Interestingly, GAPDH and cALD2 were both identified by a yeast one hybrid technique as an interaction partner of the gene coding for NADP-malate dehydrogenase (Hameister et al., 2007).

3. REGULATION OF GENE EXPRESSION VIA MODIFICATION OF TRANSCRIPTION FACTORS

Despite the growing number of identified target proteins for S-nitrosylation in different organisms, still only very few examples are known where NO directly regulates transcription factors and consequently gene expression. In *Escherichia coli*, for instance, endogenous S-nitrosylation induced by anaerobic respiration was shown to activate the transcription factor OxyR resulting in the induction of genes involved in protection of endogenous nitrosative stress. Moreover, aerobic condition-induced oxidative stress was shown to activate OxyR by oxygen-dependent thiol modifications. Interestingly, the NO- and oxygen-dependent mechanisms resulted in the induction of distinct regulons and thereby activated different sets of genes (Seth, Hausladen, Wang, & Stamler, 2012). More evidences for a direct regulation of gene expression by NO are reported in mammals. Several transcription factors (e.g. NF-κB, hypoxia-inducible factors, tumour suppressor p53 or different zinc finger transcription factors) were identified as being modulated by S-nitrosylation with a consequent effect on gene transcription (Sha & Marshall, 2012). In higher plant, the immune co-activator NPR1 is involved in NO signaling (as discussed above) and controls the transcription of SA-dependent defence genes by interacting with a TGA1 transcription factor in the nucleus. The TGA transcription factors are members of a bZIP family, and some core elements for DNA binding of these proteins were shown to be enriched in the NO-regulated genes (Palmieri et al., 2008). Two cysteine residues of TGA1 (Cys-260 and Cys-266) form an intramolecular disulphide bridge under oxidizing condition, thus preventing its interaction with NPR1 (Despres et al., 2003). Furthermore, TGA1 was reported to be S-nitrosylated and S-glutathionylated at Cys-260 and Cys-266 and this PTM increased the DNA-binding activity of TGA1 to the activation sequence-1 (as-1) element located in the promoter of several defence-related genes (Lindermayr et al., 2010). The DNA-binding activity of TGA1 was further increased in the presence of NPR1. The authors suggested that the S-nitrosylation of TGA1 might protect the protein from an oxidative modification and allows a more efficient NPR1-TGA1 interaction. The plant R2R3 MYB domain proteins are redox-regulated transcription factors containing two Cys residues that need to be reduced for their transcriptional activity (Heine, Hernandez, & Grotewold, 2004). The finding that the S-nitrosylation of Cys-53 by SNP and GSNO inhibited the DNA-binding activity of the *A. thaliana* transcription factor AtMYB2 provided the first in vitro evidence for the direct regulation of the activity of MYB transcription factors by NO (Serpa et al., 2007). Similar inhibition triggered by S-nitrosylation at the same Cys-53 position was demonstrated for AtMYB30 (Tavares et al., 2014). AtMYB30 is a positive regulator of plant defence and hypersensitive responses which activates genes related to the lipid biosynthesis pathways (Marino et al., 2013). In non-infected plants, MYB30 was shown to interact with the E3 ubiquitin ligase MIEL1, which directed it to proteasomal degradation. After bacterial infection, repression of MIEL1 resulted in an accumulation of MYB30 and the execution of the hypersensitive response to arrest bacterial growth. The study of the interplay between ubiquitination and S-nitrosylation of MYB30 in cell death pathways could provide evidences for cross-regulations of different PTMs. A new mechanism based on targeted degradation of group VII ERF transcription factors via the N-end rule pathway was recently proposed to sense NO in vivo (Gibbs et al., 2014). This small group of ERF transcription factors was found to be destabilized by NO through the N-end rule pathway of proteolysis, while they were stabilized in the lack of NO. The authors proposed a mechanism in which seed germination is regulated by group VII ERFs through the modulation of the ABA-insensitive 5 transcription factor, indicating a crosstalk between NO and ABA. Similar effect of NO on the stability of transcription factors was described in response to iron deficiency. The bHLH Fer-like Fe deficiency-induced transcription factor (FIT) controls iron uptake regulating downstream gene expressions, like ferric reductase oxidase 2 (*FRO2*) and iron-regulated transporter 1 (*IRT1*) (Meiser & Bauer, 2012). FIT was shown to be subjected to proteasomal degradation, the turnover of the protein being counteracted by NO (Meiser, Lingam, & Bauer, 2011). Moreover, FIT interacts with the ethylene-dependent transcription factor EIN3 that stabilizes it, thus providing a link between iron uptake and ethylene signaling (Lingam et al., 2011). Both signaling components, NO and ethylene, were shown to promote the production of each other and to be necessary for the upregulation of iron-acquisition genes (Garcia, Suarez, Romera, Alcantara, & Perez-Vicente, 2011). An

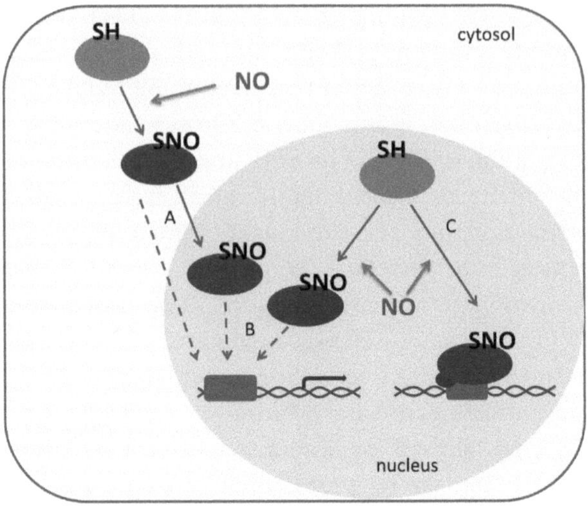

Figure 2 S-Nitrosylation-mediated regulation of gene expression. S-Nitrosylation of proteins can modulate a signaling pathway and/or results in protein translocation (A and B). Moreover, nitric oxide can directly affect DNA-binding proteins (C). Both results finally in transcriptional changes. SH, thiol group; SNO, S-nitrosothiol; dashed arrow, signaling pathways.

overview of S-nitrosylation-mediated regulation of gene expression is shown in Figure 2.

4. REGULATION OF GENE EXPRESSION VIA MODIFICATION OF CHROMATIN STRUCTURE

Modifications of DNA and histones, the core components of chromatin, affect the gene expression pattern by alteration of the chromatin structure. Important modifications are cytosine DNA methylation and PTMs, such as acetylation, methylation, phosphorylation, ubiquitinylation, sumoylation, ubiquitination and ADP-ribosylation at specific amino acid residues on the N-terminal tails of core histones (Feng & Jacobsen, 2011; Johnson et al., 2004; Zhang, Sridhar, Zhu, Kapoor, & Zhu, 2007). Genes encoding for DNA and histone-modifying enzymes and the regulation of epigenetic processes in plants are extensively reviewed by Pikaard and Mittelsten Scheid (2014).

4.1 NO-Dependent Regulation of Histone Acetylation

PTMs play an important role in modification of chromatin structure as they allow a cell to quickly reply to internal or external appeals. Acetylation of lysine (Lys) residues of histone tails plays a key role in regulation of gene expression. The transfer of acetyl groups from acetyl coenzyme A on Lys residues of histone tails is catalyzed by histone acetyltransferases. This leads to neutralization of the positive charge of the Lys residue and a decreased interaction between histones and negatively charged DNA. This results in a relaxed structure of chromatin making it accessible for transcription factors. Conversely, histone deacetylases (HDACs) remove acetyl groups and are recruited to maintain the chromatin in an inactivated and condensed state (Hollender & Liu, 2008). However, Kurdistani and Grunstein (2003) report that these enzymes are not only part of the transcriptional machinery, but also involved in other chromosomal processes such as DNA replication, repair and heterochromatin formation.

There is increasing evidence that catalytic activity of HDACs can be inhibited by redox molecules, such as NO, resulting in alteration of the chromatin structure (Figure 3). Until now, results about redox regulation of histone modifications are mainly based on research in the human/animal field. Nott, Watson, Robinson, Crepaldi, and Riccio (2008) reported that stimuli in neurons induce nuclear accumulation of NO and S-nitrosylation of many nuclear proteins. They also demonstrated that human HDAC2 is a target of the brain-derived neurotropic factor which causes NO synthesis in neurons. It was demonstrated that S-nitrosylation of HDAC2 occurs at the Cys residues Cys-262 and Cys-274 and does not inhibit the deacetylase activity of HDAC2, but causes its release from CREB-regulated gene promoter. This process induces an increase of histone acetylation at neurotrophin-dependent promoter regions and an activation of gene transcription. It was also shown that histone acetylation is inhibited and gene transcription is downregulated when Cys-262 and Cys-274 were changed to alanine. Moreover, the mutated HDAC2 did not dissociate from gene promoters even after treatment of neurons with NO donors. S-Nitrosylation of HDAC2 was also found in muscle of dystrophin-deficient MDX mice (Colussi et al., 2008). Evidence that the enzymatic activity of HDAC2 in muscle cells is impaired by NO while it remains unchanged in neurons was reported, suggesting that S-nitrosylation in muscles occurs at different Cys compared to neurons (Nott & Riccio, 2009). In addition, Feng, Jing, Fang, Gu, and Xu (2011) showed that recombinant HDAC8 is S-nitrosylated by GSNO in vitro. The enzymatic activity of this enzyme is significantly reduced by GSNO and SNO-Cys in a time- and concentration-dependent manner and can be restored by treatment with DTT. Interestingly, the NO donor SNP has no effect on

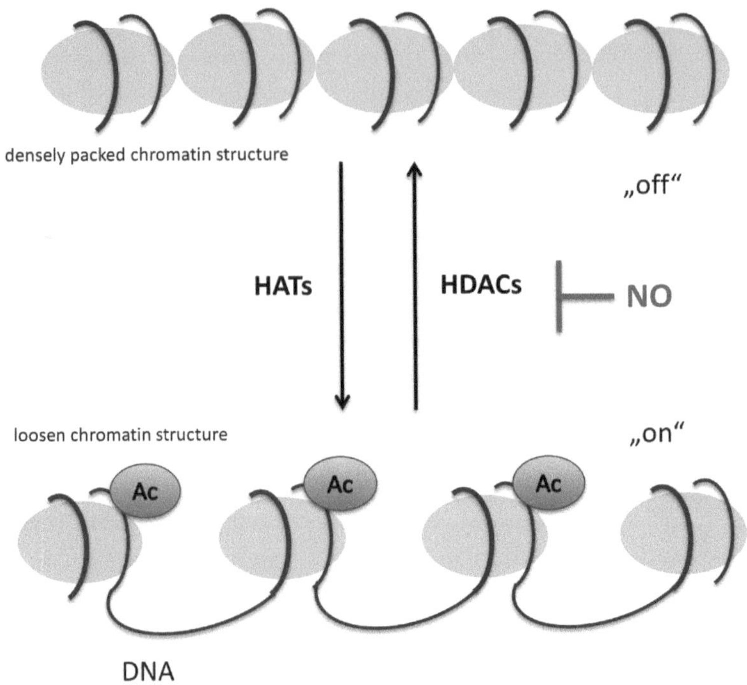

Figure 3 Histone acetylation/deacetylation alters chromatin structure. Acetylation of histones catalyzed by HATs results in opening chromatin, making it accessible for initiation of gene transcription. Conversely, deacetylation catalyzed by HDACs leads to maintain chromatin in the closed structure. Nitric oxide might affect HDACs activity, resulting in the hyperacetylation of Lys residues of histone tails and a loosen chromatin structure. HAT, histone acetyltransferase; HDAC, histone deacetylase; blue (light grey in print versions), histone complexes.

S-nitrosylation of HDAC8, suggesting that a special structure for transferring NO is required. Illi et al. (2008) also reported an indirect regulation of class II HDAC by NO. They demonstrated that NO activates PP2A (protein phosphatase 2) which binds to the pCamkIV/HDACs complex and dephosphorylates it. This process induces the transport of class II HDAC4 and 5 to the nucleus of endothelial cells and deacetylation of histones which, in turn, leads to the regulation of chromatin folding.

Three different HDAC families exist in plants. The largest and widely studied one is homologous to the yeast RPD3. This type of HDACs consists of 12 members that all possess a characteristic HDAC domain requiring zinc (Zn) ion for the catalytic activity. Based on their structure, they can be further divided into three classes. The members of the second family,

HD-tuins (HD2), have been originally found in maize and seem to be plant-specific proteins, although their homologue *cis-trans* prolyl isomerases are present in other eukaryotes (Aravind et al., 1998; Lusser, Brosch, Loidl, Haas, & Loidl, 1997). Moreover, HD2 proteins show structural differences compared to other HDAC proteins but display a sequence similarity to the FK506-binding protein. They are composed of three domains: the N-terminal domain which possesses a conserved pentapeptide MEFWG region, required for gene repression activity, a high-charged acidic domain and variable C-terminal domain (Dangl, Brosch, Haas, Loidl, & Lusser, 2001). Interestingly, most of the plants HD-tuins, analyzed until now, contain a zinc-finger motif on the C-terminus that is probably involved in protein—protein interaction (Ma, Lv, Zhang, & Yang, 2013). The third family of HDACs in plants is represented by unique NAD-dependent SIR2-like proteins. These proteins are not sensitive to any HDAC specific such as trichostatin A or sodium butyrate (Hollender & Liu, 2008). So far, little is known about the function of sirtuins in plants.

Bourque et al. (2011) demonstrated that HD2s act as negative regulator of cryptogein-induced cell death in *N. tabacum*. This elicitor is known to trigger NO production (Lamotte et al., 2004), suggesting a connection between NO-signaling and type-2 HDAC function. Accordingly, the production of NO trigger by cryptogein was strongly reduced in cells impaired in the expression of HD2s (Kulik et al., 2015). In the proteomic study of Chaki et al. (2015), 117, nuclear proteins were identified as potential target for NO. Among those, four plant-specific HD2 proteins have been demonstrated to be S-nitrosylated in vitro. So far, very little is known about NO regulation of plant HDACs. There is evidence that some members of the RPD3-like family possess highly conserved Cys residues. Notably, Cys-262 and Cys-274 of mammalian HDAC2, that are shown to be NO regulated, could be identified in some *A. thaliana* HDACs. Moreover, Liu, Zhang, Yu, Xiong, and Xia (2015) reported the redox sensitivity of a Cys residue that has been recently identified in *A. thaliana* HDAC19. Therefore, plant HDACs can be considered as interesting candidates for further studies (Mengel, Chaki, Shekariesfahlan, & Lindermayr, 2013).

4.2 NO-Dependent Regulation of Histone and DNA Methylation

Redox processes might also play a role in methylation-dependent regulation of gene expression. In *A. thaliana*, non-genic DNA methylation takes place in all three sequence contexts (CG, CHG and CHH, where H = A, T or C) on transposable elements and other repetitive DNA elements resulting in transcriptional silencing. CG-sequence gene body methylation is also found in transcribed genes (Feng & Jacobsen, 2011).

The key players for cytosine methylation in *A. thaliana* include four methyltransferases, their antagonizing demethylases, as well as chromatin remodelling ATPases and methyl cytosine-binding proteins (Furner and Matzke (2011) and references cited therein). Histone methylation can be associated with transcriptionally active or repressed regions (reviewed in Feng and Jacobsen (2011)). In *A. thaliana*, the following Lys residues on the N-terminal tail of histone H3 are subject to methylation: K4, K9, K27 and K36 (Johnson et al., 2004; Zhang et al., 2007). SET domain proteins are responsible for histone Lys methylation, of which 49 are encoded in the *A. thaliana* genome. Demethylation is catalyzed by two classes of proteins: the Lys-specific histone demethylase-like proteins (LDL) and the Jumonji C (JmjC)-domain-containing proteins (Pikaard & Mittelsten Scheid, 2014).

Until now, S-nitrosylation of DNA or histone methyltransferase or demethylase has not been reported in plants. However, based on studies on the human JmjC-domain-containing demethylase KDM3A (Hickok, Vasudevan, Antholine, & Thomas, 2013), Fe(II)-dependent plant JmjC-domain proteins might be targets for metal nitrosylation by the formation of a nitrosyl—iron complex with the non-haem Fe(II) coordinated by a 2-histidine-1-carboxylate facial triad in their catalytic pocket. Further, the application of the NO donor SNP causes DNA hypomethylation (mainly in CHG sites) in *Oryza sativa* L. spp. *Japonica* cultivars, going along with altered expression of chromatin remodelling and DNA methylation-modifying enzymes (Ou et al., 2015). To this end, DNA methylation might be regulated via differential expression of DNA methyl modifying proteins rather than by their inhibition through NO-based PTMs.

Apart from that, NO might be an issue in the supply of S-adenosylmethionine (SAM), the major methyl donor, and in the removal of the by-product inhibitor S-adenosylhomocysteine (SAH) of transmethyl reactions. DNA and histones are subjected to methylation by specific SAM-dependent methyltransferases (MTs). In each methyltransfer reaction SAH is formed, which is further converted into homocysteine (Hcy) and adenosine (Ado) by S-adenosylhomocysteine hydrolase (SAHH). The equilibrium of this reversible reaction favouring SAH synthesis (de la Haba & Cantoni, 1959; Palmer & Abeles, 1976) is driven towards hydrolysis of SAH due to removal of its products by downstream enzymes (Poulton & Butt, 1976). Methionine synthase converts Hcy to methionine, which is in turn adenylated to SAM by S-adenosylmethionine synthetase (SAMS), whereas Ado is metabolized in the adenosine salvage cycle. The levels of SAM and SAH are considered to be important regulators of cellular methylation processes. Interestingly, several proteomic studies revealed key enzymes of the methylation cycle as targets for S-nitrosylation (Abat & Deswal, 2009; Hu et al., 2015; Lindermayr et al., 2005; Puyaubert et al., 2014): Cobalamin-independent methionine synthase (MT), methionine s-adenosyltransferase (SAMS) and SAH hydrolase (SAHH). In *A. thaliana* different SAMS isoforms exist, which are differentially inhibited by protein S-nitrosylation (Lindermayr, Saalbach, Bahnweg, & Durner, 2006). Only the isoform SAMS1, but not SAMS2 or SAMS3, was reversibly inhibited by GSNO. It was demonstrated that S-nitrosylation of the Cys-114 residue of SAMS1, which is located next to the catalytic centre as part of the active site loop, is responsible for inhibition. A similar differential regulation of SAMS activity was observed in mammals. Here two genes encode different SAMS isoforms. While SAMS1A is reversibly inactivated by NO, SAMS2A is not affected (Perez-Mato, Castro, Ruiz, Corrales, & Mato, 1999). All these results suggest that NO plays a regulatory role in the synthesis of the major metyrosin-tyrosinehyl-group donor in the cell.

In *A. thaliana* two genes encode SAHH isoforms, but only SAHH1 is supposed to play a role DNA methylation processes (Rocha et al., 2005; Vriet, Hennig, & Laloi, 2015). Beside S-nitrosylation, Tyr nitration has been observed in SAHH of sunflower (*Helianthus annuus* L.). This type of modification decreased the activity of this enzyme (Chaki et al., 2009). S-Nitrosylation of *A. thaliana* SAHH1 upon cold stress was reported, but the physiological consequence of cold stress-induced S-nitrosylation of SAHH1 is not yet investigated (Puyaubert & Baudouin, 2014; Puyaubert et al., 2014). Previous studies demonstrated the importance of SAHH activity towards chromatin modifications (reviewed in Pikaard and Mittelsten Scheid (2014) and Vriet et al. (2015)): mutations in the *AtSAHH1* gene lead to reduced cytosine methylation and the release of transcriptional gene silencing (Jordan, West, Bottley, Sheikh, & Furner, 2007; Mull, Ebbs, & Bender, 2006; Rocha et al., 2005). Further, the expression of antisense RNA of SAHH in tobacco plants resulted in a loss of DNA methylation in repetitive elements (Tanaka et al., 1997) and the application of the SAHH inhibitor dihydroxypropyladenine reduces levels of DNA and histone methylation at endogenous repeats in *A. thaliana* (Baubec et al., 2010). In sum, SAHH plays an important role in chromatin modification (Baubec et al., 2010), and NO-dependent regulation of this enzyme might be an important mechanism to regulate gene expression.

5. CONCLUSION

Protein S-nitrosylation is a very important redox-dependent modification in plants and seems to be involved in regulation of many different physiological processes. However, we are just at the beginning of understanding the impact of this modification on nuclear plant proteins. Although, the presence of S-nitrosylating species in this compartment has not been demonstrated until now, evidence accumulates that S-nitrosylation of nuclear plant proteins may participate in regulation of transcription. In animals, several transcriptional regulators are regulated by this type of PTM. For instance, the regulation of various zinc-finger-containing transcription factors, including egr-1 and NFκB, are mediated by S-nitrosylation. Especially zinc-finger motifs are very sensitive to S-nitrosylation, making this class of transcription factors very interesting targets for further studies in plants, too. Moreover, NO also seems to affect chromatin structure, suggesting a regulatory function of NO in epigenetic processes. Key chromatin remodeller enzymes, such as HDACs are shown to be S-nitrosylated in the human/animal system. There is accumulating evidence that similar redox-mediated epigenetic mechanisms are present in plants as well. However, since most studies have been carried out in in vitro systems, the in vivo relevance, the physiological function as well as the exact molecular mechanisms still need to be demonstrated.

REFERENCES

Abat, J. K., & Deswal, R. (2009). Differential modulation of S-nitrosoproteome of *Brassica juncea* by low temperature: change in S-nitrosylation of Rubisco is responsible for the inactivation of its carboxylase activity. *Proteomics, 9*(18), 4368–4380. http://dx.doi.org/10.1002/pmic.200800985.

Ahlfors, R., Brosche, M., Kollist, H., & Kangasjarvi, J. (2009). Nitric oxide modulates ozone-induced cell death, hormone biosynthesis and gene expression in *Arabidopsis thaliana*. *Plant Journal, 58*(1), 1–12. http://dx.doi.org/10.1111/j.1365-313X.2008.03756.x.

Aravind, L., Koonin, E. V., Dangl, M., Lusser, A., Brosch, G., Loidl, A., et al. (1998). Second family of histone deacetylases. *Science, 280*(5367), 1167.

Astier, J., & Lindermayr, C. (2012). Nitric oxide-dependent posttranslational modification in plants: an update. *International Journal of Molecular Sciences, 13*(11), 15193–15208. http://dx.doi.org/10.3390/ijms131115193.

Badri, D. V., Loyola-Vargas, V. M., Du, J., Stermitz, F. R., Broeckling, C. D., Iglesias-Andreu, L., et al. (2008). Transcriptome analysis of *Arabidopsis* roots treated with signaling compounds: a focus on signal transduction, metabolic regulation and secretion. *New Phytologist, 179*(1), 209–223. http://dx.doi.org/10.1111/j.1469-8137.2008.02458.x.

Baubec, T., Dinh, H. Q., Pecinka, A., Rakic, B., Rozhon, W., Wohlrab, B., et al. (2010). Cooperation of multiple chromatin modifications can generate unanticipated stability of epigenetic states in *Arabidopsis*. *Plant Cell, 22*(1), 34–47. http://dx.doi.org/10.1105/tpc.109.072819.

Begara-Morales, J. C., Sanchez-Calvo, B., Luque, F., Leyva-Perez, M. O., Leterrier, M., Corpas, F. J., et al. (2014). Differential transcriptomic analysis by RNA-Seq of GSNO-responsive genes between *Arabidopsis* roots and leaves. *Plant and Cell Physiology, 55*(6), 1080–1095. http://dx.doi.org/10.1093/pcp/pcu044.

Besson-Bard, A., Astier, J., Rasul, S., Wawer, I., Dubreuil-Maurizi, C., Jeandroz, S., et al. (2009). Current view of nitric oxide-responsive genes in plants. *Plant Science, 177*(4), 302–309. http://dx.doi.org/10.1016/j.plantsci.2009.06.006.

Bourque, S., Dutartre, A., Hammoudi, V., Blanc, S., Dahan, J., Jeandroz, S., et al. (2011). Type-2 histone deacetylases as new regulators of elicitor-induced cell death in plants. *New Phytologist, 192*(1), 127–139. http://dx.doi.org/10.1111/j.1469-8137.2011.03788.x.

Camejo, D., Romero-Puertas Mdel, C., Rodriguez-Serrano, M., Sandalio, L. M., Lazaro, J. J., Jimenez, A., et al. (2013). Salinity-induced changes in S-nitrosylation of pea mitochondrial proteins. *Journal of Proteomics, 79*, 87–99. http://dx.doi.org/10.1016/j.jprot.2012.12.003.

Cantrel, C., Vazquez, T., Puyaubert, J., Reze, N., Lesch, M., Kaiser, W. M., et al. (2011). Nitric oxide participates in cold-responsive phosphosphingolipid formation and gene expression in *Arabidopsis thaliana*. *New Phytologist, 189*(2), 415–427. http://dx.doi.org/10.1111/j.1469-8137.2010.03500.x.

Chaki, M., Shekariesfahlan, A., Ageeva, A., Mengel, A., Von Toerne, C., Durner, J., et al. (2015). Identification of nuclear target proteins for S-nitrosylation in pathogen-treated *Arabidopsis thaliana* cell cultures. *Plant Science, 238*, 115–126.

Chaki, M., Valderrama, R., Fernandez Ocana, A. M., Carreras, A., Lopez-Jaramillo, J., Luque, F., et al. (2009). Protein targets of tyrosine nitration in sunflower (*Helianthus annuus* L.) hypocotyls. *Journal of Experimental Botany, 60*(15), 4221–4234. http://dx.doi.org/10.1093/jxb/erp263.

Colussi, C., Mozzetta, C., Gurtner, A., Illi, B., Rosati, J., Straino, S., et al. (2008). HDAC2 blockade by nitric oxide and histone deacetylase inhibitors reveals a common target in Duchenne muscular dystrophy treatment. *Proceedings of the National Academy of Sciences of the United States of America, 105*(49), 19183–19187. http://dx.doi.org/10.1073/pnas.0805514105.

Corpas, F. J., Leterrier, M., Valderrama, R., Airaki, M., Chaki, M., Palma, J. M., et al. (2011). Nitric oxide imbalance provokes a nitrosative response in plants under abiotic stress. *Plant Science: An International Journal of Experimental Plant Biology, 181*(5), 604–611. http://dx.doi.org/10.1016/j.plantsci.2011.04.005.

Dangl, M., Brosch, G., Haas, H., Loidl, P., & Lusser, A. (2001). Comparative analysis of HD2 type histone deacetylases in higher plants. *Planta*, 280–285.

Delledonne, M., Xia, Y., Dixon, R. A., & Lamb, C. (1998). Nitric oxide functions as a signal in plant disease resistance. *Nature, 394*(6693), 585–588. http://dx.doi.org/10.1038/29087.

Despres, C., Chubak, C., Rochon, A., Clark, R., Bethune, T., Desveaux, D., et al. (2003). The *Arabidopsis* NPR1 disease resistance protein is a novel cofactor that confers redox regulation of DNA binding activity to the basic domain/leucine zipper transcription factor TGA1. *Plant Cell, 15*(9), 2181–2191.

Durner, J., Wendehenne, D., & Klessig, D. F. (1998). Defense gene induction in tobacco by nitric oxide, cyclic GMP, and cyclic ADP-ribose. *Proceedings of the National Academy of Sciences of the United States of America, 95*(17), 10328–10333.

Espunya, M. C., De Michele, R., Gomez-Cadenas, A., & Martinez, M. C. (2012). S-Nitrosoglutathione is a component of wound- and salicylic acid-induced systemic responses in *Arabidopsis thaliana*. *Journal of Experimental Botany, 63*(8), 3219–3227. http://dx.doi.org/10.1093/jxb/ers043.

Feechan, A., Kwon, E., Yun, B. W., Wang, Y., Pallas, J. A., & Loake, G. J. (2005). A central role for S-nitrosothiols in plant disease resistance. *Proceedings of the National Academy of Sciences of the United States of America, 102*(22), 8054–8059. http://dx.doi.org/10.1073/pnas.0501456102.

Feng, J., Jing, F., Fang, H., Gu, L., & Xu, W. (2011). Expression, purification, and S-nitrosylation of recombinant histone deacetylase 8 in *Escherichia coli*. *BioScience Trends, 5*(1), 17–22. http://dx.doi.org/10.5582/bst.2011.v5.1.17.

Feng, S., & Jacobsen, S. E. (2011). Epigenetic modifications in plants: an evolutionary perspective. *Current Opinion in Plant Biology, 14*(2), 179–186. http://dx.doi.org/10.1016/j.pbi.2010.12.002.

Ferrarini, A., De Stefano, M., Baudouin, E., Pucciariello, C., Polverari, A., Puppo, A., et al. (2008). Expression of *Medicago truncatula* genes responsive to nitric oxide in pathogenic and symbiotic conditions. *Molecular Plant-Microbe Interaction, 21*(6), 781–790. http://dx.doi.org/10.1094/MPMI-21-6-0781.

Furner, I. J., & Matzke, M. (2011). Methylation and demethylation of the *Arabidopsis* genome. *Current Opinion in Plant Biology, 14*(2), 137–141. http://dx.doi.org/10.1016/j.pbi.2010.11.004.

Garcia, M. J., Suarez, V., Romera, F. J., Alcantara, E., & Perez-Vicente, R. (2011). A new model involving ethylene, nitric oxide and Fe to explain the regulation of Fe-acquisition genes in Strategy I plants. *Plant Physiology & Biochemistry, 49*(5), 537–544. http://dx.doi.org/10.1016/j.plaphy.2011.01.019.

Gibbs, D. J., Md Isa, N., Movahedi, M., Lozano-Juste, J., Mendiondo, G. M., Berckhan, S., et al. (2014). Nitric oxide sensing in plants is mediated by proteolytic control of group VII ERF transcription factors. *Molecular Cell, 53*(3), 369–379. http://dx.doi.org/10.1016/j.molcel.2013.12.020.

Grun, S., Lindermayr, C., Sell, S., & Durner, J. (2006). Nitric oxide and gene regulation in plants. *Journal of Experimental Botany, 57*(3), 507–516. http://dx.doi.org/10.1093/jxb/erj053.

de la Haba, G., & Cantoni, G. L. (1959). The enzymatic synthesis of s-adenosyl-l-homocysteine from adenosine and homocysteine. *Journal of Biological Chemistry, 234*(3), 603–608.

Hameister, S., Becker, B., Holtgrefe, S., Strodtkotter, I., Linke, V., Backhausen, J. E., et al. (2007). Transcriptional regulation of NADP-dependent malate dehydrogenase: comparative genetics and identification of DNA-binding proteins. *Journal of Molecular Evolution, 65*(4), 437–455. http://dx.doi.org/10.1007/s00239-007-9025-9.

Hara, M. R., Agrawal, N., Kim, S. F., Cascio, M. B., Fujimuro, M., Ozeki, Y., et al. (2005). S-nitrosylated GAPDH initiates apoptotic cell death by nuclear translocation following Siah1 binding. *Nature Cell Biology, 7*(7), 665–674. http://dx.doi.org/10.1038/ncb1268.

Heine, G. F., Hernandez, J. M., & Grotewold, E. (2004). Two cysteines in plant R2R3 MYB domains participate in REDOX-dependent DNA binding. *Journal of Biological Chemistry, 279*(36), 37878–37885. http://dx.doi.org/10.1074/jbc.M405166200.

Henry, E., Fung, N., Liu, J., Drakakaki, G., & Coaker, G. (2015). Beyond glycolysis: GAPDHs are multi-functional enzymes involved in regulation of ROS, autophagy, and plant immune responses. *PLoS Genetics, 11*(4), e1005199. http://dx.doi.org/10.1371/journal.pgen.1005199.

Hickok, J. R., Vasudevan, D., Antholine, W. E., & Thomas, D. D. (2013). Nitric oxide modifies global histone methylation by inhibiting jumonji C domain-containing demethylases. *Journal of Biological Chemistry, 288*(22), 16004–16015. http://dx.doi.org/10.1074/jbc.M112.432294.

Hollender, C., & Liu, Z. (2008). Histone deacetylase genes in *Arabidopsis* development. *Journal of Integrative Plant Biology, 50*(7), 875–885. http://dx.doi.org/10.1111/j.1744-7909.2008.00704.x.

Holtgrefe, S., Gohlke, J., Starmann, J., Druce, S., Klocke, S., Altmann, B., et al. (2008). Regulation of plant cytosolic glyceraldehyde 3-phosphate dehydrogenase isoforms by thiol modifications. *Physiologia Plantarum, 133*(2), 211–228. http://dx.doi.org/10.1111/j.1399-3054.2008.01066.x.

Hu, J., Huang, X., Chen, L., Sun, X., Lu, C., Zhang, L., et al. (2015). Site-specific nitrosoproteomic identification of endogenously S-nitrosylated proteins in *Arabidopsis*. *Plant Physiology, 167*(4), 1731–1746. http://dx.doi.org/10.1104/pp.15.00026.

Huang, X., von Rad, U., & Durner, J. (2002). Nitric oxide induces transcriptional activation of the nitric oxide-tolerant alternative oxidase in *Arabidopsis* suspension cells. *Planta, 215*(6), 914–923. http://dx.doi.org/10.1007/s00425-002-0828-z.

Illi, B., Dello Russo, C., Colussi, C., Rosati, J., Pallaoro, M., Spallotta, F., et al. (2008). Nitric oxide modulates chromatin folding in human endothelial cells via protein phosphatase 2A activation and class II histone deacetylases nuclear shuttling. *Circulation Research, 102*(1), 51–58. http://dx.doi.org/10.1161/CIRCRESAHA.107.157305.

Johnson, L., Mollah, S., Garcia, B. A., Muratore, T. L., Shabanowitz, J., Hunt, D. F., et al. (2004). Mass spectrometry analysis of *Arabidopsis* histone H3 reveals distinct combinations of post-translational modifications. *Nucleic Acids Research, 32*(22), 6511–6518. http://dx.doi.org/10.1093/nar/gkh992.

Jordan, N. D., West, J. P., Bottley, A., Sheikh, M., & Furner, I. (2007). Transcript profiling of the hypomethylated hog1 mutant of *Arabidopsis*. *Plant Molecular Biology, 65*(5), 571–586. http://dx.doi.org/10.1007/s11103-007-9221-4.

Kornberg, M. D., Sen, N., Hara, M. R., Juluri, K. R., Nguyen, J. V., Snowman, A. M., et al. (2010). GAPDH mediates nitrosylation of nuclear proteins. *Nature Cell Biology, 12*(11), 1094–1100. http://dx.doi.org/10.1038/ncb2114.

Kovacs, I., Durner, J., & Lindermayr, C. (2015). Crosstalk between nitric oxide and glutathione is required for NONEXPRESSOR OF PATHOGENESIS-RELATED GENES 1 (NPR1)-dependent defense signaling in *Arabidopsis thaliana*. *New Phytologist*. http://dx.doi.org/10.1111/nph.13502.

Kulik, A., Noirot, E., Grandperret, V., Bourque, S., Fromentin, J., Salloignon, P., et al. (2015). Interplays between nitric oxide and reactive oxygen species in cryptogein signalling. *Plant Cell and Environment, 38*(2), 331–348. http://dx.doi.org/10.1111/pce.12295.

Kurdistani, S. K., & Grunstein, M. (2003). Histone acetylation and deacetylation in yeast. *Nature Reviews Molecular Cell Biology, 4*(4), 276–284.

Lamotte, O., Gould, K., Lecourieux, D., Sequeira-Legrand, A., Lebrun-Garcia, A., Durner, J., et al. (2004). Analysis of nitric oxide signaling functions in tobacco cells challenged by the elicitor cryptogein. *Plant Physiology, 135*(1), 516–529. http://dx.doi.org/10.1104/pp.104.038968.

Lee, U., Wie, C., Fernandez, B. O., Feelisch, M., & Vierling, E. (2008). Modulation of nitrosative stress by S-nitrosoglutathione reductase is critical for thermotolerance and plant growth in *Arabidopsis*. *Plant Cell, 20*(3), 786–802. http://dx.doi.org/10.1105/tpc.107.052647.

Lin, A., Wang, Y., Tang, J., Xue, P., Li, C., Liu, L., et al. (2012). Nitric oxide and protein S-nitrosylation are integral to hydrogen peroxide-induced leaf cell death in rice. *Plant Physiology, 158*(1), 451–464. http://dx.doi.org/10.1104/pp.111.184531.

van der Linde, K., Gutsche, N., Leffers, H. M., Lindermayr, C., Muller, B., Holtgrefe, S., et al. (2011). Regulation of plant cytosolic aldolase functions by redox-modifications. *Plant Physiology & Biochemistry, 49*(9), 946–957. http://dx.doi.org/10.1016/j.plaphy.2011.06.009.

Lindermayr, C., Saalbach, G., Bahnweg, G., & Durner, J. (2006). Differential inhibition of *Arabidopsis* methionine adenosyltransferases by protein S-nitrosylation. *Journal of Biological Chemistry, 281*(7), 4285–4291. http://dx.doi.org/10.1074/jbc.M511635200.

Lindermayr, C., Saalbach, G., & Durner, J. (2005). Proteomic identification of S-nitrosylated proteins in *Arabidopsis*. *Plant Physiology, 137*(3), 921–930. http://dx.doi.org/10.1104/pp.104.058719.

Lindermayr, C., Sell, S., Muller, B., Leister, D., & Durner, J. (2010). Redox regulation of the NPR1-TGA1 system of *Arabidopsis thaliana* by nitric oxide. *Plant Cell, 22*(8), 2894–2907. http://dx.doi.org/10.1105/tpc.109.066464.

Lingam, S., Mohrbacher, J., Brumbarova, T., Potuschak, T., Fink-Straube, C., Blondet, E., et al. (2011). Interaction between the bHLH transcription factor FIT and ETHYLENE INSENSITIVE3/ETHYLENE INSENSITIVE3-LIKE1 reveals molecular linkage between the regulation of iron acquisition and ethylene signaling in *Arabidopsis*. *Plant Cell, 23*(5), 1815–1829. http://dx.doi.org/10.1105/tpc.111.084715.

Liu, P., Zhang, H., Yu, B., Xiong, L., & Xia, Y. (2015). Proteomic identification of early salicylate- and flg22-responsive redox-sensitive proteins in *Arabidopsis*. *Scientific Reports, 5*, 8625. http://dx.doi.org/10.1038/srep08625. http://www.nature.com/articles/srep08625#supplementary-information.

Lusser, A., Brosch, G., Loidl, A., Haas, H., & Loidl, P. (1997). Identification of Maize histone deacetylase HD2 as an acidic nucleolar phosphoprotein. *Science, 277*(5322), 88–91. http://dx.doi.org/10.1126/science.277.5322.88.

Ma, X., Lv, S., Zhang, C., & Yang, C. (2013). Histone deacetylases and their functions in plants. *Plant Cell Reports, 32*(4), 465–478. http://dx.doi.org/10.1007/s00299-013-1393-6.

Marino, D., Froidure, S., Canonne, J., Ben Khaled, S., Khafif, M., Pouzet, C., et al. (2013). *Arabidopsis* ubiquitin ligase MIEL1 mediates degradation of the transcription factor MYB30 weakening plant defence. *Nature Communications, 4*, 1476. http://dx.doi.org/10.1038/ncomms2479.

Martinez-Ruiz, A., Cadenas, S., & Lamas, S. (2011). Nitric oxide signaling: classical, less classical, and nonclassical mechanisms. *Free Radical Biology & Medicine, 51*(1), 17–29. http://dx.doi.org/10.1016/j.freeradbiomed.2011.04.010.

Meiser, J., & Bauer, P. (2012). Looking for the hub in Fe signaling. *Plant Signaling & Behavior, 7*(6), 688–690. http://dx.doi.org/10.4161/psb.20150.

Meiser, J., Lingam, S., & Bauer, P. (2011). Posttranslational regulation of the iron deficiency basic helix-loop-helix transcription factor FIT is affected by iron and nitric oxide. *Plant Physiology, 157*(4), 2154–2166. http://dx.doi.org/10.1104/pp.111.183285.

Mengel, A., Chaki, M., Shekariesfahlan, A., & Lindermayr, C. (2013). Effect of nitric oxide on gene transcription—S-nitrosylation of nuclear proteins. *Frontiers of Plant Science, 4*, 293. http://dx.doi.org/10.3389/fpls.2013.00293.

Mou, Z., Fan, W., & Dong, X. (2003). Inducers of plant systemic acquired resistance regulate NPR1 function through redox changes. *Cell, 113*(7), 935–944.

Mull, L., Ebbs, M. L., & Bender, J. (2006). A histone methylation-dependent DNA methylation pathway is uniquely impaired by deficiency in *Arabidopsis* S-adenosylhomocysteine hydrolase. *Genetics, 174*(3), 1161–1171. http://dx.doi.org/10.1534/genetics.106.063974.

Neill, S. J., Desikan, R., & Hancock, J. T. (2003). Nitric oxide signalling in plants. *New Phytologist, 159*(1), 11–35. http://dx.doi.org/10.1046/j.1469-8137.2003.00804.x.

Nott, A., & Riccio, A. (2009). Nitric oxide-mediated epigenetic mechanisms in developing neurons. *Cell Cycle, 8*(5), 725–730. http://dx.doi.org/10.4161/cc.8.5.7805.

Nott, A., Watson, P. M., Robinson, J. D., Crepaldi, L., & Riccio, A. (2008). S-Nitrosylation of histone deacetylase 2 induces chromatin remodelling in neurons. *Nature, 455*(7211), 411–415. http://dx.doi.org/10.1038/nature07238.

Ou, X., Zhuang, T., Yin, W., Miao, Y., Wang, B., Zhang, Y., et al. (2015). DNA methylation changes induced in Rice by exposure to high concentrations of the nitric oxide modulator, sodium nitroprusside. *Plant Molecular Biology Reporter.* http://dx.doi.org/10.1007/s11105-014-0843-9.

Pajerowska-Mukhtar, K. M., Emerine, D. K., & Mukhtar, M. S. (2013). Tell me more: roles of NPRs in plant immunity. *Trends in Plant Science, 18*(7), 402–411. http://dx.doi.org/10.1016/j.tplants.2013.04.004.

Palmer, J. L., & Abeles, R. H. (1976). Mechanism for enzymatic thioether formation. Mechanism of action of S-adenosylhomocysteinase. *Journal of Biological Chemistry, 251*(18), 5817–5819.

Palmieri, M. C., Sell, S., Huang, X., Scherf, M., Werner, T., Durner, J., et al. (2008). Nitric oxide-responsive genes and promoters in *Arabidopsis thaliana*: a bioinformatics approach. *Journal of Experimental Botany, 59*(2), 177–186. http://dx.doi.org/10.1093/jxb/erm345.

Parani, M., Rudrabhatla, S., Myers, R., Weirich, H., Smith, B., Leaman, D. W., et al. (2004). Microarray analysis of nitric oxide responsive transcripts in *Arabidopsis*. *Plant Biotechnology Journal, 2*(4), 359–366. http://dx.doi.org/10.1111/j.1467-7652.2004.00085.x.

Perez-Mato, I., Castro, C., Ruiz, F. A., Corrales, F. J., & Mato, J. M. (1999). Methionine adenosyltransferase S-nitrosylation is regulated by the basic and acidic amino acids surrounding the target thiol. *Journal of Biological Chemistry, 274*(24), 17075–17079.

Pikaard, C. S., & Mittelsten Scheid, O. (2014). Epigenetic regulation in plants. *Cold Spring Harbor Perspectives in Biology, 6*(12). http://dx.doi.org/10.1101/cshperspect.a019315.

Polverari, A., Molesini, B., Pezzotti, M., Buonaurio, R., Marte, M., & Delledonne, M. (2003). Nitric oxide-mediated transcriptional changes in *Arabidopsis thaliana. Molecular Plant-Microbe Interaction, 16*(12), 1094–1105. http://dx.doi.org/10.1094/mpmi.2003.16.12.1094.

Poulton, J. E., & Butt, V. S. (1976). Purification and properties of S-adenosyl-l-homocysteine hydrolase from leaves of spinach beet. *Archives of Biochemistry and Biophysics, 172*(1), 135–142. http://dx.doi.org/10.1016/0003-9861(76)90058-8.

Puyaubert, J., & Baudouin, E. (2014). New clues for a cold case: nitric oxide response to low temperature. *Plant Cell and Environment, 37*(12), 2623–2630. http://dx.doi.org/10.1111/pce.12329.

Puyaubert, J., Fares, A., Reze, N., Peltier, J. B., & Baudouin, E. (2014). Identification of endogenously S-nitrosylated proteins in *Arabidopsis* plantlets: effect of cold stress on cysteine nitrosylation level. *Plant Science, 215-216*, 150–156. http://dx.doi.org/10.1016/j.plantsci.2013.10.014.

Rocha, P. S., Sheikh, M., Melchiorre, R., Fagard, M., Boutet, S., Loach, R., et al. (2005). The *Arabidopsis* HOMOLOGY-DEPENDENT GENE SILENCING1 gene codes for an S-adenosyl-L-homocysteine hydrolase required for DNA methylation-dependent gene silencing. *Plant Cell, 17*(2), 404–417. http://dx.doi.org/10.1105/tpc.104.028332.

Rusterucci, C., Espunya, M. C., Diaz, M., Chabannes, M., & Martinez, M. C. (2007). S-nitrosoglutathione reductase affords protection against pathogens in *Arabidopsis*, both locally and systemically. *Plant Physiology, 143*(3), 1282–1292. http://dx.doi.org/10.1104/pp.106.091686.

Saez, D. E., & Slebe, J. C. (2000). Subcellular localization of aldolase B. *Journal of Cellular Biochemistry, 78*(1), 62–72.

Schutze, K., Harter, K., & Chaban, C. (2008). Post-translational regulation of plant bZIP factors. *Trends in Plant Science, 13*(5), 247–255. http://dx.doi.org/10.1016/j.tplants.2008.03.002.

Serpa, V., Vernal, J., Lamattina, L., Grotewold, E., Cassia, R., & Terenzi, H. (2007). Inhibition of AtMYB2 DNA-binding by nitric oxide involves cysteine S-nitrosylation. *Biochemical and Biophysical Research, 361*(4), 1048–1053. http://dx.doi.org/10.1016/j.bbrc.2007.07.133.

Seth, D., Hausladen, A., Wang, Y. J., & Stamler, J. S. (2012). Endogenous protein S-Nitrosylation in *E. coli*: regulation by OxyR. *Science, 336*(6080), 470–473. http://dx.doi.org/10.1126/science.1215643.

Sha, Y., & Marshall, H. E. (2012). S-nitrosylation in the regulation of gene transcription. *Biochimica et Biophysica Acta, 1820*(6), 701–711. http://dx.doi.org/10.1016/j.bbagen.2011.05.008.

Tada, Y., Spoel, S. H., Pajerowska-Mukhtar, K., Mou, Z., Song, J., Wang, C., et al. (2008). Plant immunity requires conformational changes [corrected] of NPR1 via S-nitrosylation and thioredoxins. *Science, 321*(5891), 952–956. http://dx.doi.org/10.1126/science.1156970.

Tanaka, H., Masuta, C., Uehara, K., Kataoka, J., Koiwai, A., & Noma, M. (1997). Morphological changes and hypomethylation of DNA in transgenic tobacco expressing antisense RNA of the S-adenosyl-l-homocysteine hydrolase gene. *Plant Molecular Biology, 35*(6), 981–986. http://dx.doi.org/10.1023/A:1005896711321.

Tanou, G., Filippou, P., Belghazi, M., Job, D., Diamantidis, G., Fotopoulos, V., et al. (2012). Oxidative and nitrosative-based signaling and associated post-translational modifications orchestrate the acclimation of citrus plants to salinity stress. *Plant Journal, 72*(4), 585–599. http://dx.doi.org/10.1111/j.1365-313X.2012.05100.x.

Tanou, G., Job, C., Rajjou, L., Arc, E., Belghazi, M., Diamantidis, G., et al. (2009). Proteomics reveals the overlapping roles of hydrogen peroxide and nitric oxide in the acclimation of citrus plants to salinity. *Plant Journal, 60*(5), 795–804. http://dx.doi.org/10.1111/j.1365-313X.2009.04000.x.

Tavares, C. P., Vernal, J., Delena, R. A., Lamattina, L., Cassia, R., & Terenzi, H. (2014). S-nitrosylation influences the structure and DNA binding activity of AtMYB30 transcription factor from *Arabidopsis thaliana. Biochimica et Biophysica Acta, 1844*(4), 810–817. http://dx.doi.org/10.1016/j.bbapap.2014.02.015.

Thomashow, M. F. (2010). Molecular basis of plant cold acclimation: insights gained from studying the CBF cold response pathway. *Plant Physiology, 154*(2), 571–577. http://dx.doi.org/10.1104/pp.110.161794.

Toledo, J. C., Jr., & Augusto, O. (2012). Connecting the chemical and biological properties of nitric oxide. *Chemical Research in Toxicology, 25*(5), 975–989. http://dx.doi.org/10.1021/tx300042g.

Tristan, C., Shahani, N., Sedlak, T. W., & Sawa, A. (2011). The diverse functions of GAPDH: views from different subcellular compartments. *Cell Signal, 23*(2), 317–323. http://dx.doi.org/10.1016/j.cellsig.2010.08.003.

Vescovi, M., Zaffagnini, M., Festa, M., Trost, P., Lo Schiavo, F., & Costa, A. (2013). Nuclear accumulation of cytosolic glyceraldehyde-3-phosphate dehydrogenase in cadmium-stressed *Arabidopsis* roots. *Plant Physiology, 162*(1), 333–346. http://dx.doi.org/10.1104/pp.113.215194.

Vriet, C., Hennig, L., & Laloi, C. (2015). Stress-induced chromatin changes in plants: of memories, metabolites and crop improvement. *Cellular and Molecular Life Sciences, 72*(7), 1261–1273. http://dx.doi.org/10.1007/s00018-014-1792-z.

Wang, C., El-Shetehy, M., Shine, M. B., Yu, K., Navarre, D., Wendehenne, D., et al. (2014). Free radicals mediate systemic acquired resistance. *Cell Reports,* 7(2), 348–355. http://dx.doi.org/10.1016/j.celrep.2014.03.032.

Wawer, I., Bucholc, M., Astier, J., Anielska-Mazur, A., Dahan, J., Kulik, A., et al. (2010). Regulation of *Nicotiana tabacum* osmotic stress-activated protein kinase and its cellular partner GAPDH by nitric oxide in response to salinity. *Biochemical Journal, 429*(1), 73–83. http://dx.doi.org/10.1042/BJ20100492.

Xu, S., Guerra, D., Lee, U., & Vierling, E. (2013). S-nitrosoglutathione reductases are low-copy number, cysteine-rich proteins in plants that control multiple developmental and defense responses in *Arabidopsis*. *Frontiers of Plant Science, 4*, 430. http://dx.doi.org/10.3389/fpls.2013.00430.

Yu, M., Lamattina, L., Spoel, S. H., & Loake, G. J. (2014). Nitric oxide function in plant biology: a redox cue in deconvolution. *New Phytologist, 202*(4), 1142–1156. http://dx.doi.org/10.1111/nph.12739.

Zago, E., Morsa, S., Dat, J. F., Alard, P., Ferrarini, A., Inze, D., et al. (2006). Nitric oxide- and hydrogen peroxide-responsive gene regulation during cell death induction in tobacco. *Plant Physiology, 141*(2), 404–411. http://dx.doi.org/10.1104/pp.106.078444.

Zeidler, D., Zahringer, U., Gerber, I., Dubery, I., Hartung, T., Bors, W., et al. (2004). Innate immunity in *Arabidopsis thaliana*: lipopolysaccharides activate nitric oxide synthase (NOS) and induce defense genes. *Proceedings of the National Academy of Sciences of the United States of America, 101*(44), 15811–15816. http://dx.doi.org/10.1073/pnas.0404536101.

Zhang, K., Sridhar, V. V., Zhu, J., Kapoor, A., & Zhu, J. K. (2007). Distinctive core histone post-translational modification patterns in *Arabidopsis thaliana*. *PLoS One, 2*(11), e1210. http://dx.doi.org/10.1371/journal.pone.0001210.

Zhao, M. G., Chen, L., Zhang, L. L., & Zhang, W. H. (2009). Nitric reductase-dependent nitric oxide production is involved in cold acclimation and freezing tolerance in *Arabidopsis*. *Plant Physiology, 151*(2), 755–767. http://dx.doi.org/10.1104/pp.109.140996.

CHAPTER THREE

Auxin and Nitric Oxide: A Counterbalanced Partnership Ensures the Redox Cue Control Required for Determining Root Growth Pattern

Natalia Correa-Aragunde, Noelia Foresi, Lorenzo Lamattina[1]
Instituto de Investigaciones Biológicas, Consejo Nacional de Investigaciones Científicas y Técnicas (CONICET), Universidad Nacional de Mar del Plata (UNMdP), Mar del Plata, Argentina
[1]Corresponding author: E-mail: lolama@mdp.edu.ar

Contents

Abstract

Auxin is the main hormone that controls growth and developmental processes in plants. In the last decade, many studies confirmed the interplay between reactive oxygen species (ROS) and reactive nitrogen species (RNS) in the auxin signaling transduction pathway. Nitric oxide (NO) is an RNS induced by, and required for the auxin-mediated lateral root formation and inhibition of the primary root elongation. Auxin induces both ROS and NO that, in turn, modulates reductases and peroxidases activities resulting in the attenuation of auxin signaling. As a consequence, a balance between ROS and NO concentrations appears to be essential for the control of the auxin action during root growth and developmental processes. In this review, we summarize and discuss the recent progress in the understanding of signaling and interacting components participating of the auxin-, NO- and ROS-modulated redox balance determining root growth pattern.

Advances in Botanical Research, Volume 77
ISSN 0065-2296
http://dx.doi.org/10.1016/bs.abr.2015.10.006

1. INTRODUCTION

Plant growth and developmental (PGD) processes are coordinated by the interplay of molecular basis that are under the control of master hormones. Auxin is probably the best characterized phytohormone associated to the regulation of PGD (Salehin, Bagchi, & Estelle, 2015). The integration and balance of endogenous hormones and external stimuli perceived by plant cells trigger the expression of a correct PGD program directed to manage adaptive responses for the particular growing situations. The precise control of the cell redox status is crucial for the success of the PGD programs at any stage of the plant life cycle. Thus, plants reach different cellular homeostatic conditions by integrating a complex network of many interacting partners. In this scenario, auxin appears to be closely associated to an appropriate balance of reactive oxygen species (ROS) and reactive nitrogen species (RNS) with the aim of reaching a correct homeostatic redox control required for executing a specific PGD program (Figure 1; Tognetti, Mühlenbock, & Van Breusegem, 2012).

Nitric oxide (NO) is an RNS paramagnetic small gas with neutral charge, displaying a high affinity for transition metals (Fe, Cu, Zn) and high reactivity with the superoxide radical ($O_2^{\bullet -}$) and molecular O_2 (Stamler, Singel, & Loscalzo, 1992). NO is soluble in both hydrophobic and hydrophilic environments, although the first preferred. The chemical and physical characteristics of NO allow it to cross biological membranes and attain a relatively easy intercellular spread (Subczynski, Lomnicka, & Hyde, 1996). Its short half-life

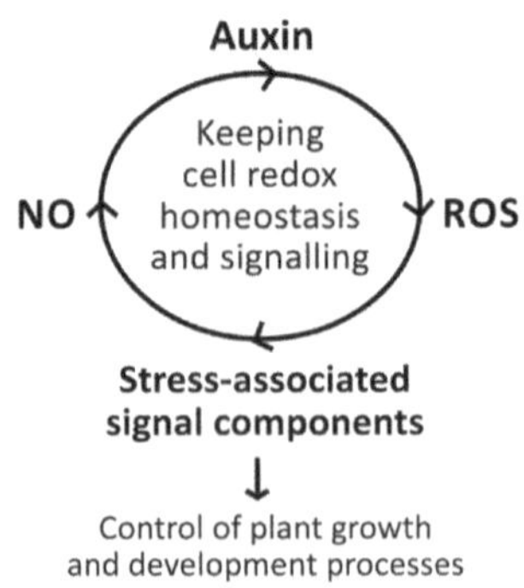

Figure 1 Simplified scheme showing the close relationships between auxin, reactive oxygen species (ROS) and nitric oxide (NO) and the balance required for keeping cell redox homeostasis, and their linkage with pathways regulating plant growth and developmental processes associated to stress-signaling components.

(less than 1 s) and its high chemical reactivity give NO characteristics of a potent biological effector. Indeed, the NO activity regulating gene expression and the biological activity of key proteins and enzymes make NO a central component involved in the homeostasis of cell physiology, participating in cell responses to various endogenous and exogenous stimuli. However, NO is a free radical and, at high concentrations, it is itself harmful for the cell integrity and viability (Lamattina, García-Mata, Graziano, & Pagnussat, 2003). At high concentrations and under certain circumstances, NO and its derived RNS such as peroxynitrite ($ONOO^-$), nitrogen dioxide ($\bullet NO_2$), dinitrogen trioxide (N_2O_3) and S-nitrosoglutathione (GSNO) which can mediate post-translational modifications of different biomolecules cause cell damages in a process named nitrosative stress (Corpas & Barroso, 2013). This process has been described in macrophages when killing bacteria or tumour cells (Nathan & Xie, 1994). However, as stated above, NO reacts rapidly with $O_2^{\bullet -}$, the first product of the NADPH oxidase (NOX) activity, in a fast reaction that generates the oxidant RNS $ONOO^-$. In this reaction, NO acts as scavenger of $O_2^{\bullet -}$, the substrate for the formation of the powerful ROS compounds H_2O_2 and hydroxyl radical ($\bullet OH$). Thus, in this way, NO functions as a protective agent breaking oxidative chain reactions and limiting cell injuries when challenged by a powerful oxidative stress.

One of the first reports in the studies of NO action in plant physiology pointed out the potent antioxidant capacity of NO under circumstances of strong oxidative stresses like those produced by the methylviologen (MV) herbicides diquat and paraquat (Beligni & Lamattina, 1999, 2002). MV rapidly breaks the electron chain transport in chloroplasts allowing O_2 being the receptor of free electrons, generating huge amounts of superoxide $O_2^{\bullet -}$. The NO acts like a potent $O_2^{\bullet -}$ scavenger since the affinity of NO for $O_2^{\bullet -}$ in aqueous solution is $k \approx 3.7 \times 10^7 M^{-1} s^{-1}$ (Stamler et al., 1992), several times higher than for reactions between NO and other radicals. The presence of NO allows MV-treated plants survive to the toxic effects generated by huge amounts of $O_2^{\bullet -}$ (Beligni & Lamattina, 1999). NO also protects potato plants from the hypersensitive response (HR) and cell death provoked by the infection with the pathogen *Phytophthora infestans* (Beligni & Lamattina, 2001; Laxalt, Beligni, & Lamattina, 1997) and other plant species challenged by abiotic stresses like exposure to UV irradiance, to heavy metals and salt stress (Laspina, Groppa, Tomaro, & Benavides, 2005; Shi, Ding, Wang, & Wei, 2007; Tossi, Amenta, Lamattina, & Cassia, 2011).

Under strong oxidative stress conditions, the increase of endogenous NO concentration through exogenous applications of NO donors like sodium nitroprusside, N-acetyl S-penicillamine or GSNO protects plants from the exacerbated ROS production. More recently, the endogenous production of NO due to the expression of nitric oxide synthase (NOS) from the unicellular green algae *Ostreococcus tauri* was shown to protect Arabidopsis plants from the oxidative damage generated by abiotic stresses (Foresi et al., 2015).

In the last 10 years, many reports have documented the close relationship between auxins and NO in PGD, and mainly in root growth processes (Correa-Aragunde, Graziano, Chevalier, & Lamattina, 2006; Correa-Aragunde, Graziano, & Lamattina, 2004; Lombardo, Graziano, Polacco, & Lamattina, 2006; Lombardo & Lamattina, 2012; Pagnussat, Simontacchi, Puntarulo, & Lamattina, 2002). In all of them, it was unambiguously demonstrated the requirement of NO for the success of the auxin action leading to normal root growth and development. In this review, we will discuss the molecular partners underpinning the NO function in controlling the cell redox homeostasis, attenuating/counterbalancing an auxin-promoted specific PDG.

2. INDOLE ACETIC ACID INDUCES OXIDATIVE STRESS AND NO PRODUCTION

As stated, auxin is probably the master phytohormone controlling growing processes in plant roots. However, phytotoxic effects due to herbicide actions of auxin at high concentrations have thoroughly been reported (Grossmann, Kwiatkowski, & Tresch, 2001; Hansen & Grossmann, 2000). The induction of ethylene and abscisic acid biosynthesis pathways as well as the accumulation of higher concentrations of H_2O_2 occurs after the auxin action. At high auxin concentrations, all these effects induce growth inhibition and cell death.

Many experimental evidence confirmed the close linkage between the production of different ROS, like $O_2^{\bullet -}$, •OH and H_2O_2, peroxidase activity, cell wall loosening and root gravitropism as a requisite of the auxin-modulated root growth processes (Joo et al., 2005; Kukavica et al., 2009; Liszkay, van der Zalm, & Schopfer, 2004; Yamaguchi & Sharp, 2010). It was shown that auxin induces NOX activity during root gravitropism (Joo, Bae, & Lee, 2001). According to these reports, the Arabidopsis mutant in the NOX gene, homologue to the catalytic subunit of mammalian

NADPH oxidase gp91^{phox}, named root hair defective 2 (*rhd2*) is impaired in the generation of ROS and in the elongation of the primary root and root hairs (Foreman et al., 2003). Additional evidence indicates that ROS alters auxin signaling, since the induction of ROS by application of ozone reduces the expression of AUX/IAA repressors and of auxin-response genes (Blomster, 2011). It was also proposed that the oxidative stress following the increase of auxin concentration is linked to the attenuation of the auxin signaling (Blomster, 2011; Peer, Cheng, & Murphy, 2013). It was found that auxin-induced ROS accumulation results in auxin oxidation (oxIAA) and its biological inactivation, since oxIAA cannot be exported from the cell. A third player, flavonoids, accumulated after auxin stimulus and is thought to scavenge ROS and to prevent the formation of oxIAA (Maloney et al., 2014).

At this point, the fact that auxin induces $O_2^{\bullet-}$ generation through the activation of NOX, but, at the same time, promotes the production of NO through the activation of nitrate reductase, and an NO synthase-like activity (Figure 2; Flores et al., 2008; Mounier, Pervent, Ljung, Gojon, & Nacry, 2014; Vidal, Moyano, Riveras, Contreras-López, & Gutiérrez, 2013) suggests that NO could, like flavonoids, attenuate the damage that could originate an exacerbated and uncontrolled increase of intracellular ROS concentration. Moreover, NO also controls $O_2^{\bullet-}$ generation by inhibition of NOX by S-nitrosylation reported during HR and regulating excessive cell death (Figure 2; Yun et al., 2011). A precise balance between auxin, NO and ROS concentrations assures the redox balance and homeostasis required for the control of PGD processes. For achieving this balance, due to the presence of ROS and RNS compounds and the stress-like characteristics associated to growth and development, many components involved in stress metabolism are required in those processes (Figure 2; Bartoli, Casalongué, Simontacchi, Marquez-Garcia, & Foyer, 2013; Iglesias et al., 2014; Xia et al., 2015).

3. THE COUNTERBALANCE BETWEEN NO AND ROS OPERATES DOWNSTREAM AUXIN AND IS CRITIC FOR DETERMINING ROOT ARCHITECTURE

The execution of a PGD program is reminiscent of a stress situation that is accompanied by increased concentration of ROS (Schmidt & Schippers, 2015). How cells deal with these ROS and use them in its own benefit to favour cell functions and growth processes is a matter of

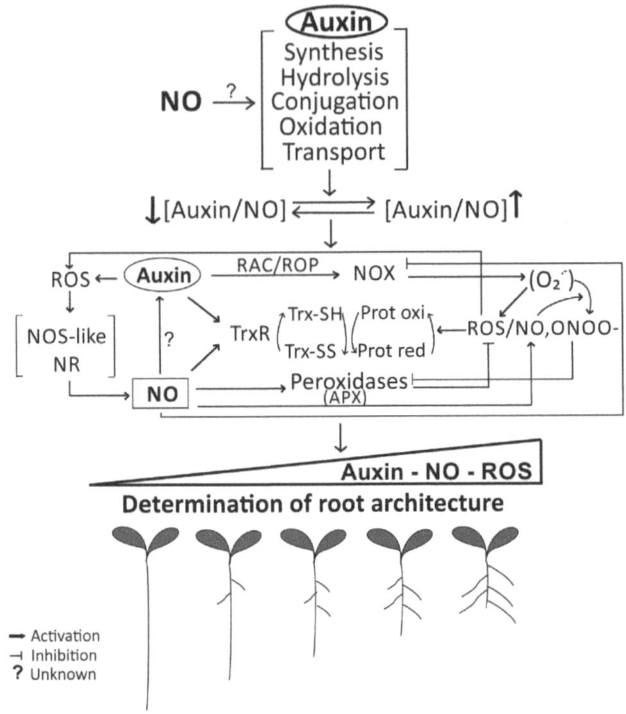

Figure 2 Interplay between auxin and nitric oxide (NO) controlling the counterbalance of redox molecules involved in root growth and development. Auxin levels in roots are mainly determined by synthesis in shoots and basipetal transport to the roots. Auxin hydrolysis, oxidation and conjugation can diminish the active form of the hormone. In roots, auxin induces $O_2^{\cdot-}$ formation through the activation of NOX via RAC/ROP GTPases. Later, $O_2^{\cdot-}$ interacts with reactive nitrogen species molecules like NO, generating $ONOO^-$. In parallel, reactive oxygen species (ROS) are able to increase NO levels in roots by the activation of nitrate reductase (NR) and nitric oxide synthase (NOS)-like enzyme, and in a negative feedback pathway, NO can inhibit NOX by S-nitrosylation. At the same time, auxin and NO activate thioredoxin reductase (TrxR) aimed to control excessive ROS-mediated oxidation of proteins. NO is able to activate the antioxidant ascorbate peroxidase (APX), thus contributing to the control of ROS, even if RNS can also inhibit APX through $ONOO^-$-mediated Tyr nitration. Finally, the dynamic balance and distribution of auxin, ROS and NO contribute to the determination of root growth-associated processes in plants.

research. First of all, cells have developed sophisticated antioxidant machinery directed to minimize the irreversible oxidation of key cellular components like proteins, lipids and nucleic acids. This machinery includes several peroxidases that detoxify H_2O_2. For instance, ascorbate peroxidase (APX)-deficient Arabidopsis displays growth suppression, altered stomatal response and the induction of the stress-related proteins like heat shock proteins (Pnueli, Liang, Rozenberg, & Mittler, 2003). Notably, APX enzymes are present in all subcellular compartments (Mittler, Vanderauwera, Gollery, & Van Breusegem, 2004). In this context, the cytosolic isoform of APX was shown to be positively regulated by NO through the S-nitrosylation of the Cys-32 residue (Begara-Morales et al., 2013; Correa-Aragunde, Foresi, Delledonne, & Lamattina, 2013; Yang et al., 2015). In a very exquisite menu of redox-based chemistry, APX is the target of several oxidative post-translational modifications that result in biochemical changes of the protein that influence its enzymatic activity. These control points of the APX enzymatic activity include Tyr nitration, S-nitrosylation, metal nitrosylation, carbonylation, Cys oxidation and thioredoxin-mediated Cys reduction; each could be sensing a specific oxidative cell state (Correa-Aragunde, Foresi, et al., 2015).

Root system is important for a wide variety of processes, such as nutrient and water uptake, anchoring and storage. The root architecture is determined by a complex network of hormone signaling pathways and environmental stimuli (Osmont, Sibout, & Hardtke, 2007). Regarding the interplay between auxin and NO, blocking the auxin transport by the treatment with naphthylphthalamic acid or scavenging NO with cPTIO results in an almost complete inhibition of lateral root formation. In contrast, the increase of either auxin or NO concentrations results in the inhibition of the primary root growth and activation of the founder cells in the pericycle to enter in cell division and generates new primordia of lateral roots (Boerjan et al., 1995; Correa-Aragunde et al., 2006, 2004; Himanen et al., 2002). The regulation of the process orchestrated by auxin and NO requires a precise control of the redox balance involving the concerted action of many redox-regulated proteins (Figure 2). This includes the simultaneous and synchronizing activity of molecules exerting actions in opposite directions to dynamically counterbalance their effects. For instance, even if a general feature of the auxin action is the NOX-dependent increase of $O_2^{\bullet-}$ and H_2O_2 concentration required for cell elongation, cell division and differentiation, auxin and NO activate, at the same time, the thioredoxin reductase (TrxR) activity leading to the attenuation of the oxidative-mediated damage of cellular components. Figure 2 describes some ways by which cells keep redox homeostasis upon stimulation by auxin, involving the production of NO and activation of reductases and peroxidases (Bashandy et al., 2010; Correa-Aragunde, Cejudo, Lamattina, 2015). Auxin activates NOX through RAC/ROP GTPases mediating auxin-responsive gene expression

(Tao, Cheung, & Wu, 2002). At the same time, NO inhibits NOX activity through S-nitrosylation (Yun et al., 2011), activates APX by S-nitrosylation of Cys-32 (Begara-Morales et al., 2013; Yang et al., 2015) and can also irreversible inhibit APX through the formation of $ONOO^-$ and the nitration of Tyr5 and Tyr235 (Begara-Morales et al., 2013), or through metal nitrosylation (Clark, Durner, Navarre, & Klessig, 2000). Furthermore, reduction of Cys residues in APX by Trx causes activity inhibition of the enzyme (Gelhaye et al., 2006). All these post-translational modifications of APX contribute to the fine control and regulation of H_2O_2 concentration, and it was not fully studied yet at the molecular level in a dynamic and changing scenario, for instance, when occurring several post-translational modifications at the same time. How NO is affecting the balance of the auxin metabolism, from the biosynthesis of the hormone until its oxidation, conjugation and transport, is completely unknown and deserves a deep study (Figure 2).

4. REDOX REGULATION OF AUXIN PERCEPTION AND SIGNALING

The auxin receptors TIR1/AFB are F-box proteins functioning in a ubiquitin ligase SCF complex to target AUX/IAA repressors for degradation (Dharmasiri, Dharmasiri, & Estelle, 2005; Kepinski & Leyser, 2005). Recent studies indicate that the redox state of the cell also influences auxin perception by TIR1/AFB. It was shown that NO activates the auxin signaling through the S-nitrosylation of TIR1/AFB (Figure 3; Terrile et al., 2012), increasing the sensitivity of the auxin pathway at low auxin concentration (Terrile et al., 2012). Recently, it was proposed a mechanism that regulates the stability of the auxin receptor TIR1/AFB and the response to auxin (Yu et al., 2015). Authors postulate that when TIR1/AFB is dissembled from the complex formed with the ubiquitinating subunits CUL1 and ASK1, it becomes more stable and resistant to auxin stimulus. In contrast, when assembled to CUL1 and ASK1, TIR1/AFB is susceptible to ubiquitination and degradation. Whether S-nitrosylation of TIR1/AFB plays a role for its preference to assemble or not to CUL1 and/or ASK1 and influences the stability of TIR1/AFB, still remains to be elucidated.

In this scenario, whereas it was shown that the activation of the auxin receptor TIR1/AFB by S-nitrosylation leads to enhance the polyubiquitination and degradation of AUX/IAA corepressors via the proteasome and activation of auxin-responsive genes (Terrile et al., 2012,

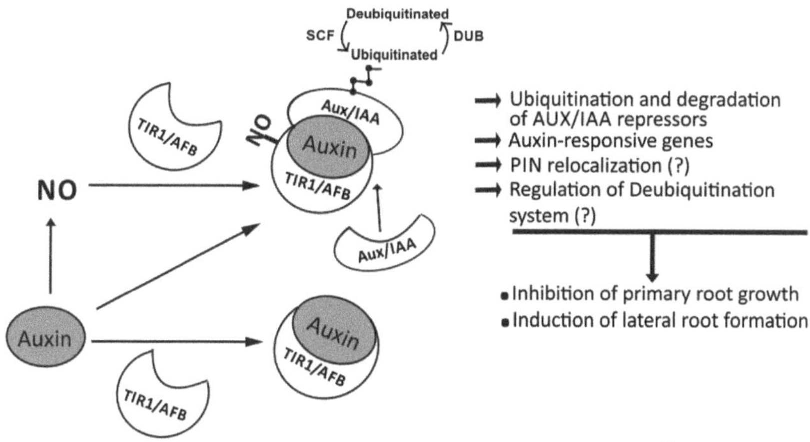

Figure 3 Proposed model of S-nitrosylation of the auxin receptor TIR1/AFB and its effects on the activation of auxin-induced responses in roots. Cys-140 and Cys-480 were shown to be S-nitrosylated residues in the TIR1/AFB and positively influence root growth in Arabidopsis (Terrile et al., 2012). This redox-based modification enhances the efficiency of TIR1/AFB interaction with AUX/IAAs repressors, facilitating their degradation and promoting auxin signaling during root growth in Arabidopsis seedlings. The effect of NO on the auxin transporters PIN and on the regulation of the deubiquitination enzyme DUB is still unknown.

Figure 3), there is no report on the regulation of auxin signaling mediated by the activation/inhibition of deubiquitinating (DUB) enzymes. In animal cells, the regulation of DUB enzymes through their inactivation/degradation promoted by H_2O_2 was recently demonstrated (Kulathu et al., 2013; Lee, Baek, Soetandyo, & Ye, 2013). Thus, putting the actors together, auxin induces the increase of H_2O_2 concentration through both the activation of NOX (Figure 2) and partial inhibition of APX1 mediated by the TrxR-Trx-dependent denitrosylation of the enzyme (Correa-Aragunde, Foresi, et al., 2015). At the same time, the increase of the H_2O_2 concentration induced by auxin could degrade/inhibit DUB enzymes, thus contributing to sustaining the polyubiquitination of AUX/IAA corepressors and their degradation, sensitizing the auxin signaling response (Figure 3). In a recent article, it was nicely probed that the progression of lateral root development correlated with the rate of turnover of IAA14, an AUX/IAA corepressor (Guseman et al., 2015). Authors concluded that AUX/IAA degradation initiates and synchronizes the auxin actions in developmental transition processes.

Notably, in animal cells DUB enzymes have been implicated in signaling pathways associated with growth factors (Inui et al., 2011). It deserves, therefore, to investigate whether DUB enzymes are regulated by redox chemistry in plants, and how it impacts on many polyubiquitination-based plant hormone actions on growth parameters. In addition, it would also be interesting to search for any NO-mediated action on the reactive Cys residue located in the catalytic site of the caspase activity belonging to DUB enzymes. Indeed, a yet unexplored mechanism that could be mediated by NO, resulting in the S-nitrosylation of the key Cys residue in DUB enzymes that might prevent the irreversible H_2O_2-mediated oxidation/degradation of DUB enzymes, was already proposed by Clague (2013).

In plants, DUB enzymes have been linked with intracellular trafficking and auxin transport (Isono et al., 2010). A redox regulation mediated by the balance of H_2O_2/NO affecting the activity of DUB enzymes and, as a consequence, the ubiquitination process would probably influence auxin signaling. This represents another yet unexplored molecular mechanism that might contribute to the plant cell decisions integrating PGD and stress tolerance responses.

5. CONCLUDING REMARKS AND PERSPECTIVES

Since long time ago, it is known that changes in redox balance belong to the many cell responses observed after auxin treatment. In the last two decades, the identification of the auxin receptor TIR1 and the gene family of auxin transporters solved one of the major enigmas in plant biology. The milestones for constructing a model that explains the auxin perception and signaling were postulated. However, secondary messengers and signal molecules linking to the redox homeostasis of the cell play a role modulating auxin responses. The knowledge of the complex interaction of all redox players in the auxin-involved molecular machinery during PGD processes is and will be a continuous challenge for plant biologists. The understanding of the interplay between ROS, RNS and auxin signaling may give a more accuracy viewing of the complexity of hormone signal transduction in plants. This fundamental knowledge is essential for the success of managing genetic programs and the improvement of commercial crops.

REFERENCES

Bartoli, C. G., Casalongué, C., Simontacchi, M., Marquez-Garcia, B., & Foyer, C. H. (2013). Interactions between hormone and redox signalling pathways in the control of growth and cross tolerance to stress. *Environmental and Experimental Botany, 94*, 73–88.

Bashandy, T., Guilleminot, J., Vernoux, T., Caparros-Ruiz, D., Ljung, K., Meyer, Y., et al. (2010). Interplay between the NADP-linked thioredoxin and glutathione systems in Arabidopsis auxin signaling. *Plant Cell, 22*, 376–391.

Begara-Morales, J. C., Sánchez-Calvo, B., Chaki, M., Valderrama, R., Mata-Pérez, C., López-Jaramillo, J., et al. (2013). Dual regulation of cytosolic ascorbate peroxidase (APX) by tyrosine nitration and S-nitrosylation. *Journal of Experimental Botany, 65*, 527–538.

Beligni, M. V., & Lamattina, L. (1999). Nitric oxide protects against cellular damage produced by methylviologen herbicides in potato plants. *Nitric Oxide, 3*, 199–208.

Beligni, M. V., & Lamattina, L. (2001). Nitric oxide in plants: the history is just beginning. *Plant, Cell and Environment, 24*, 267–278.

Beligni, M. V., & Lamattina, L. (2002). Nitric oxide interferes with plant photo-oxidative stress by detoxifying reactive oxygen species. *Plant, Cell and Environment, 25*, 737–748.

Blomster, T. (2011). Apoplastic reactive oxygen species transiently decrease auxin signaling and cause stress-induced morphogenic response in Arabidopsis. *Plant Physiology, 157*, 1866–1883.

Boerjan, W., Cervera, M. T., Delarue, M., Beeckman, T., Dewitte, W., Bellini, C., et al. (1995). Superroot, a recessive mutation in Arabidopsis, confers auxin overproduction. *Plant Cell, 7*, 1405–1419.

Clague, M. J. (2013). Biochemistry: oxidation controls the DUB step. *Nature, 497*, 49–50.

Clark, D., Durner, J., Navarre, D. A., & Klessig, D. F. (2000). Nitric oxide inhibition of tobacco catalase and ascorbate peroxidase. *Molecular Plant–Microbe Interactions, 13*, 1380–1384.

Corpas, F. J., & Barroso, J. B. (2013). Nitro-oxidative stress vs oxidative or nitrosative stress in higher plants. *New Phytologist, 199*, 633–635.

Correa-Aragunde, N., Cejudo, F. J., & Lamattina, L. (2015). Nitric oxide is required for the auxin-induced activation of NADPH-dependent thioredoxin reductase and protein denitrosylation during root growth responses in arabidopsis. *Annals of botany, 116*, 695–702.

Correa-Aragunde, N., Foresi, N., Delledonne, M., & Lamattina, L. (2013). Auxin induces redox regulation of ascorbate peroxidase 1 activity by S-nitrosylation/denitrosylation balance resulting in changes of root growth pattern in Arabidopsis. *Journal of Experimental Botany, 64*, 3339–3349.

Correa-Aragunde, N., Foresi, N., & Lamattina, L. (2015). Nitric oxide is a ubiquitous signal for maintaining redox balance in plant cells: regulation of ascorbate peroxidase as a case study. *Journal of Experimental Botany, 66*, 2913–2921.

Correa-Aragunde, N., Graziano, M., Chevalier, C., & Lamattina, L. (2006). Nitric oxide modulates the expression of cell cycle regulatory genes during lateral root formation in tomato. *Journal of Experimental Botany, 57*, 581–588.

Correa-Aragunde, N., Graziano, M., & Lamattina, L. (2004). Nitric oxide plays a central role in determining lateral root development in tomato. *Planta, 218*, 900–905.

Dharmasiri, N., Dharmasiri, S., & Estelle, M. (2005). The F-box protein TIR1 is an auxin receptor. *Nature, 435*, 441–445.

Flores, T., Todd, C. D., Tovar-Mendez, A., Dhanoa, P. K., Correa-Aragunde, N., Hoyos, M. E., et al. (2008). Arginase-negative mutants of Arabidopsis exhibit increased nitric oxide signaling in root development. *Plant Physiology, 147*, 1936–1946.

Foreman, J., Demidchik, V., Bothwell, J. H. F., Mylona, P., Miedema, H., Torres, M. A., et al. (2003). Reactive oxygen species produced by NADPH oxidase regulate plant cell growth. *Nature, 422*, 442–446.

Foresi, N., Mayta, M. L., Lodeyro, A. F., Scuffi, D., Correa-Aragunde, N., García-Mata, C., Casalongué, C., Carrillo, N., & Lamattina, L. (2015). Expression of the tetrahydrofolate-dependent nitric oxide synthase from the green alga Ostreococcus tauri increases tolerance to abiotic stresses and influences stomatal development in Arabidopsis. *The Plant Journal, 82*, 806–821.

Gelhaye, E., Navrot, N., Macdonald, I. K., Rouhier, N., Raven, E. L., & Jacquot, J.-P. (2006). Ascorbate peroxidase-thioredoxin interaction. *Photosynthesis Research, 89*, 193–200.

Grossmann, K., Kwiatkowski, J., & Tresch, S. (2001). Auxin herbicides induce H_2O_2 overproduction and tissue damage in cleavers (*Galium aparine* L.). *Journal of Experimental Botany, 52*, 1811–1816.

Guseman, J. M., Hellmuth, A., Lanctot, A., Feldman, T. P., Moss, B. L., Klavins, E., et al. (2015). Auxin-induced degradation dynamics set the pace for lateral root development. *Development, 89*, 1–5.

Hansen, H., & Grossmann, K. (2000). Auxin-induced ethylene triggers abscisic acid biosynthesis and growth inhibition. *Plant Physiology, 124*, 1437–1448.

Himanen, K., Boucheron, E., Vanneste, S., Engler, J. D. A., Inzé, D., & Beeckman, T. (2002). Auxin-mediated cell cycle activation during early lateral root initiation. *Plant Cell, 14*, 2339–2351.

Iglesias, M. J., Terrile, M. C., Windels, D., Lombardo, M. C., Bartoli, C. G., Vazquez, F., et al. (2014). MiR393 regulation of auxin signaling and redox-related components during acclimation to salinity in Arabidopsis. *PLoS One, 9*, e107678.

Inui, M., Manfrin, A., Mamidi, A., Martello, G., Morsut, L., Soligo, S., et al. (2011). USP15 is a deubiquitylating enzyme for receptor-activated SMADs. *Nature Cell Biology, 13*, 1368–1375.

Isono, E., Katsiarimpa, A., Müller, I. K., Anzenberger, F., Stierhof, Y.-D., Geldner, N., et al. (2010). The deubiquitinating enzyme AMSH3 is required for intracellular trafficking and vacuole biogenesis in *Arabidopsis thaliana*. *Plant Cell, 22*, 1826–1837.

Joo, J. H., Bae, Y. S., & Lee, J. S. (2001). Role of auxin-induced reactive oxygen species in root gravitropism. *Plant Physiology, 126*, 1055–1060.

Joo, J. H., Yoo, H. J., Hwang, I., Lee, J. S., Nam, K. H., & Bae, Y. S. (2005). Auxin-induced reactive oxygen species production requires the activation of phosphatidylinositol 3-kinase. *FEBS Letters, 579*, 1243–1248.

Kepinski, S., & Leyser, O. (2005). The Arabidopsis F-box protein TIR1 is an auxin receptor. *Nature, 435*, 446–451.

Kukavica, B., Mojović, M., Vuccinić, Ž., Maksimović, V., Takahama, U., & Jovanović, S. V. (2009). Generation of hydroxyl radical in isolated pea root cell wall, and the role of cell wall-bound peroxidase, Mn-SOD and phenolics in their production. *Plant and Cell Physiology, 50*, 304–317.

Kulathu, Y., Garcia, F. J., Mevissen, T. E. T., Busch, M., Arnaudo, N., Carroll, K. S., et al. (2013). Regulation of A20 and other OTU deubiquitinases by reversible oxidation. *Nature Communications, 4*, 1569.

Lamattina, L., García-Mata, C., Graziano, M., & Pagnussat, G. (2003). Nitric oxide: the versatility of an extensive signal molecule. *Annual Review of Plant Biology, 54*, 109–136.

Laspina, N. V., Groppa, M. D., Tomaro, M. L., & Benavides, M. P. (2005). Nitric oxide protects sunflower leaves against Cd-induced oxidative stress. *Plant Science, 169*, 323–330.

Laxalt, A. M., Beligni, M. V., & Lamattina, L. (1997). Nitric oxide preserves the level of chlorophyll in potato leaves infected by *Phytophthora infestans*. *European Journal of Plant Patholology, 103*, 643–651.

Lee, J.-G., Baek, K., Soetandyo, N., & Ye, Y. (2013). Reversible inactivation of deubiquitinases by reactive oxygen species in vitro and in cells. *Nature Communications, 4*, 1568.

Liszkay, A., van der Zalm, E., & Schopfer, P. (2004). Production of reactive oxygen intermediates (O2−, H_2O_2, and OH) by maize roots and their role in wall loosening and elongation growth. *Plant Physiology, 136*, 3114–3123.

Lombardo, M. C., Graziano, M., Polacco, J. C., & Lamattina, L. (2006). Nitric oxide functions as a positive regulator of root hair development. *Plant Signaling and Behavior, 1*, 28–33.

Lombardo, M. C., & Lamattina, L. (2012). Nitric oxide is essential for vesicle formation and trafficking in Arabidopsis root hair growth. *Journal of Experimental Botany, 63*, 4875–4885.

Maloney, G. S., DiNapoli, K. T., & Muday, G. K. (2014). The anthocyanin reduced tomato mutant demonstrates the role of flavonols in tomato lateral root and root hair development. *Plant physiology, 166*, 614–631.

Mittler, R., Vanderauwera, S., Gollery, M., & Van Breusegem, F. (2004). Reactive oxygen gene network of plants. *Trends in Plant Science, 9*, 490–498.

Mounier, E., Pervent, M., Ljung, K., Gojon, A., & Nacry, P. (2014). Auxin-mediated nitrate signalling by NRT1. 1 participates in the adaptive response of Arabidopsis root architecture to the spatial heterogeneity of nitrate availability. *Plant, Cell and Environment, 37*, 162–174.

Nathan, C., & Xie, Q. W. (1994). Regulation of biosynthesis of nitric oxide. *Journal of Biological Chemistry, 269*, 13725–13728.

Osmont, K. S., Sibout, R., & Hardtke, C. S. (2007). Hidden branches: developments in root system architecture. *Annual Review of Plant Biology, 58*, 93–113.

Pagnussat, G. C., Simontacchi, M., Puntarulo, S., & Lamattina, L. (2002). Nitric oxide is required for root organogenesis. *Plant Physiology, 129*, 954–956.

Peer, W. A., Cheng, Y., & Murphy, A. S. (2013). Evidence of oxidative attenuation of auxin signalling. *Journal of Experimental Botany, 64*, 2629–2639.

Pnueli, L., Liang, H., Rozenberg, M., & Mittler, R. (2003). Growth suppression, altered stomatal responses, and augmented induction of heat shock proteins in cytosolic ascorbate peroxidase (Apx1)-deficient Arabidopsis plants. *Plant Journal, 34*, 187–203.

Salehin, M., Bagchi, R., & Estelle, M. (2015). SCFTIR1/AFB-based auxin perception: mechanism and role in plant growth and development. *Plant Cell, 27*, 9–19.

Schmidt, R., & Schippers, J. H. M. (2015). ROS-mediated redox signaling during cell differentiation in plants. *Biochimica et Biophysica Acta, 1850*, 1497–1508.

Shi, Q., Ding, F., Wang, X., & Wei, M. (2007). Exogenous nitric oxide protect cucumber roots against oxidative stress induced by salt stress. *Plant Physiology and Biochemistry, 45*, 542–550.

Stamler, J. S., Singel, D. J., & Loscalzo, J. (1992). Biochemistry nitric oxide and redox-activated forms. *Science, 258*, 1898–1902.

Subczynski, W. K., Lomnicka, M., & Hyde, J. S. (1996). Permeability of nitric oxide through lipid bilayer membranes. *Free Radical Research, 24*, 343–349.

Tao, L., Cheung, A. Y., & Wu, H. (2002). Plant Rac-Like GTPases are activated by auxin and mediate auxin-responsive gene expression. *Plant Cell, 14*, 2745–2760.

Terrile, M. C., París, R., Calderón-Villalobos, L. I., Iglesias, M. J., Lamattina, L., Estelle, M., et al. (2012). Nitric oxide influences auxin signaling through S-nitrosylation of the Arabidopsis TRANSPORT INHIBITOR RESPONSE 1 auxin receptor. *Plant Journal, 70*, 492–500.

Tognetti, V. B., Mühlenbock, P. E. R., & Van Breusegem, F. (2012). Stress homeostasis–the redox and auxin perspective. *Plant, Cell and Environment, 35*, 321–333.

Tossi, V., Amenta, M., Lamattina, L., & Cassia, R. (2011). Nitric oxide enhances plant ultraviolet-B protection up-regulating gene expression of the phenylpropanoid biosynthetic pathway. *Plant, Cell and Environment, 34*, 909–921.

Vidal, E. A., Moyano, T. C., Riveras, E., Contreras-López, O., & Gutiérrez, R. (2013). Systems approaches map regulatory networks downstream of the auxin receptor AFB3 in the nitrate response of *Arabidopsis thaliana* roots. *Proceedings of the National Academy of Sciences of the United States ofAmerica, 110*, 12840–12845.

Xia, X.-J., Zhou, Y.-H., Shi, K., Zhou, J., Foyer, C. H., & Yu, J.-Q. (2015). Interplay between reactive oxygen species and hormones in the control of plant development and stress tolerance. *Journal of Experimental Botany, 66*, 2839–2856.

Yamaguchi, M., & Sharp, R. E. (2010). Complexity and coordination of root growth at low water potentials: recent advances from transcriptomic and proteomic analyses. *Plant, Cell and Environment, 33*, 590–603.

Yang, H., Mu, J., Chen, L., Feng, J., Hu, J., Li, L., et al. (2015). S-Nitrosylation positively regulates ascorbate peroxidase activity during plant stress responses. *Plant Physiology, 167*, 1604–1615.

Yu, H., Zhang, Y., Moss, B. L., Bargmann, B. O. R., Wang, R., Prigge, M., et al. (2015). Untethering the TIR1 auxin receptor from the SCF complex increases its stability and inhibits auxin response. *Nature Plants, 1*, 3.

Yun, B.-W., Feechan, A., Yin, M., Saidi, N. B. B., Le Bihan, T., Yu, M., et al. (2011). S-nitrosylation of NADPH oxidase regulates cell death in plant immunity. *Nature, 478*, 264–268.

CHAPTER FOUR

Control of Nitrogen Assimilation in Plants through *S*-nitrosothiols

Lucas Frungillo*, Steven H. Spoel§, Ione Salgado*,[1]
*Department of Plant Biology, Institute of Biology, University of Campinas, Campinas/SP, Brazil
§Institute of Molecular Plant Sciences, School of Biological Sciences, University of Edinburgh, Edinburgh, UK
[1]Corresponding author: E-mail: ionesalgado10@gmail.com

Contents

Abstract

The inorganic ion nitrate is the primary source of nitrogen for land plants, and the availability of this nutrient in the soil represents a bottleneck in crop yield. To assimilate nitrate, plants employ a variety of transporters and reductases expressed in different tissues and organs to transport and catalyse the sequential reduction of assimilates. Nitrate assimilation is a high-energy consuming process subject to tight metabolic control, which is not yet fully understood. Recently, nitrate assimilation was demonstrated to be regulated by a feedback mechanism involving the free radical nitric oxide (NO). NO primarily acts through covalent attachment to thiol groups of Cys residues, causing *S*-nitrosylation, a reversible post-translational protein modification. Previous evidence has indicated that *S*-nitrosylation feedback regulates nitrate transporters and reductases in a novel mechanism involving the production and scavenging of NO. In this review, we will discuss recent advances in the field of nitrate assimilation, focussing on the interplay between this process and NO-mediated redox signaling pathways in plants.

Advances in Botanical Research, Volume 77
ISSN 0065-2296
http://dx.doi.org/10.1016/bs.abr.2015.10.011

1. INTRODUCTION

Nitrogen (N) is an essential nutrient required for plant growth, development and yield. To meet nutritional needs, plants have evolved a highly specialized transport system for uptake of N in various forms available in soil. N can be acquired through the roots as inorganic ions (i.e. nitrate, nitrite, ammonium and dinitrogen) and organic molecules (i.e. urea and amino acids), and its availability largely differs depending on the region and type of soil (Crawford, 1995). Among the different N forms available to plants, nitrate (NO_3^-) is the most abundant source for annual crops (Ruffel, Gojon, & Lejay, 2014). Millions of tons of nitrate-containing fertilizers are applied annually as an agricultural practice to ensure adequate N supply to support plant productivity (Crawford & Glass, 1998). Despite enormous agricultural fertilization programs, crops are frequently challenged with N deprivation, reflecting the high mobility of nitrate ions leached from the soil, eventually leading to river eutrophication (Howarth & Marino, 2006). Thus, an understanding of how plants respond to and assimilate available nitrate is essential to increase N use efficiency, avoid detrimental environmental impacts and prevent economic losses.

Recently, crosstalk between the nitrate assimilatory pathway and nitric oxide (NO)-mediated redox signaling has been revealed in the model plant *Arabidopsis thaliana* (Frungillo, Skelly, Loake, Spoel, & Salgado, 2014). NO is a free radical that plays key roles in various physiological processes during plant growth, development and defence against environmental cues (Groß, Durner, & Gaupels, 2013; Salgado, Martínez, Oliveira, & Frungillo, 2013; Yu, Lamattina, Spoel, & Loake, 2014). The broad range of effects of NO or related molecules in plants primarily reflects the effects of this molecule on gene expression and post-translational regulation of proteins (Corpas et al., 2008; Lozano-Juste, Colom-Moreno, & León, 2011; Spoel, Tada, & Loake, 2010; Malik, Hussain, Yun, Spoel, & Loake, 2011; Yu et al., 2014). Despite its importance in plant biology, it remains unclear how NO homeostasis and signaling specificity are achieved.

In this review, we discuss how NO homeostasis might control the uptake and reduction of nitrate in plants. We also critically examine potential links between the NO-mediated control of nitrate assimilation and other metabolic processes that facilitate prompt responses to environmental and cellular fluctuations in N status and mediate adjustments in growth and development accordingly.

2. NITRATE UPTAKE AND TRANSPORT

To achieve N homeostasis and sustain development and growth, land plants are equipped with a complex apparatus to scavenge nitrate from the soil (reviewed by Krapp et al., 2014; Ruffel et al., 2014). Four gene families in *A. thaliana* encode nitrate transporters: nitrate transporter 1/peptide transporter (NRT1/PTR), nitrate transporter 2 (NRT2), chloride channels (CLC) and slow chloride channel 1 homologues (SLAC1/SLAH). Most members of the NRT1/PTR family were initially named according to the first identified substrate. However, several members of this group transport more than one substrate, and the analysis of sequence homologies showed no correlation with substrate selectivity (reviewed by Krapp et al., 2014). Recent efforts to develop a practical, straightforward and unified nomenclature for NRT1/PTR proteins have suggested a new acronym, NPF (NRT1 PTR Family), to designate this group of plant proteins (Léran et al., 2014). Biochemical and phylogenetic analyses have also led to the identification of eight different clades within the NPF family (NPF1 to NPF8) (Léran et al., 2014). Thus, for the sake of clarity, when citing a member of the NPF family (the new proposed nomenclature), the respective former name (NRT) is also provided. Currently, among the 73 genes grouped in these gene families, 24 genes have been characterized and related to nitrate transport, comprising influx and/or efflux throughout different cellular compartments in plants (Krapp et al., 2014). Increased attention has been directed to NPF6.3/NRT1.1 and NRT2.1, as these proteins play key roles in nitrate influx and signaling in the roots.

Nitrate is actively taken up through the roots via specific transporter systems differentially recruited, depending on the ion availability in soil (Tsay, Chiu, Tsai, Ho, & Hsu, 2007; Wang, Hsu, & Tsay, 2012). The concentration of nitrate in the soil largely varies from less than 1−70 mM (Crawford, 1995). Hence, to ensure adequate N acquisition in plants, the nitrate transport system in roots has evolved to cope with large variations in nitrate availability by adjusting its sensitivity to nitrate (Dechorgnat et al., 2011; Tsay et al., 2007; Wang et al., 2012). Nitrate uptake through the roots is based on the activity of the high-affinity transport system (HATS, *Km* in the micromolar range) and low-affinity transport system (LATS, *Km* in the millimolar range) (Wang et al., 2012) (Figure 1). For instance, at high availability, nitrate uptake is performed through the LATS, in which the main effector is the product of *NPF6.3/NRT1.1* gene expression (Ho, Lin,

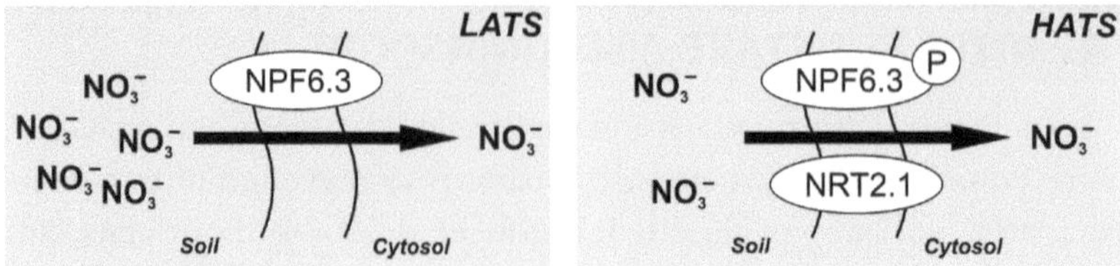

Figure 1 Systems of nitrate (NO_3^-) transport in plant roots. The recruitment of transporters and post-translational state might differ depending on nitrate availability in the soil. Low-affinity transport system (LATS): under nitrate abundance, nitrate transport primarily relies on NPF6.3/NRT1.1 activity (left panel). High-affinity transport system (HATS): under nitrate shortage, NPF6.3/NRT1.1 exhibits increased affinity to nitrate through phosphorylation (P) at Tyr101, and the pure high-affinity NRT2.1 is recruited (right panel).

Hu, & Tsay, 2009; Liu, Huang, & Tsay, 1999). In contrast, low nitrate availability induces the HATS by activating *NRT2.1* expression and phosphorylating the NPF6.3/NRT1.1 transporter, thereby increasing its affinity to nitrate (Tsay et al., 2007; Wang et al., 2012). Thus, while NRT2.1 is a pure high-affinity transporter, NPF6.3/NRT1.1 is a dual-affinity transporter, involved in nitrate uptake through both LATS and HATS, depending on its post-translational state.

Recently, two crystallographic studies have shed light on how phosphorylation at the Thr101 residue of NPF6.3/NRT1.1 alters nitrate transport (Parker & Newstead, 2014; Sun et al., 2014). Both studies examined the inward-facing conformational state of the NPF6.3/NRT1.1 protein, comprising 12 transmembrane helices. Together, these studies have indicated that the phosphorylation of NPF6.3/NRT1.1 induces a dimer-to-monomer switch, thereby increasing protein flexibility and ultimately, reducing the *Km* to nitrate (Parker & Newstead, 2014; Sun et al., 2014). However, it remains unknown whether this conformational modification increases the binding affinity for nitrate or merely improves transport capacity through accelerating nitrate shuttling.

Several lines of evidence have indicated a role for NPF6.3/NRT1.1 in nitrate assimilation far beyond the acquisition of this nutrient through the roots. It has been suggested that the post-translational modification of NPF6.3/NRT1.1 not only controls transport capacity, but is also critical for the plant response to nitrate availability (Ho et al., 2009). Plants rapidly trigger expression of nitrate assimilatory pathway genes in response to nitrate supply, a process called the nitrate primary response (NPR) (Ho et al., 2009; Krouk et al., 2010; Wang, Xing, Wang, Tran, & Crawford, 2009). Notably,

NPR is observed in mutants impaired in the initial reductive steps of nitrate assimilation, indicating that the signal to this response is nitrate itself (Ruffel et al., 2011; Wang et al., 2004; Wilkinson & Crawford, 1993). NPR is characterized by two distinct levels of gene induction, depending on the nitrate concentration: at nitrate availabilities higher than 1 mM, the induction of nitrate-responsive genes was approximately twice as strong as induction at nitrate levels below the millimolar range (Ho et al., 2009; Hu, Wang, & Tsay, 2009). The *npf6.3/nrt1.1* mutant exhibits trace expression of *NRT2.1*, a HATS component marker gene for NPR. Interestingly, replacing Thr101 with an Asp, thereby mimicking phosphorylation, suppressed upregulation of the *NPF6.3/NRT1.1* gene at nitrate availabilities above millimolar concentrations. Conversely, replacing the Thr101 residue with an Ala induced a stronger response to nitrate, regardless of nitrate concentration (Ho et al., 2009). Taken together, these data suggest that rather than simply acting as a transporter, NPF6.3/NRT1.1 functions as both a receptor and transporter, or transceptor, to sense nitrate availability and coordinate plant responses through critical phosphorylative events. In accordance with the observation that NPR is triggered as rapidly as 3 min post application of nitrate, the transport activity of NPF6.3/NRT1.1 can be decoupled from its sensing activity (Ho et al., 2009), suggesting that NPF6.3/NRT1.1 is responsible for signaling nitrate availability.

NPF7.3/NRT1.5 and NPF7.2/NRT1.8, members of the NPF7 subfamily, have been implicated in nitrate loading in the xylem through the control of the efflux and influx of nitrate, respectively, from xylem vessels (Chen, Lv, Li, Yi, & Gong, 2012; Li et al., 2010; Lin et al., 2008). The observation that *NPF6.3/NRT1.1* is co-expressed with *NPF7.3/NRT1.5* and *NPF7.2/NRT1.8* in mature parts of the roots, including the endodermis and stele (Huang et al., 2004; Remans et al., 2006), suggested a role for NPF6.3/NRT1.1 in nitrate translocation to the shoots (Léran et al., 2013). Accordingly, phenotypic analysis indicated that *npf6.3/nrt1.1* mutants display normal nitrate uptake through the roots (Muños et al., 2004), but exhibited an unexpected delay in nitrate distribution to the shoots (Léran et al., 2013). Furthermore, after loading *Xenopus* oocytes expressing *AtNPF6.3/NRT1.1* with ^{15}N-labelled nitrate and subsequently measuring the appearance of ^{15}N-labelled nitrate in the incubation buffer, Léran et al. (2013) demonstrated that NPF6.3/NRT1.1 mediates nitrate efflux, even in the absence of a favourable gradient. These results suggest that the NPF6.3/NRT1.1 transporter functions synergistically with NPF7.2/NRT1.8 for nitrate translocation through the plant. However, the molecular mechanism determining whether NPF6.3/NRT1.1 acts in the influx or efflux of nitrate, and the role of this bidirectional transport in root nitrate uptake remains unknown.

In addition to a role in transporting and sensing nitrate, NPF6.3/NRT1.1 has also been implicated in shaping root architecture (Remans et al., 2006; Krouk et al., 2010). Formation of lateral roots from the primary root is an important mechanism through which plants forage for water and nutrients. Lateral roots are initiated as mitotically active cells in the pericycle of primary roots that protrude through the epidermis after a few days. Lateral root formation in nitrate-rich patches is strikingly associated with lateral root elongation. The MADS box gene *ANR1* encodes a transcription factor and is required for lateral root elongation in Arabidopsis (Remans et al., 2006). Interestingly, *npf6.3/nrt1.1* mutant plants displayed reduced *ANR1* levels and, accordingly, reduced root colonization, a phenotype that restrains plant growth. Additionally, NPF6.3/NRT1.1 transports the hormone auxin from the developing lateral root, thereby negatively impacting lateral root elongation by reducing the auxin concentration (Krouk et al., 2010). Collectively, these findings clearly indicate that NPF6.3/NRT1.1 is a key element in orchestrating biochemical and morphological responses in plants through nitrate transport and sensing.

3. NITRATE ASSIMILATION

Incorporation of N atoms from inorganic ion nitrate into the carbon skeleton to form N-containing organic molecules, such as amino acids, proteins and nucleotides, is one of the most energy-consuming biochemical pathways in nature (Crawford, 1995). Initiated through N scavenging from the soil, the nitrate assimilatory pathway comprises several redox reactions that together consume 12 ATPs per N atom assimilated (Bloom, Sukrapanna, & Warner, 1992). In plants, once taken up through the root, nitrate is transported to the leaves where this molecule is stored in vacuoles or effectively assimilated into organic compounds. The first reductive step in nitrate assimilation is its reduction to nitrite (NO_2^-) through the activity of the cytosolic enzyme nitrate reductase (NR). This reaction involves the transfer of two electrons donated from NADPH or NADH, depending on the NR isoform (Warner & Kleinhofs, 1992). In Arabidopsis, NR is encoded by two structural genes, *NIA1* and *NIA2*, of which NIA2 accounts for the majority of NR activity in shoots (Wilkinson & Crawford, 1993).

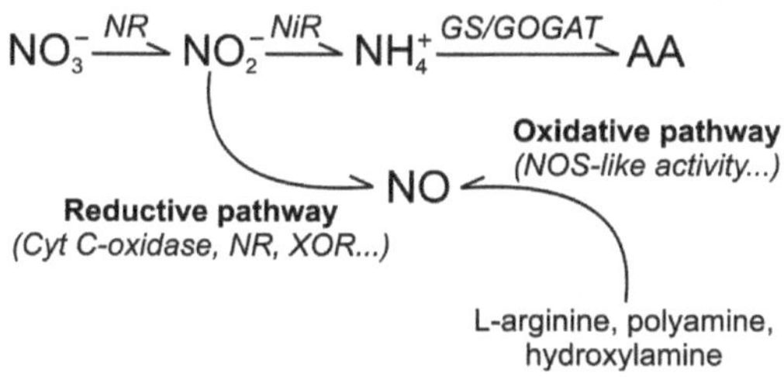

Figure 2 Proposed links between nitrate (NO_3^-) assimilation and nitric oxide (NO) production in plants. As depicted in the schematic representation, the substrates for either the reductive or oxidative pathways in NO production are directly or indirectly provided through nitrate assimilation. NO_2^-, nitrite; NH_4^+, ammonium; Cyt C-oxidase, mitochondrial inner membrane-bounded cytochrome c oxidase; NR, nitrate reductase; XOR, peroxisomal xanthine oxidoreductase; NiR, nitrite reductase; GS/GOGAT, glutamine synthetase/glutamine oxoglutarate aminotransferase; AA, amino acids; NOS, nitric oxide synthase.

Due to high reactivity and potential toxicity, nitrite is promptly removed from the cells (Zheng, Wisedchaisri, & Gonen, 2013) or transported to chloroplasts in the leaves or plastids in the roots for further reduction. Nitrite is then reduced to ammonium through nitrite reductase at the expense of six electrons donated from reduced ferredoxin (Fd) (Crawford, 1995). The final process in nitrate assimilation is the conversion of ammonia into amino acids in chloroplasts and plastids through a two-step pathway (Figure 2). First, glutamine synthetase incorporates ammonia into glutamate, forming glutamine. Sequentially, glutamine oxoglutarate aminotransferase transfers the amide group of glutamine to the organic acid oxoglutarate to yield two glutamate molecules. After this step, several transaminases mediate the production of other amino acids through transamination reactions (Lea, 1993).

Due to the requirement of the C skeleton to incorporate the N atom from nitrate into organic molecules, it is likely that nitrate assimilation is closely associated with photosynthetic processes in plants. Indeed, several lines of evidence have supported the idea that plant C status is associated with nitrate assimilation through a stimulatory effect in nitrate uptake, at least in the short term (Ruffel et al., 2014). For example, while examining non-nodulated soya bean plants (*Glycine max*), Delhon, Gojon, Tillard, and Passama (1996) showed that the diurnal variation of nitrate transporter expression was associated with plant C status. Interestingly, feeding experiments with glucose and sucrose as a source of C were effective in stimulating nitrate transporter expression and nitrate uptake in soya beans and Arabidopsis plants (Delhon et al., 1996; Kiba et al., 2012; Lejay et al., 1999, 2003).

Thus, it has been suggested that C is initially assimilated in the leaves and subsequently transported to roots where it stimulates nitrate uptake (Delhon et al., 1996; Lejay et al., 1999). In addition to physiological interplay, photosynthesis and nitrate assimilation are also associated at the biochemical level. Maintenance of the pool of reduced Fd, necessary for the nitrite to ammonia reduction, is achieved through chloroplastic electron transport in photosystem I during photosynthesis in the leaves and the oxidative pentose phosphate pathway (OPPP) in the roots (Crawford, 1995). Conversely, nitrate supply and sensing regulates the expression of OPPP-related genes (Wang, Okamoto, Xing, & Crawford, 2003; Wang et al., 2004, 2009), indicating that nitrate assimilation also impacts C metabolism. Taken together, these data indicate a complex C:N regulatory network that profoundly impacts plant metabolism. Although the underlying molecular mechanisms are not yet fully understood, fluctuations in the rate of photosynthesis or nitrate assimilation might be sensed by each other.

4. LINKS BETWEEN NITRATE ASSIMILATION AND NITRIC OXIDE FORMATION

According to the role of NR in the nitrate assimilation pathway, NR-double-deficient (*nia1nia2*) *A. thaliana* plants are unable to reduce nitrate to nitrite and consequently show a drastic reduction in leaf amino acid levels, reduced biomass and typical pale leaves (Frungillo et al., 2014; Oliveira, Justino, Sodek, & Salgado, 2009; Wilkinson & Crawford, 1993). Additionally, *nia1nia2* plants display severe disease susceptibility when challenged with the avirulent strain *Pseudomonas syringae* pv. *maculicola* (Modolo et al., 2006; Oliveira et al., 2009). Interestingly, feeding experiments with the end products of the N assimilation pathway, such as L-arginine or L-glutamine, effectively restored the *nia1nia2* leaf amino acid content to levels compatible with wild-type (WT), but failed to rescue disease resistance (Oliveira et al., 2009). Still, *nia1nia2* mutants are defective in seed germination, seedling establishment, seedling development, secondary metabolite synthesis, drought stress adaptation and floral induction (Lozano-Juste & León, 2010; Santos-Filho et al., 2012; Seligman, Saviani, Oliveira, Pinto-Maglio, & Salgado, 2008). These data indicate that in addition to a role in N homeostasis, the nitrate assimilation pathway is required for proper development and responses to biotic and abiotic stimuli.

Several studies have demonstrated that *nia1nia2* mutant plants show reduced NO production and emission (Frungillo et al., 2014; Lozano-Juste &

León, 2010; Modolo, Augusto, Almeida, Magalhaes, & Salgado, 2005; Modolo et al., 2006; Oliveira et al., 2009), indicating an association between nitrate assimilation and NO production. The active redox molecule NO is a free radical with signaling action in all living organisms. Particularly in plants, NO has been implicated in growth and development and responses to biotic and abiotic cues (reviewed by Salgado et al., 2013). Despite the relevance of NO signaling in plant biology, the synthesis of NO remains a matter of debate (Figure 2). In mammals, NO is synthesized through a family of nitric oxide synthase (NOS) enzymes that catalyse oxidation of the guanidine nitrogen of L-arginine to produce L-citrulline and NO in a reaction dependent on molecular oxygen (O_2) and NADPH (Alderton, Cooper, & Knowles, 2001). Several lines of evidence have demonstrated NOS activity in plants. For example, NO production, estimated through the oxidation of L-arginine to L-citrulline in plant extracts, has been reported (Delledonne, Xia, Dixon, & Lamb, 1998; Durner, Wendehenne, & Klessig, 1998; Gupta, Fernie, Kaiser, & van Dongen, 2011; Modolo, Cunha, Braga, & Salgado, 2002). Importantly, human NOS inhibitors suppress NO production through L-arginine-dependent NOS activity in several plant species (Corpas et al., 2006; Delledonne, Zeier, Marocco, & Lamb, 2001; Durner et al., 1998). Using a genomic approach, Foresi et al. (2010) identified an enzyme with 45% similarity to human NOS in the unicellular green algae *Ostreococcus tauri*. Molecular characterization revealed that OtNOS shares structural and kinetics similarities to human NOS. *Escherichia coli* transformed with recombinant OtNOS displayed increased NO production in response to L-arginine treatment and oxidative challenge (Foresi et al., 2010). Intriguingly, however, genomic analyses have not identified homologues of mammalian NOS or OtNOS in higher plants. Instead, it has been suggested that NO synthesis in plants occurs through polyamine and hydroxylamine oxidation (Arasimowicz-Jelonek, Floryszak-Wieczorek, & Kubiś, 2009; Groppa, Rosales, Iannone, & Benavides, 2008; Rümer, Kapuganti, & Kaiser, 2009). Despite evidence showing that NO production through these sources is physiologically relevant, the precise mechanism underlying these pathways has not been resolved (Figure 2).

In addition to the above oxidative routes, NO can be synthesized through at least four different nitrite reductive pathways (Figure 2). In tobacco (*Nicotiana tabacum*) roots, nitrite was reduced to NO through the activity of the membrane-bound nitrite:NO reductase (Ni:NOR). Ni:NOR-mediated NO production plays a role in mycorrhizal fungus interactions in a nitrate-dependent manner (Moche et al., 2010).

At the cellular level, under low O_2 tensions, nitrite accumulates through NR activation and the partial inhibition of plastidic nitrite reduction (Botrel, Magne, & Kaiser, 1996; Kaiser & Förster, 1989). Under these conditions, nitrite can be reduced to NO through the activity of peroxisomal xanthine oxidoreductase (XOR), mitochondrial inner membrane-bounded cytochrome *c* oxidase (Cyt C-oxidase) and cytosolic NR. The nitrite to NO reduction via XOR occurs at the expense of NADH or xanthine as reducing agents (Figure 2), potentially representing the local interaction of NO with reactive oxygen intermediates (ROIs) (Cantu-Medellin & Kelley, 2013; Godber et al., 2000).

Oxygen, the final electron acceptor in the mitochondrial respiratory chain, can be partially reduced through electron leakage, resulting in the generation of the superoxide anion (O_2^-) and representing a significant mechanism of ROI production in mitochondria, particularly under oxygen shortage (Schmoldt, Benthe, & Haberland, 1975). Interestingly, under hypoxia, nitrite acts as an alternative electron acceptor in the respiratory chain in the mitochondria of mammals (Castello, David, McClure, Crook, & Poyton, 2006; Kozlov, Staniek, & Nohl, 1999) and plants (Planchet, Jagadis Gupta, Sonoda, & Kaiser, 2005; Wulff, Oliveira, Saviani, & Salgado, 2009). The mitochondrial reduction of nitrite to NO alleviates the stress induced under low oxygen tension via the flow of electrons through the mitochondrial respiratory chain and the maintenance of needful ATP generation (Oliveira, Salgado, & Sodek, 2013; Planchet et al., 2005). Moreover, reduction of nitrite to NO through mitochondrial electron transport was significant during the incompatible interaction of *A. thaliana* with *P. syringae* (Modolo et al., 2005).

In addition to a major role in nitrate assimilation, NR has also been implicated in the reduction of nitrite to NO (Rockel, Strube, Rockel, Wildt, & Kaiser, 2002; Yamasaki & Sakihama, 2000) (Figure 2). However, the reduction of nitrite to NO through NR catalysis is dependent on high concentrations of nitrite and low oxygen tensions (Planchet et al., 2005; Rockel et al., 2002) and this mechanism might be physiologically relevant only under specific conditions. In presence of nitrite, the rate of in vitro NO production in leaf homogenates of *A. thaliana* plants defective in the two structurally related NR genes (*NIA1* and *NIA2*) was similar to that in WT plants (Modolo et al., 2005). Importantly, this NR-independent reduction of nitrite to NO was abolished using inhibitors of mitochondrial respiration, suggesting that NO production might be derived from electron leakage in the mitochondrial respiratory chain (Modolo et al., 2005). The production of NO in the nitrogen-fixing nodules of *Medicago truncatula* might also result from a two-step mechanism involving NR followed by mitochondrial electron transport (Horchani et al., 2011). Thus, in contrast to direct involvement in nitrite-to-NO catalysis, the primary role of NR in NO homeostasis is the production of nitrite, i.e., providing the substrate for NO production (Modolo et al., 2005; Salgado et al., 2013).

5. REDOX SIGNALING BY NO THROUGH PROTEIN MODIFICATION

NO induces *S*-nitrosylation, a redox-based, post-translational modification that alters protein behaviour in a wide range of situations (reviewed by Groβ et al., 2013; Spoel et al., 2010; Salgado et al., 2013; Yu et al., 2014). Effective post-translational redox signaling relies on the specific and reversible nature of these redox-based modifications. Based on these features, *S*-nitrosylation is pivotal for the molecular transfer of NO bioactivity (Figure 3). *S*-nitrosylation involves the addition of an NO moiety to specific biologically active thiol groups of cysteine residues, forming protein *S*-nitrosothiols (protein-SNO) (Besson-Bard, Pugin, & Wendehenne, 2008). *S*-nitrosylation alters the localization, activity and function of a wide variety of proteins in plants (Spoel & Loake, 2011; Salgado et al., 2013; Yu et al., 2014). Reaction of NO with Cys residues is dependent on O_2 and might be limited to aerated and hydrophobic microenvironments (Besson-Bard et al., 2008; Liu, Miller, Joshi, Thomas, & Lancaster, 1998). Alternatively,

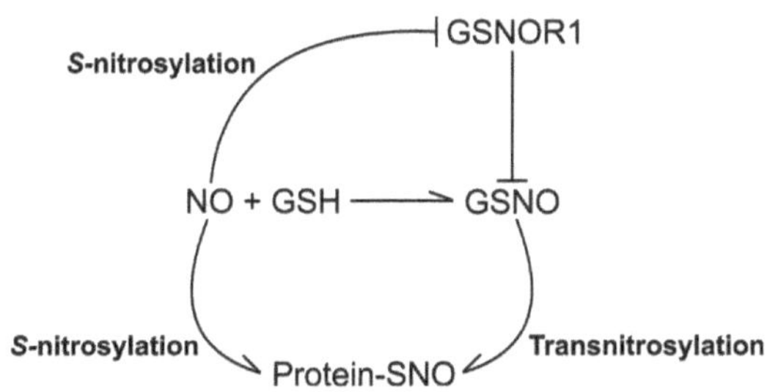

Figure 3 Kinetics of nitric oxide (NO) signaling through *S*-nitrosylation. In cellular systems, NO might react with glutathione (GSH) to form *S*-nitrosoglutathione (GSNO). GSNO levels are controlled through the enzymatic activity of GSNO reductase 1 (GSNOR1). NO regulates its own bioavailability through the control of GSNOR1 activity through inhibitory *S*-nitrosylation. Both NO and GSNO form *S*-nitrosothiols (protein-SNO), and although they overlap, recent studies have indicated that NO and GSNO target different sets of proteins for *S*-nitrosylation.

NO reacts with glutathione (GSH), a major antioxidant in cells, forming *S*-nitrosoglutathione (GSNO) (Liu et al., 2001). GSNO acts as an NO carrier, increasing the half-life of this compound in biological systems (Lindermayr, Saalbach, & Durner, 2005; Stamler, Singel, & Loscalzo, 1992), and has been implicated in NO signaling through the formation of protein-SNO. By acting as an *S*-nitrosylation agent, GSNO transfers an NO moiety to thiol groups in proteins through *trans*-nitrosylation (Marino & Gladyshev, 2010) (Figure 3).

Important advances in the field of NO-mediated redox signaling have been achieved through the recognition of enzymatic control of protein-SNO. The intracellular level of GSNO is controlled through the evolutionary conserved cytosolic enzyme *S*-nitrosoglutathione reductase (GSNOR1) (Feechan et al., 2005; Liu et al., 2001). GSNOR1 primarily reduces GSNO to oxidized glutathione and NH_4^+, thereby reducing the likelihood of protein *S*-nitrosylation (Chen et al., 2009; Frungillo et al., 2013; Lee, Wie, Fernandez, Feelisch, & Vierling, 2008; Liu et al., 2001) (Figure 3). *GSNOR1*-deficient plants exhibit increased global levels of protein-SNO, which negatively correlate with pathogen resistance (Feechan et al., 2005; Kneeshaw, Gelineau, Tada, Loake, & Spoel, 2014; Yun et al., 2011), herbicide resistance (Chen et al., 2009) and heat acclimation (Lee et al., 2008), revealing the biological relevance of GSNOR1 in controlling plant responses to environmental cues. Recently, new layers of complexity have been demonstrated in NO signaling in plants. While GSNOR1 globally reduces protein-SNO formation through a reduction of the intracellular GSNO pool, characterization of the thioredoxin (TRX) system suggests a function for these oxidoreductases in the direct control of protein-SNO. Pharmacological and genetic inhibition of the TRX system resulted in marked cellular accumulation of protein-SNO (Correa-Aragunde, Cejudo, & Lamattina, 2015; Kneeshaw et al., 2014). Moreover, the pathogen-inducible TRX*h*5 enzyme acted as a direct protein-SNO reductase in vitro. Subsequent genetic experiments with different (S)NO mutants revealed that TRX*h*5 selectively discriminates between protein-SNO substrates in vivo, representing the first demonstration of specificity in the reversion of protein-SNO.

In addition to a role in protein *S*-nitrosylation, NO reacts with O_2^- through rate-limiting diffusion to yield the potent oxidant peroxynitrite ($ONOO^-$). In biological systems, $ONOO^-$ permanently reacts with Tyr residues in proteins to form nitrotyrosine, a process referred to as Tyr-nitration (Radi, 2004). Tyr-nitration plays a crucial role in hypersensitive responses and abiotic stress responses (Chaki et al., 2011; Romero-Puertas et al., 2007). In vitro and in vivo evidence has indicated that the nitrotyrosine level is controlled through the $ONOO^-$ detoxification activity of peroxiredoxin II E (PrxII E) in plant cells (Romero-Puertas et al., 2007). Interestingly, PrxII E has been identified as an *S*-nitrosylated target. *S*-nitrosylation of PrxII E in plants challenged with avirulent pathogens inhibited the $ONOO^-$ detoxifying activity of this enzyme, leading to a marked increase in nitrotyrosine content and resulting in nitrosative stress (Romero-Puertas et al., 2007).

6. THE ROLE OF NO IN NITRATE ASSIMILATION PATHWAYS

Despite increasing knowledge concerning the multiple pathways involved in NO homeostasis, coordination among NO synthesis and scavenging has only recently been revealed. Frungillo et al. (2014) have proposed that NO controls self-generation and scavenging through the control of nitrate assimilation pathways and GSNOR1 activity (Figures 3 and 4).

Previous genetic and biochemical analysis of nitrate uptake in the roots of *A. thaliana* plants demonstrated that the high-affinity transporter NRT2.1, a marker component of HATS, is upregulated at the transcriptional level under nitrate starvation (Lejay et al., 1999). As discussed earlier, the transcriptional regulation of NRT2.1 in response to nitrate availability enables plants to circumvent shortages in nitrate supply and ensure adequate nitrate uptake. Remarkably, the NR-double mutant plant, *nia1nia2*, exhibited highly elevated *NRT2.1* expression even when nitrate was adequately supplied (Lejay et al., 1999). Although these observations strongly suggest feedback repression of nitrate uptake (Lejay et al., 1999; Muños et al., 2004), the

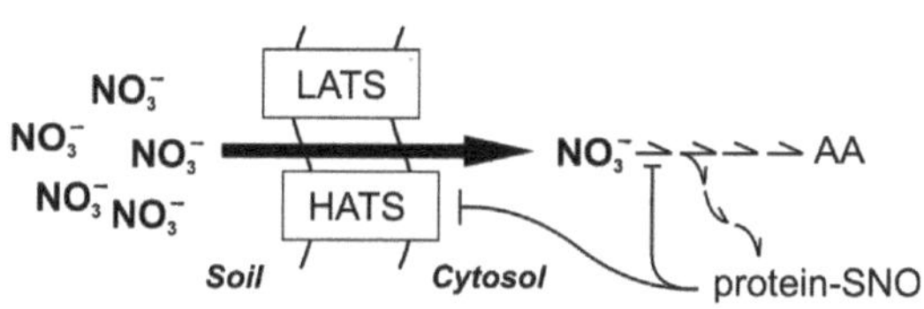

Figure 4 Proposed control of nitrate (NO_3^-) assimilation through *S*-nitrosothiols (protein-SNO) in plants. Protein-SNO is produced as a consequence of nitrate assimilation. Evidence has indicated that protein-SNO feedback regulates nitrate assimilation through the inhibition of transporters and reductases. AA, amino acids; LATS, low-affinity transport system; HATS, high-affinity transport system.

identity of the metabolite implicated in the control of nitrate assimilation and the molecular mechanism remained largely unknown. Recently, it was proposed that NO fine-tunes both uptake and reduction of nitrate by controlling its own bioavailability (Frungillo et al., 2014). Gene expression analysis in roots of *nitric oxide overproducer1* (*nox1*) and *gsnor1* mutants revealed that *NRT2.1* expression is repressed in these genotypes compared to WT plants. Consistently, the pharmacological treatment of WT roots with GSNO or the NO donor, diethylamine NONOate (DEA/NO), was effective in suppressing *NRT2.1* gene expression. Importantly, whereas *NRT2.1* expression was affected by NO and GSNO, the expression of NPF6.3/*NRT1.1* remained unchanged (Frungillo et al., 2014). These observations suggested a switch from high- to low-affinity nitrate transport modulated by NO/GSNO, which might impact nitrate uptake through the roots (Frungillo et al., 2014) (Figure 4). Considering that NPF6.3/NRT1.1 plays a key role in sensing nitrate availability and controlling the switch between LATS and HATS, it is tempting to speculate that NO signaling is involved in the nitrate-sensing activity of NPF6.3/NRT1.1. Still, in addition to transcriptional control, recent studies have suggested that NRT2.1 is also subjected to post-translational modification (Laugier et al., 2012). However, the molecular mechanism underlying the control of NRT2.1 activity remains unclear. Thus, it could be fruitful to determine whether NO or related molecules exert a role in directly controlling the activity of NRTs.

The analysis of in vitro enzymatic activity revealed reduced NR activity in *gsnor1* mutant plants compared with WT plants (Frungillo et al., 2014). Conversely, plants with presumably reduced protein-SNO levels resulting from the overexpression of *GSNOR1* under the constitutive 35S promoter showed increased NR activity. Thus, SNO has been suggested to negatively regulate nitrate uptake and reduction through a feedback mechanism and consequently impair plant growth (Figure 4). Accordingly, while the leaf area and biomass accumulation were markedly decreased in *gsnor1* plants, plant vigour increased in *GSNOR1*-overexpressing plant lines. Remarkably, when *gsnor1* plants were fed with L-glutamine, the primary end product of nitrate assimilation, plant vigour was rescued to WT levels, suggesting that the (S)NO-mediated feedback mechanism of nitrate assimilation might significantly undermine plant growth. Collectively, these data indicate that the nitrate assimilation pathway is regulated through NO signaling (Frungillo et al., 2014). However, how appropriate adjustments in nitrate assimilation are achieved was only revealed in studies of GSNOR1 activity.

In feeding experiments with controlled nitrate availability, GSNOR1 activity was inversely associated with nitrate supply, indicating crosstalk between nitrate assimilation and NO signaling (Frungillo et al., 2014). Previous studies have suggested that plant GSNOR1 might be the target of post-translational modification, which could impact the activity of this enzyme (Lee et al., 2008) (Figure 3). Indeed, measurements of in vitro enzymatic activity in the presence of different NO donors and by products of the nitrate assimilation pathway indicated that GSNOR1 could be directly inhibited by NO. Application of the well-established biotin switch technique on plants transformed with epitope-tagged 35S::FLAG-GSNOR1 in WT and *nox1* backgrounds showed that GSNOR1 is the target of inhibitory *S*-nitrosylation in vitro and in vivo, thereby preventing GSNO catabolism (Frungillo et al., 2014).

Studies indicating that GSNOR1 activity is directly inhibited through NO-mediated post-translational modification are intriguing, as intuitively an increase in the denitrosylating activity of GSNOR1 would be expected during NO production. This apparent discrepancy can be explained by the fact that the redox-active molecules NO and GSNO control partially overlapping, yet significantly different subsets of protein-SNO (Forrester, Foster, Benhar, & Stamler, 2009; Frungillo et al., 2014; Kneeshaw et al., 2014). Indeed, the observation that GSNOR1 is a target of *S*-nitrosylation is an elucidative example of disparate subsets of protein-SNO.

Inhibitory *S*-nitrosylation of GSNOR1 might represent a molecular mechanism through which NO bioavailability is controlled and nitrate assimilation is adjusted according to N demand (Frungillo et al., 2014). Additionally, GSNOR1 activity could be slightly inhibited through the in vitro addition of high levels of $ONOO^-$ (Frungillo et al., 2014), suggesting that different NO-mediated mechanisms might control GSNOR1 activity at the post-translational level. However, additional studies are needed to determine the biological relevance of this process.

Evidence suggests that NR activity might also be subjected to NO-mediated redox control at the post-transcriptional level. NR is a homodimer cytoplasmic enzyme involved in different regulatory strategies (reviewed by Campbell, 1999; Xiong, Fu, Yang, Zhu, & Tao, 2012). Although variations in NR mRNA levels have been reported, particularly under stress situations (Lea, Leydecker, Quilleré, Meyer, & Lillo, 2006), NR has long been known to be regulated through reversible phosphorylation (Crawford, 1995). The recruitment of 14-3-3 proteins to phosphorylated Ser534 of NIA2 inhibits NR activity and promotes protein degradation (Kaiser & Huber,

2001; Moorhead et al., 1996; Su, Huber, & Crawford, 1996). However, phosphorylation at Ser627 through the mitogen-activated protein kinase 6 (MPK6) stimulates NIA2 activity. This mechanism for the regulation of NR activity plays a role in NO production under oxidative stress (Wang, Du, Li, Ren, & Song, 2010). Additionally, *AtSIZ1* positively regulates Arabidopsis NR through small ubiquitin-related modifier (SUMO) proteins via its E3 SUMO ligase activity. Consistent with NR activation through *AtSIZ1*, mutant *siz1-2* plants displayed reduced NO production and a dwarf phenotype (Park, Song, & Seo, 2011). Furthermore, evidence suggests that NR activity might also be subject to NO-mediated redox control at the post-transcriptional level. Intriguingly, this NO-mediated redox control of NR activity might be associated with nitrate supply. The roots of tomato plants (*Solanum lycocarpum*) exposed to different NO donors in nutrient solution containing high nitrate levels (5 mM) showed marked inhibition of NR activity. Conversely, NO stimulated NR activity in plants fed with low nitrate (0.5 mM), and this effect was reversed after NO removal from the medium (Jin, Du, Zhang, Lin, & Tang, 2009). These results suggest that NR is the target of labile NO-mediated modification, such as *S*-nitrosylation. Thus, it might be interesting to investigate whether NR is indeed *S*-nitrosylated and if so, it might be worthwhile to determine the potential crosstalk between different NR post-translational modifications.

NO might also exert indirect effects on N assimilation through metabolically interconnected pathways. For example, sunflower (*Helianthus annuus* L.) plants subjected to high temperature stress showed downregulated *GSNOR1* gene expression and activity, finely correlated with an increase in protein-SNO and Tyr-nitration levels (Chaki et al., 2011). Chaki et al. (2011) proposed that protein-SNO act as NO reservoirs during heat stress to mediate the generation of $ONOO^-$. Nitroproteome analysis identified 22 proteins as targets of Tyr-nitration under temperature stress, including ferredoxin-NADP oxidoreductase (FNR). FNR catalyses the electron transfer from reduced Fd to $NADP^+$ during the final step of photosynthesis (Carrillo & Ceccarelli, 2003; Chitnis, 2001). In vitro assays indicated that FNR activity was inhibited through the $ONOO^-$ donor SIN-1, suggesting a role for NO derivatives in controlling photosynthesis (Chaki et al., 2011). Together with the fact that in vitro GSNOR1 activity is also inhibited in the presence of $ONOO^-$ (Frungillo et al., 2014) and photosynthesis is closely associated with N assimilation (see below), it is tempting to speculate that $ONOO^-$-mediated redox signaling impacts N assimilation through the post-transcriptional control of FNR and GSNOR1 activities.

One of the well-established targets of NO is the enzyme aconitase. Aconitase is involved in the stereoisomerization of citrate to isocitrate in the cytosol and mitochondrial matrix. In both animals and plants, aconitase is inhibited through NO in a reversible manner (Gardner, Costantino, Szabó, & Salzman, 1997; Navarre, Wendehenne, Durner, Noad, & Klessig, 2000). In addition to the involvement of this enzyme in the citric acid cycle and cellular energy metabolism, the regulation of aconitase activity might also be a key for the provision of the C skeleton to amino acid biosynthesis (Igamberdiev & Gardeström, 2003). The roots of Arabidopsis plants under hypoxia show NO production through an NR-dependent pathway, resulting in the significant inhibition of aconitase activity and a consequent increase in citrate levels (Gupta et al., 2012). Considering that NR activity and amino acid levels are markedly increased under hypoxia, the NO-dependent inhibition of aconitase leads to a shift towards amino acid biosynthesis (Gupta et al., 2012).

Taken together, the multiple roles of NO-mediated signaling in N metabolism, as discussed in this review, suggest that specificity during plant responses to environmental cues might be achieved through a balance between the synthesis and scavenging of NO and related molecules in a stimulus-specific manner. Importantly, different sources of NO, newly described pathways of NO degradation in plants and the molecular associations of these features should be addressed in future studies of plant NO-mediated redox systems.

7. CONCLUSIONS AND FUTURE REMARKS

Particularly in plants, NO synthesis is achieved through the operation of multiple oxi-reductive routes. These different pathways for NO synthesis have represented a trammel in the genetic manipulation of NO signaling in plants. Attempts to identify the primary source of NO in plants frequently generate discrepant results, and together with the fact that different mechanisms for NO production occur in distinct subcellular sites, have indicated that NO homeostasis depends on the specificity of the stimulus and the triggered cellular response. Alternatively, the molecular mechanisms underlying NO scavenging have only recently been revealed (Frungillo et al., 2014; Kneeshaw et al., 2014) and might lead to significant advances in the NO research field.

Recently, a novel NO-mediated feedback mechanism for the fine-tuning of nitrate assimilation has been proposed (Frungillo et al., 2014). Importantly, the redox control of GSNOR1 activity through *S*-nitrosylation has been suggested as a point of convergence in the control of nitrate assimilation and NO signaling in plants. Investigation of the exact site of *S*-nitrosylation in GSNOR1 should be the next step towards understanding the role of the post-transcriptional control of this enzyme. These data might provide information concerning possible different strategies to control GSNOR1 activity at protein level.

Although the interdependency between nitrate assimilation and photosynthesis has been firmly established, the underlying molecular mechanisms remain poorly understood. Considering that key photosynthetic proteins are targets of NO, it is tempting to speculate that the coordination of C and N metabolism is mediated through NO. Future studies focusing on the NO-mediated post-translational modification of proteins in N and C metabolism might therefore be of interest.

Recent advances in the redox control of plant metabolism, particularly those concerning the NO-mediated post-translational modification of key enzymes, might foster future efforts to improve N use efficiency in agriculture and reduce the cost and environmental impact of fertilization.

ACKNOWLEDGEMENTS

The authors would like to thank Dr David Wendehenne for kindly inviting this review. This work was financially supported through a PhD studentship from the Fundação de Amparo à Pesquisa do Estado de São Paulo (FAPESP) to L.F., a Royal Society University Research Fellowship (UF090321) to S.H.S. and research grants from the Conselho Nacional de Desenvolvimento Científico e Tecnológico (CNPq) to I.S. (no. 311860/2013-3).

REFERENCES

Alderton, W. K., Cooper, C. E., & Knowles, R. G. (2001). Nitric oxide synthases: structure, function and inhibition. *Biochemical Journal, 357*, 593–615.

Arasimowicz-Jelonek, M., Floryszak-Wieczorek, J., & Kubiś, J. (2009). Interaction between polyamine and nitric oxide signaling in adaptive responses to drought in cucumber. *Journal of Plant Growth Regulation, 28*, 177–186.

Besson-Bard, A., Pugin, A., & Wendehenne, D. (2008). New insights into nitric oxide signaling in plants. *Annual Review of Plant Biology, 59*, 21–39.

Bloom, A. J., Sukrapanna, S. S., & Warner, R. L. (1992). Root respiration associated with ammonium and nitrate absorption and assimilation by barley. *Plant Physiology, 99*, 1294–1301.

Botrel, A., Magne, C., & Kaiser, W. M. (1996). Nitrate reduction, nitrite reduction and ammonium assimilation in barley roots in response to anoxia. *Plant Physiology & Biochemistry, 34*, 607–760.

Campbell, W. H. (1999). Nitrate reductase structure, function and regulation: bridging the gap between biochemistry and physiology. *Annual Review of Plant Physiology and Plant Molecular Biology, 50*, 277–303.

Cantu-Medellin, N., & Kelley, E. E. (2013). Xanthine oxidoreductase-catalyzed reduction of nitrite to nitric oxide: insights regarding where, when and how. *Nitric Oxide, 34*, 19–26.

Carrillo, N., & Ceccarelli, E. A. (2003). Open questions in ferredoxin-NADP+ reductase catalytic mechanism. *European Journal of Biochemistry, 270*, 1900–1915.

Castello, P. R., David, P. S., McClure, T., Crook, Z., & Poyton, R. O. (2006). Mitochondrial cytochrome oxidase produces nitric oxide under hypoxic conditions: implications for oxygen sensing and hypoxic signaling in eukaryotes. *Cell Metabolism, 3*, 277–287.

Chaki, M., Valderrama, R., Fernández-Ocaña, A. M., Carreras, A., Gómez-Rodríguez, M. V., López-Jaramillo, J., et al. (2011). High temperature triggers the metabolism of *S*-nitrosothiols in sunflower mediating a process of nitrosative stress which provokes the inhibition of ferredoxin-NADP reductase by tyrosine nitration. *Plant Cell and Environment, 34*, 1803–1818.

Chen, C.-Z., Lv, X.-F., Li, J.-Y., Yi, H.-Y., & Gong, J.-M. (2012). *Arabidopsis* NRT1.5 is another essential component in the regulation of nitrate reallocation and stress tolerance. *Plant Physiology, 159*, 1582–1590.

Chen, R., Sun, S., Wang, C., Li, Y., Liang, Y., An, F., et al. (2009). The *Arabidopsis* PARAQUAT RESISTANT2 gene encodes an *S*-nitrosoglutathione reductase that is a key regulator of cell death. *Cell Research, 19*, 1377–1387.

Chitnis, P. R. (2001). Photosystem I: function and physiology. *Annual Review of Plant Physiology and Plant Molecular Biology, 52*, 593–626.

Corpas, F. J., Barroso, J. B., Carreras, A., Valderrama, R., Palma, J. M., León, A. M., et al. (2006). Constitutive arginine-dependent nitric oxide synthase activity in different organs of pea seedlings during plant development. *Planta, 224*, 246–254.

Corpas, F. J., Chaki, M., Fernández-Ocaña, A., Valderrama, R., Palma, J. M., Carreras, A., et al. (2008). Metabolism of reactive nitrogen species in pea plants under abiotic stress conditions. *Plant and Cell Physiology, 49*, 1711–1722.

Correa-Aragunde, N., Cejudo, F. J., & Lamattina, L. (2015). Nitric oxide is required for the auxin-induced activation of NADPH-dependent thioredoxin reductase and protein denitrosylation during root growth responses in arabidopsis. *Annals of Botany, 116*, 695–702.

Crawford, N. M. (1995). Nitrate: nutrient and signal for plant growth. *Plant Cell, 7*, 859–868.

Crawford, N. M., & Glass, A. D. (1998). Molecular and physiological aspects of nitrate uptake in plants. *Trends in Plant Science, 3*, 389–395.

Dechorgnat, J., Nguyen, C. T., Armengaud, P., Jossier, M., Diatloff, E., Filleur, S., et al. (2011). From the soil to the seeds: the long journey of nitrate in plants. *Journal of Experimental Botany, 62*, 1349–1359.

Delhon, P., Gojon, A., Tillard, P., & Passama, L. (1996). Diurnal regulation of NO − 3 uptake in soybean plants IV. Dependence on current photosynthesis and sugar availability to the roots. *Journal of Experimental Botany, 47*, 893–900.

Delledonne, M., Xia, Y., Dixon, R. A., & Lamb, C. (1998). Nitric oxide functions as a signal in plant disease resistance. *Nature, 394*, 585–588.

Delledonne, M., Zeier, J., Marocco, A., & Lamb, C. (2001). Signal interactions between nitric oxide and reactive oxygen intermediates in the plant hypersensitive disease resistance response. *Proceedings of the National Academy of Sciences of the United States of America, 98*, 13454–13459.

Durner, J., Wendehenne, D., & Klessig, D. F. (1998). Defense gene induction in tobacco by nitric oxide, cyclic GMP, and cyclic ADP-ribose. *Proceedings of the National Academy of Sciences of the United States of America, 95*, 10328–10333.

Feechan, A., Kwon, E., Yun, B. W., Wang, Y., Pallas, J. A., & Loake, G. J. (2005). A central role for *S*-nitrosothiols in plant disease resistance. *Proceedings of the National Academy of Sciences of the United States of America, 102*, 8054–8059.

Foresi, N., Correa-Aragunde, N., Parisi, G., Caló, G., Salerno, G., & Lamattina, L. (2010). Characterization of a nitric oxide synthase from the plant kingdom: NO generation from the green alga *Ostreococcus tauri* is light irradiance and growth phase dependent. *Plant Cell, 22*, 3816–3830.

Forrester, M. T., Foster, M. W., Benhar, M., & Stamler, J. S. (2009). Detection of protein *S*-nitrosylation with the biotin-switch technique. *Free Radical Biology & Medicine, 46*, 119–126.

Frungillo, L., de Oliveira, J. F., Saviani, E. E., Oliveira, H. C., Martínez, M. C., & Salgado, I. (2013). Modulation of mitochondrial activity by *S*-nitrosoglutathione reductase in *Arabidopsis thaliana* transgenic cell lines. *Biochimica et Biophysica Acta—Bioenergetics, 1827*, 239–247.

Frungillo, L., Skelly, M. J., Loake, G. J., Spoel, S. H., & Salgado, I. (2014). *S*-nitrosothiols regulate nitric oxide production and storage in plants through the nitrogen assimilation pathway. *Nature Communications, 5*, 5401.

Gardner, P. R., Costantino, G., Szabó, C., & Salzman, A. L. (1997). Nitric oxide sensitivity of the aconitases. *Journal of Biological Chemistry, 272*, 25071–25076.

Godber, B. L., Doel, J. J., Sapkota, G. P., Blake, D. R., Stevens, C. R., Eisenthal, R., et al. (2000). Reduction of nitrite to nitric oxide catalyzed by xanthine oxidoreductase. *Journal of Biological Chemistry, 275*, 7757–7763.

Groß, F., Durner, J., & Gaupels, F. (2013). Nitric oxide, antioxidants and prooxidants in plant defence responses. *Frontiers of Plant Science, 4*, 419.

Groppa, M. D., Rosales, E. P., Iannone, M. F., & Benavides, M. P. (2008). Nitric oxide, polyamines and Cd-induced phytotoxicity in wheat roots. *Phytochemistry, 69*, 2609–2615.

Gupta, K. J., Fernie, A. R., Kaiser, W. M., & van Dongen, J. T. (2011). On the origins of nitric oxide. *Trends in Plant Science, 16*, 160–168.

Gupta, K. J., Shah, J. K., Brotman, Y., Jahnke, K., Willmitzer, L., Kaiser, W. M., et al. (2012). Inhibition of aconitase by nitric oxide leads to induction of the alternative oxidase and to a shift of metabolism towards biosynthesis of amino acids. *Journal of Experimental Botany, 63*, 1773–1784.

Ho, C.-H., Lin, S.-H., Hu, H.-C., & Tsay, Y.-F. (2009). CHL1 functions as a nitrate sensor in plants. *Cell, 138*, 1184–1194.

Horchani, F., Prévot, M., Boscari, A., Evangelisti, E., Meilhoc, E., Bruand, C., et al. (2011). Both plant and bacterial nitrate reductases contribute to nitric oxide production in *Medicago truncatula* nitrogen-fixing nodules. *Plant Physiology, 155*, 1023–1036.

Howarth, R. W., & Marino, R. (2006). Nitrogen as the limiting nutrient for eutrophication in coastal marine ecosystems: evolving views over three decades. *Limnology and Oceanography, 51*, 364–376.

Hu, H.-C., Wang, Y.-Y., & Tsay, Y.-F. (2009). AtCIPK8, a CBL-interacting protein kinase, regulates the low-affinity phase of the primary nitrate response. *Plant Journal, 57*, 264–278.

Huang, X., Stettmaier, K., Michel, C., Hutzler, P., Mueller, M. J., Durner, J., et al. (2004). Nitric oxide is induced by wounding and influences jasmonic acid signaling in *Arabidopsis thaliana*. *Planta, 218*, 938–946.

Igamberdiev, A. U., & Gardeström, P. (2003). Regulation of NAD- and NADP-dependent isocitrate dehydrogenases by reduction levels of pyridine nucleotides in mitochondria and cytosol of pea leaves. *Biochimica et Biophysica Acta, 1606*, 117–125.

Jin, C. W., Du, S. T., Zhang, Y. S., Lin, X. Y., & Tang, C. X. (2009). Differential regulatory role of nitric oxide in mediating nitrate reductase activity in roots of tomato (*Solanum lycocarpum*). *Annals of Botany, 104*, 9–17.

Kaiser, W. M., & Förster, J. (1989). Low CO(2) prevents nitrate reduction in leaves. *Plant Physiology, 91*, 970–974.

Kaiser, W. M., & Huber, S. C. (2001). Post-translational regulation of nitrate reductase: mechanism, physiological relevance and environmental triggers. *Journal of Experimental Botany, 52*, 1981–1989.

Kiba, T., Feria-Bourrellier, A. B., Lafouge, F., Lezhneva, L., Boutet-Mercey, S., Orsel, M., et al. (2012). The *Arabidopsis* nitrate transporter NRT2.4 plays a double role in roots and shoots of nitrogen-starved plants. *Plant Cell, 24*, 245–258.

Kneeshaw, S., Gelineau, S., Tada, Y., Loake, G. J., & Spoel, S. H. (2014). Selective protein denitrosylation activity of thioredoxin-h5 modulates plant immunity. *Molecular Cell, 56*, 153–162.

Kozlov, A. V., Staniek, K., & Nohl, H. (1999). Nitrite reductase activity is a novel function of mammalian mitochondria. *FEBS Letters, 454*, 127–130.

Krapp, A., David, L. C., Chardin, C., Girin, T., Marmagne, A., Anne-Sophie, L., et al. (2014). Nitrate transport and signalling in *Arabidopsis*. *Journal of Experimental Botany, 65*, 789–798.

Krouk, G., Lacombe, B., Bielach, A., Perrine-Walker, F., Malinska, K., Mounier, E., et al. (2010). Nitrate-regulated auxin transport by NRT1.1 defines a mechanism for nutrient sensing in plants. *Developmental Cell, 18*, 927–937.

Laugier, E., Bouguyon, E., Mauriès, A., Tillard, P., Gojon, A., Lejay, L., et al. (2012). Regulation of high-affinity nitrate uptake in roots of *Arabidopsis* depends predominantly on posttranscriptional control of the NRT2.1/NAR2.1 transport system. *Plant Physiology, 158*, 1067–1078.

Lea, P. J. (1993). *Plant Biochemistry and Molecular Biology*, 155–180.

Lea, U. S., Leydecker, M.-T., Quilleré, I., Meyer, C., & Lillo, C. (2006). Posttranslational regulation of nitrate reductase strongly affects the levels of free amino acids and nitrate, whereas transcriptional regulation has only minor influence. *Plant Physiology, 140*, 1085–1094.

Lee, U., Wie, C., Fernandez, B. O., Feelisch, M., & Vierling, E. (2008). Modulation of nitrosative stress by *S*-nitrosoglutathione reductase is critical for thermotolerance and plant growth in Arabidopsis. *Plant Cell, 20*, 786–802.

Lejay, L., Gansel, X., Cerezo, M., Tillard, P., Müller, C., Krapp, A., et al. (2003). Regulation of root ion transporters by photosynthesis: functional importance and relation with hexokinase. *Plant Cell, 15*, 2218–2232.

Lejay, L., Tillard, P., Lepetit, M., Olive, F. D., Filleur, S., Daniel-Vedele, F., et al. (1999). Molecular and functional regulation of two NO3- uptake systems by N- and C-status of *Arabidopsis* plants. *Plant Journal, 18*, 509–519.

Léran, S., Muños, S., Brachet, C., Tillard, P., Gojon, A., Lacombe, B., et al. (2013). *Arabidopsis* NRT1.1 is a bidirectional transporter involved in root-to-shoot nitrate translocation. *Molecular Plant, 6*, 1984–1987.

Léran, S., Varala, K., Boyer, J. C., Chiurazzi, M., Crawford, N., Daniel-Vedele, F., et al. (2014). A unified nomenclature of nitrate transporter 1/peptide transporter family members in plants. *Trends in Plant Science, 19*, 5–9.

Li, J.-Y., Fu, Y. L., Pike, S. M., Bao, J., Tian, W., Zhang, Y., et al. (2010). The *Arabidopsis* nitrate transporter NRT1.8 functions in nitrate removal from the xylem sap and mediates cadmium tolerance. *Plant Cell, 22*, 1633–1646.

Lin, S.-H., Kuo, H. F., Canivenc, G., Lin, C. S., Lepetit, M., Hsu, P. K., et al. (2008). Mutation of the *Arabidopsis* NRT1.5 nitrate transporter causes defective root-to-shoot nitrate transport. *Plant Cell, 20*, 2514–2528.

Lindermayr, C., Saalbach, G., & Durner, J. (2005). Proteomic identification of *S*-nitrosylated proteins in *Arabidopsis*. *Plant Physiology, 137*, 921–930.

Liu, K., Huang, C., & Tsay, Y. (1999). CHL1 is a dual-affinity nitrate transporter of *Arabidopsis*. *Plant Cell, 11*, 865–874.

Liu, L., Hausladen, A., Zeng, M., Que, L., Heitman, J., & Stamler, J. S. (2001). A metabolic enzyme for *S*-nitrosothiol conserved from bacteria to humans. *Nature, 410*, 490–494.

Liu, X., Miller, M. J., Joshi, M. S., Thomas, D. D., & Lancaster, J. R. (1998). Accelerated reaction of nitric oxide with O_2 within the hydrophobic interior of biological membranes. *Proceedings of the National Academy of Sciences of the United States of America, 95*, 2175–2179.
Lozano-Juste, J., Colom-Moreno, R., & León, J. (2011). In vivo protein tyrosine nitration in *Arabidopsis thaliana. Journal of Experimental Botany, 62*, 3501–3517.
Lozano-Juste, J., & León, J. (2010). Enhanced abscisic acid-mediated responses in nia1nia2-noa1-2 triple mutant impaired in NIA/NR- and AtNOA1-dependent nitric oxide biosynthesis in Arabidopsis. *Plant Physiology, 152*, 891–903.
Malik, S. I., Hussain, A., Yun, B.-W., Spoel, S. H., & Loake, G. J. (2011). GSNOR-mediated de-nitrosylation in the plant defence response. *Plant Science, 181*, 540–544.
Marino, S. M., & Gladyshev, V. N. (2010). Structural analysis of cysteine *S*-nitrosylation: a modified acid-based motif and the emerging role of trans-nitrosylation. *Journal of Molecular Biology, 395*, 844–859.
Moche, M., Stremlau, S., Hecht, L., Göbel, C., Feussner, I., Stöhr, C., et al. (2010). Effect of nitrate supply and mycorrhizal inoculation on characteristics of tobacco root plasma membrane vesicles. *Planta, 231*, 425–436.
Modolo, L. V., Augusto, O., Almeida, I. M. G., Magalhaes, J. R., & Salgado, I. (2005). Nitrite as the major source of nitric oxide production by *Arabidopsis thaliana* in response to *Pseudomonas syringae. FEBS Letters, 579*, 3814–3820.
Modolo, L. V., Augusto, O., Almeida, I. M. G., Pinto-Maglio, C. A. F., Oliveira, H. C., & Seligman, K. (2006). Decreased arginine and nitrite levels in nitrate reductase-deficient *Arabidopsis thaliana* plants impair nitric oxide synthesis and the hypersensitive response to *Pseudomonas syringae. Plant Science, 171*, 34–40.
Modolo, L. V., Cunha, F. Q., Braga, M. R., & Salgado, I. (2002). Nitric oxide synthase-mediated phytoalexin accumulation in soybean cotyledons in response to the *Diaporthe phaseolorum* f. sp. *meridionalis* elicitor. *Plant Physiology, 130*, 1288–1297.
Moorhead, G., Douglas, P., Morrice, N., Scarabel, M., Aitken, A., MacKintosh, C., et al. (1996). Phosphorylated nitrate reductase from spinach leaves is inhibited by 14-3-3 proteins and activated by fusicoccin. *Current Biology, 6*, 1104–1113.
Muños, S., Cazettes, C., Fizames, C., Gaymard, F., Tillard, P., Lepetit, M., et al. (2004). Transcript profiling in the chl1-5 mutant of *Arabidopsis* reveals a role of the nitrate transporter NRT1.1 in the regulation of another nitrate transporter, NRT2.1. *Plant Cell, 16*, 2433–2447.
Navarre, D. A., Wendehenne, D., Durner, J., Noad, R., & Klessig, D. F. (2000). Nitric oxide modulates the activity of tobacco aconitase. *Plant Physiology, 122*, 573–582.
Oliveira, H. C., Justino, G. C., Sodek, L., & Salgado, I. (2009). Amino acid recovery does not prevent susceptibility to *Pseudomonas syringae* in nitrate reductase double-deficient *Arabidopsis thaliana* plants. *Plant Science, 176*, 105–111.
Oliveira, H. C., Salgado, I., & Sodek, L. (2013). Involvement of nitrite in the nitrate-mediated modulation of fermentative metabolism and nitric oxide production of soybean roots during hypoxia. *Planta, 237*, 255–264.
Park, B. S., Song, J. T., & Seo, H. S. (2011). *Arabidopsis* nitrate reductase activity is stimulated by the E3 SUMO ligase AtSIZ1. *Nature Communications, 2*, 400.
Parker, J. L., & Newstead, S. (2014). Molecular basis of nitrate uptake by the plant nitrate transporter NRT1.1. *Nature, 507*, 68–72.
Planchet, E., Jagadis Gupta, K., Sonoda, M., & Kaiser, W. M. (2005). Nitric oxide emission from tobacco leaves and cell suspensions: rate limiting factors and evidence for the involvement of mitochondrial electron transport. *Plant Journal, 41*, 732–743.
Radi, R. (2004). Nitric oxide, oxidants, and protein tyrosine nitration. *Proceedings of the National Academy of Sciences of the United States of America, 101*, 4003–4008.
Remans, T., Nacry, P., Pervent, M., Filleur, S., Diatloff, E., Mounier, E., et al. (2006). The *Arabidopsis* NRT1.1 transporter participates in the signaling pathway triggering root

colonization of nitrate-rich patches. *Proceedings of the National Academy of Sciences of the United States of America, 103*, 19206–19211.

Rockel, P., Strube, F., Rockel, A., Wildt, J., & Kaiser, W. M. (2002). Regulation of nitric oxide (NO) production by plant nitrate reductase in vivo and in vitro. *Journal of Experimental Botany, 53*, 103–110.

Romero-Puertas, M. C., Laxa, M., Mattè, A., Zaninotto, F., Finkemeier, I., Jones, A. M., et al. (2007). *S*-nitrosylation of peroxiredoxin II E promotes peroxynitrite-mediated tyrosine nitration. *Plant Cell, 19*, 4120–4130.

Ruffel, S., Gojon, A., & Lejay, L. (2014). Signal interactions in the regulation of root nitrate uptake. *Journal of Experimental Botany, 65*, 5509–5517.

Ruffel, S., Krouk, G., Ristova, D., Shasha, D., Birnbaum, K. D., & Coruzzi, G. M. (2011). Nitrogen economics of root foraging: transitive closure of the nitrate-cytokinin relay and distinct systemic signaling for N supply vs. demand. *Proceedings of the National Academy of Sciences of the United States of America, 108*, 18524–18529.

Rümer, S., Kapuganti, J. G., & Kaiser, W. M. (2009). Oxidation of hydroxylamines to NO by plant cells. *Plant Signaling & Behavior, 4*, 853–855.

Salgado, I., Martínez, M. C., Oliveira, H. C., & Frungillo, L. (2013). Nitric oxide signaling and homeostasis in plants: a focus on nitrate reductase and *S*-nitrosoglutathione reductase in stress-related responses. *Brazilian Journal of Botany, 36*, 89–98.

Santos-Filho, P. R., Vitor, S. C., Frungillo, L., Saviani, E. E., Oliveira, H. C., Salgado, I., et al. (2012). Nitrate reductase- and nitric oxide-dependent activation of sinapoylglucose:malate sinapoyltransferase in leaves of *Arabidopsis thaliana*. *Plant and Cell Physiology, 53*, 1607–1616.

Schmoldt, A., Benthe, H. F., & Haberland, G. (1975). Digitoxin metabolism by rat liver microsomes. *Biochemical Pharmacology, 24*, 1639–1641.

Seligman, K., Saviani, E. E., Oliveira, H. C., Pinto-Maglio, C. A. F., & Salgado, I. (2008). Floral transition and nitric oxide emission during flower development in *Arabidopsis thaliana* is affected in nitrate reductase-deficient plants. *Plant and Cell Physiology, 49*, 1112–1121.

Spoel, S. H., & Loake, G. J. (2011). Redox-based protein modifications: the missing link in plant immune signalling. *Current Opinion in Plant Biology, 14*, 358–364.

Spoel, S. H., Tada, Y., & Loake, G. J. (2010). Post-translational protein modification as a tool for transcription reprogramming. *New Phytologist, 186*, 333–339.

Stamler, J. S., Singel, D. J., & Loscalzo, J. (1992). Biochemistry of nitric oxide and its redox-activated forms. *Science, 258*, 1898–1902.

Su, W., Huber, S. C., & Crawford, N. M. (1996). Identification in vitro of a post-translational regulatory site in the hinge 1 region of *Arabidopsis* nitrate reductase. *Plant Cell, 8*, 519–527.

Sun, J., Bankston, J. R., Payandeh, J., Hinds, T. R., Zagotta, W. N., & Zheng, N. (2014). Crystal structure of the plant dual-affinity nitrate transporter NRT1.1. *Nature, 507*, 73–77.

Tsay, Y.-F., Chiu, C.-C., Tsai, C.-B., Ho, C.-H., & Hsu, P.-K. (2007). Nitrate transporters and peptide transporters. *FEBS Letters, 581*, 2290–2300.

Wang, P., Du, Y., Li, Y., Ren, D., & Song, C.-P. (2010). Hydrogen peroxide-mediated activation of MAP kinase 6 modulates nitric oxide biosynthesis and signal transduction in *Arabidopsis*. *Plant Cell, 22*, 2981–2998.

Wang, R., Okamoto, M., Xing, X., & Crawford, N. M. (2003). Microarray analysis of the nitrate response in *Arabidopsis* roots and shoots reveals over 1000 rapidly responding genes and new linkages to glucose, trehalose-6-phosphate, iron, and sulfate metabolism. *Plant Physiology, 132*, 556–567.

Wang, R., Tischner, R., Gutiérrez, R. A., Hoffman, M., Xing, X., Chen, M., et al. (2004). Genomic analysis of the nitrate response using a nitrate reductase-null mutant of *Arabidopsis*. *Plant Physiology, 136*, 2512–2522.

Wang, R., Xing, X., Wang, Y., Tran, A., & Crawford, N. M. (2009). A genetic screen for nitrate regulatory mutants captures the nitrate transporter gene NRT1.1. *Plant Physiology, 151*, 472–478.

Wang, Y.-Y., Hsu, P.-K., & Tsay, Y.-F. (2012). Uptake, allocation and signaling of nitrate. *Trends in Plant Science, 17*, 458–467.

Warner, R. L., & Kleinhofs, A. (1992). Genetics and molecular biology of nitrate metabolism in higher plants. *Physiologia Plantarum, 85*, 245–252.

Wilkinson, J. Q., & Crawford, N. M. (1993). Identification and characterization of a chlorate-resistant mutant of *Arabidopsis thaliana* with mutations in both nitrate reductase structural genes NIA1 and NIA2. *Molecular & General Genetics, 239*, 289–297.

Wulff, A., Oliveira, H. C., Saviani, E. E., & Salgado, I. (2009). Nitrite reduction and superoxide-dependent nitric oxide degradation by *Arabidopsis* mitochondria: influence of external NAD(P)H dehydrogenases and alternative oxidase in the control of nitric oxide levels. *Nitric Oxide, 21*, 132–139.

Xiong, J., Fu, G., Yang, Y., Zhu, C., & Tao, L. (2012). Tungstate: is it really a specific nitrate reductase inhibitor in plant nitric oxide research? *Journal of Experimental Botany, 63*, 33–41.

Yamasaki, H., & Sakihama, Y. (2000). Simultaneous production of nitric oxide and peroxynitrite by plant nitrate reductase: in vitro evidence for the NR-dependent formation of active nitrogen species. *FEBS Letters, 468*, 89–92.

Yu, M., Lamattina, L., Spoel, S. H., & Loake, G. J. (2014). Nitric oxide function in plant biology: a redox cue in deconvolution. *New Phytologist, 202*, 1142–1156.

Yun, B.-W., Feechan, A., Yin, M., Saidi, N. B., Le Bihan, T., Yu, M., et al. (2011). *S*-nitrosylation of NADPH oxidase regulates cell death in plant immunity. *Nature, 478*, 264–268.

Zheng, H., Wisedchaisri, G., & Gonen, T. (2013). Crystal structure of a nitrate/nitrite exchanger. *Nature, 497*, 647–651.

CHAPTER FIVE

Functional Implications of *S*-Nitrosothiols under Nitrooxidative Stress Induced by Abiotic Conditions

Francisco J. Corpas*[,1], Mounira Chaki[§], Juan C. Begara-Morales[§], Raquel Valderrama[§], Beatriz Sánchez-Calvo[§], Juan B. Barroso[§]

*Group of Antioxidants, Free Radicals and Nitric Oxide in Biotechnology, Food and Agriculture, Department of Biochemistry, Cell and Molecular Biology of Plants, Estación Experimental del Zaidín, CSIC, Granada, Spain

[§]Group of Biochemistry and Cell Signaling in Nitric Oxide, Department of Biochemistry and Molecular Biology, University of Jaén, Jaén, Spain

[1]Corresponding author: E-mail: javier.corpas@eez.csic.es

Contents

Abstract

S-Nitrosothiols (SNOs) is a family of nitric oxide (NO)-derived molecules resulting from NO binding to sulphhydryl (-SH) groups present in specific cysteine residues of peptides and proteins. This redox modification is reversible and is regarded as a cellular mechanism to extend the physiological function of NO. SNOs are classified as having either low or high molecular mass, with *S*-nitrosoglutathione and proteins being the most commonly studied, respectively. This article provides an overview of the importance of SNOs, especially in stress situations where NO content appears to be unaffected or even downregulated, with SNOs being responsible for the generation of nitrooxidative stress under these stress conditions.

Advances in Botanical Research, Volume 77
ISSN 0065-2296
http://dx.doi.org/10.1016/bs.abr.2015.10.004

List of Abbreviations

GAPC1 Cytoplasmic glyceraldehyde-3-phosphate dehydrogenase
GSH Reduced glutathione
GSNO *S*-Nitrosoglutathione
NO Nitric oxide
$ONOO^-$ Peroxynitrite
PTMs Posttranslational modifications
SNOs *S*-Nitrosothiols

1. INTRODUCTION

The nitric oxide (NO) metabolism in higher plants has become a new line of research since NO was demonstrated to be involved in the regulation of a wide range of physiological processes and also in the mechanism of response to adverse environmental conditions (Airaki et al., 2015; Baudouin, 2011; Besson-Bard, Pugin, & Wendehenne, 2008; Chaki, Shekariesfahlan et al., 2015; Corpas & Barroso, 2015; Corpas et al., 2011; Scheler, Durner, & Astier, 2013; Shapiro, 2005). Enzymatic NO generation is still the subject of intense research in order to determine how, where and when NO is produced in a specific organ under specific physiological and stress conditions (Corpas, Palma, del Río, & Barroso, 2009; Fröhlich & Durner, 2011; del Río, Corpas, & Barroso, 2004; Wojtaszek, 2000). Nevertheless, significant advances have been made in relation to the mechanisms involved in NO activity at the biochemical and physiological level. NO can interact directly with metal centres as NO has a high affinity for ferrous iron as well as other free radicals. It also reacts, for example, with superoxide to generate peroxynitrite ($ONOO^-$) which is a strong oxidant molecule (Calcerrada, Peluffo, & Radi, 2011). However, many of the biochemical functions displayed by NO involve thorough modifications in the main macromolecules, which include proteins, nucleic acids and lipids (Hess, Matsumoto, Kim, Marshall, & Stamler, 2005; Lamotte et al., 2015; Pfeiffer, Mayer, & Hemmens, 1999; Sánchez-Calvo, Barroso, & Corpas, 2013; Souza, Peluffo, & Radi, 2008; Trostchansky, Bonilla, González-Perilli, & Rubbo, 2013).

There is considerably more literature and research on NO biochemistry in animal systems than in plants, with available information being, in many cases, the starting point for developing specific research on plant cells. Thus, NO can form a covalent bond with sulphydryl (-SH) groups present in specific cysteine residues of peptides and proteins to generate a family of NO-derived molecules designated as *S*-nitrosothiols (SNOs). In animals, the importance of SNOs in neuronal, vascular and immune systems as mechanisms of signaling and protection has been the subject of intensive study in the fields of health and disease (Foster, McMahon, & Stamler, 2003; Foster, Pawloski, Singel, & Stamler, 2005; Gaston et al., 1998; Kevil & Patel, 2010). Moreover, in prokaryotic cells, SNOs have also attracted the attention of researchers due to possible involvement in their own pathogenesis (Laver et al., 2013). In plant systems, less is known about these molecules although there has been a special focus on mediated posttranslational modifications (PTMs). However, these PTMs, especially in proteins, are designated as *S*-nitrosylation, although *S*-nitrosation would be a more appropriate term (Hogg, 2002; Smith & Marletta, 2012).

2. BIOCHEMISTRY OF SNOs

SNOs result from the interaction between NO and a thiol. This process requires one-electron oxidation and can take place either throughout NO autoxidation to N_2O_3 or the addition of NO to a glutathionyl radical formed during this reaction (Broniowska, Diers, & Hogg, 2013; Broniowska & Hogg, 2012; Couturier, Chibani, Jacquot, & Rouhier, 2013; Smith & Marletta, 2012). Depending on the mass of the molecule containing the thiol group, SNOs can be referred to as low molecular mass (LMM) or high molecular mass (HMM) SNOs. Although *S*-nitrosoglutathione (GSNO) is considered to be the most abundant LMM SNO, this group includes other molecules such as *S*-nitrosocysteine and *S*-nitrosocysteinylglycine, which have been the subject of less study in the field of plant research (Corpas et al., 2013). On the other hand, when NO binding to sulphydryl (-SH) groups presents in specific cysteine residues of proteins, they are called HMM SNOs. In biological systems, SNOs are more stable than NO although their stability depends on the presence of trace metal ions (specifically copper and iron) which enhance their degradation (Askew, Barnett, McAninly, & Williams, 1995; Vanin, Malenkova, & Serezhenkov, 1997). In addition, the presence of reducing agents, such as thiols and ascorbate, can boost metal-ion-dependent decay (Dicks, Beloso, & Williams, 1997).

SNOs could mediate three main biological reactions: (1) NO release, (2) transnitrosylation, and (3) *S*-thiolation (Hogg, 2002; Zhang & Hogg, 2005). NO release is produced in the presence of metal ions such as copper and iron and reducing agents ($2RSNO + Cu^{2+} \rightarrow RSSR + Cu^{+} + 2NO$ or $RSNO + Cu^{+} \rightarrow RSN^{-} + NO + Cu^{2+}$). Trans-nitrosylation involves the transfer of the nitroso functional group from a nitrosothiol to a thiol

(RSNO + R′SH ⇆ RSH + R′SNO). Finally, *S*-thiolation involves a nucleophilic attack on RSNO sulphur by a thiolate anion, resulting in a disulphide and nitroxyl anion as products ($RSNO + R'S^- \rightarrow RSSR + NO^-$). In plant systems, the study of SNOs has focused on NO release and transnitrosylation but, to our knowledge, no information is available on *S*-thiolation. Figure 1 shows a simple model of the SNO metabolism and its potential interactions with other molecules in cells constituting the cellular pool of SNOs. The reactions involved, including *S*-nitrosylation and *S*-transnitrosylation, are underlined.

Plants research has concentrated mainly on identifying the protein targets of SNOs (Astier, Kulik, et al., 2012). Initially, these studies were carried out using NO donors, mainly GSNO, and a significant number of potential targets were identified using proteomic approaches. A pioneer study using *Arabidopsis thaliana* identified up to 63 proteins from cell cultures and 52 proteins from leaves (Lindermayr, Saalbach, & Durner, 2005). Later studies have confirmed and expanded the number of endogenous *S*-nitrosylated proteins under physiological and adverse conditions (Astier, Besson-Bard, et al., 2012; Begara-Morales et al., 2013, 2015; Begara-Morales, Sanchez-Calvo, Chaki, et al., 2014; Chaki, Shekariesfahlan et al., 2015; Chaki, Alvarez de Morales, et al., 2015; Fares, Nespoulous, Rossignol, & Peltier, 2014; Fares, Rossignol, & Peltier, 2011; Kato, Takemoto, & Kawakita, 2013; Tanou et al., 2012; Yu, Yun, Spoel, & Loake, 2012). Table 1 summarizes the *S*-nitrosylated proteins recognized in different plant species where the Cys affected and their effects

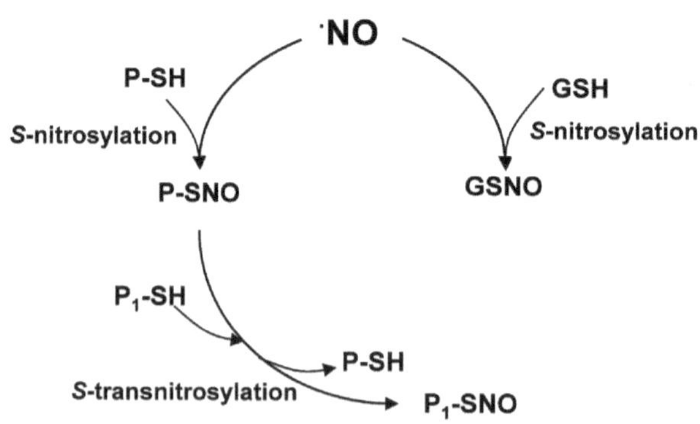

Figure 1 Simple model of *S*-nitrosothiols (SNOs) metabolism in cells. The interaction between a specific sulphydryl (-SH) group present in a protein (P) or a peptide, i.e. glutathione and nitric oxide (NO), enables SNOs to be generated by a process of *S*-nitrosylation. SNOs can also transfer NO to the sulphydryl (-SH) group of other proteins (P_1-SH) through a process of *S*-transnitrosylation between proteins. *S*-nitrosoglutathion (GSNO) is considered to be the most abundant low molecular mass SNO in higher plants.

Table 1 List of *S*-Nitros(yl)ated proteins in Higher Plants Where It Has Been Identified the *S*-Nitrosylated Cys Residue(s), and the Effect of This Posttranslational Modification on Their Function

Protein	Plant Species	Subcellular Localization	Effects	Identified *S*-Nitros(yl)ated Cys	References
Methionine adenosyltransferase	*Arabidopsis thaliana*	Cytoplasm	Inhibition	114[a]	Lindermayr, Saalbach, Bahnweg, and Durner (2006)
Cytosolic NAD-dependent glyceraldehyde 3-P dehydrogena (GAPDH)	*A. thaliana*	Cytoplasm	Inhibition	155, 159[a] 149[a,f]	Holtgrefe et al. (2008) Zaffagnini et al. (2013)
Histidine-containing phosphotransfer protein 1 AHP1	*A. thaliana*	Cytoplasm	Inhibition	115[a]	Feng et al. (2013)
Rubisco large subunit	*A. thaliana, Kalanchoe pinnata, Brassica juncea*	Chloroplast	Inhibition carboxylase activity	175[b]	Fares et al. (2011), Abat, Mattoo, and Deswal (2008) and Abat and Deswal (2009)
Cytosolic ascorbate peroxidase (APX)	*A. thaliana, Pisum sativum*	Cytoplasm	Increased activity	32[b]	Fares et al. (2011), Begara-Morales, Sánchez-Calvo, Chaki, et al. (2014), and Yang et al. (2015)
Dehydroascrobate reductase (DHAR)	*A. thaliana, Solanum tuberosum*	Cytoplasm	Unknown	20, 147[a,c]	Fares et al. (2011) and Kato et al. (2013)
Monodehydroascorbate reductase (MDAR)	*P. sativum*	Peroxisome	Inhibition	68[e]	Begara-Morales et al. (2015)
Glutathione reductase	*P. sativum*	Chloroplast Cytoplasm	No effect	BS[d]	Begara-Morales et al. (2015)

(*Continued*)

Table 1 List of *S*-Nitros(yl)ated proteins in Higher Plants Where It Has Been Identified the *S*-Nitrosylated Cys Residue(s), and the Effect of This Posttranslational Modification on Their Function—cont'd

Protein	Plant Species	Subcellular Localization	Effects	Identified *S*-Nitros(yl)ated Cys	References
Photosystem I apoprotein A2	*A. thaliana*	Chloroplast (thylakoid membrane)	Unknown	559[b]	Fares et al. (2011)
Tubulin β	*A. thaliana*	Cytoskeleton	Unknown	56[b]	Lindermayr et al. (2005) and Fares et al. (2011)
Auxin receptor (TIR1)	*A. thaliana*		Increased activity	140[c]	Terrile et al. (2012)
Transcription factor MYB2	*A. thaliana*	Nucleus	Inhibition	53[d]	Serpa et al. (2007)
Transcription factor NPR1	*A. thaliana*	Nucleus	Inhibition	156[c]	Tada et al. (2008)
Transcription factor TGA1	*A. thaliana*	Nucleus	Activation	172, 287, 260[a]	Lindermayr, Sell, Müller, Leister, and Durner (2010)
Transcription factor AtMYB3	*A. thaliana*	Nucleus	Inhibition DNA binding	49, 53[f]	Tavares et al. (2014)
Calnexin (CNX1)	*A. thaliana*	Endoplasmic reticulum membrane	Unknown	108[b]	Fares et al. (2011)
Elongation factor EF1-a	*A. thaliana*, *B. juncea*	Nucleus	Unknown	87[b]	Lindermayr et al. (2005), Tanou et al. (2009), and Fares et al. (2011)
Vacuolar ATPase subunit		Vacuole	Unknown	201[b]	Fares et al. (2011)
Phosphate transporter (PHT3; 1)	*A. thaliana*	Mitochondrial inner membrane	Unknown	104[b]	Fares et al. (2011)

Adenylate translocater	*A. thaliana*		Unknown	130[b]	Fares et al. (2011)
NADPH oxidase	*A. thaliana*	Cytoplasmic membrane	Inhibition	890[a,c]	Yun et al. (2011)
Peroxiredoxin II E	*A. thaliana*	Chloroplast (stroma)	Inhibition	121[a]	Romero-Puertas et al. (2007)
Peroxiredoxin II F	*P. sativum*	Mitochondrium	Decrease peroxidase activity	58[c]	Camejo et al. (2015)
Metacaspase AtMC9	*A. thaliana*	Apoplast	Inhibition	147[c]	Belenghi et al. (2007)
Phytochelatin (PC2, PC3 and PC4)	*A. thaliana*	Cytoplasm/vacuole	No effect	2[a]	Elviri et al. (2010)
Cysteine protease RD21	*A. thaliana*	Cytoplasm	Unknown	233, 342[b]	Fares et al. (2011)
NADP-dependent isocitrate dehydrogenase	*A. thaliana*	Cytoplasm	Unknown	75, 269, 363[b]	Fares et al. (2011)
Ferredoxin-NADP reductase	*A. thaliana*, *P. sativum*	Chloroplast (thylakoid membrane)	Unknown	108[e]	Holzmeister et al. (2011) and Begara-Morales et al. (2013)
CDC48 (cell division cycle 48)	*Nicotiana tabacum*	Cytoplasm	Inhibition	526[a]	Astier, Besson-Bard, et al. (2012)
Phenylalanine ammonia-lyase 2 (PAL2)	*Populus* × *canescens*		No effect	557[e]	Vanzo et al. (2014)
Sucrose nonfermenting 1 (SNF1)-related protein kinase 2.6 (SnRK2.6)	*A. thaliana*	Guard cells	Inhibition	137[f]	Wang et al. (2015)

[a]Mass spectrometric techniques.
[b]Combination of Biotin Switch and labelling with isotope-coded affinity tags.
[c]Combination of Biotin Switch and Site-Directed Mutagenesis.
[d]Biotin Switch.
[e]In silico identification.
[f]Site-Directed Mutagenesis.

on protein function/activity have been identified. The identification of new protein candidates is still underway, and these advances are based on the analysis of specific *S*-nitrosylated proteins in the different subcellular compartments (chloroplasts, mitochondria, nucleus, peroxisomes, etc.) and also on the use of new mass spectrometric technologies.

As protein *S*-nitrosylation is a reversible process, special attention has been devoted to understanding the mechanism of nitrosylation/denitrosylation in order to determine whether this process is enzymatically regulated. In the case of GSNO, considered to be the most abundant LMM SNO, the enzyme GSNO reductase is regarded as the key player in plant denitrosylation (Leterrier et al., 2011; Malik, Hussain, Yun, Spoel, & Loake, 2011). However, in the case of HMM SNOs, the situation is not so clear-cut. It has been reported that cytoplasmic glyceraldehyde-3-phosphate dehydrogenase (GAPC1) from *A. thaliana* is nitrosylated and causes activity to be inhibited. However, this process is reversible, and both nitrosylation and denitrosylation of purified GAPC1 have been reported to correlate with the (GSH)/(GSNO) ratio (Zaffagnini et al., 2013).

Very recently, a comparative large-scale, site-specific proteomic analysis in *Arabidopsis* wild-type and the mutant *gsnor1-3*, characterized to have a higher level of GSNO, has allowed to identify 926 proteins which expand the data set of theses target proteins (Hu et al., 2015).

3. ROLE OF GSNO AS CELLULAR SIGNAL

In plant systems, glutathione (GSH) is the major nonprotein thiol which contains antioxidant properties (Noctor, Queval, Mhamdi, Chaouch, & Foyer, 2011). Consequently, GSNO is considered a major LMM SNO in plant systems as well as an NO vehicle over long distances, which is important in relation to redox signaling mechanisms (Airaki et al., 2011; Malik et al., 2011). At the experimental level, exogenous GSNO has been used as an NO donor in order to identify potential protein targets in higher plants and, as mentioned above, this facilitated the identification of a set of potential proteins to be *S*-nitrosylated.

Similarly, NO donors have been used in medium- and large-scale transcriptomic analyses, with the aid of cDNA amplified fragment length polymorphism and microarray technology to identify NO-responsive genes mainly in the organs (leaves and roots) and cell cultures of *A. thaliana* (Ahlfors, Brosche, Kollist, & Kangasjarvi, 2009; Badri et al., 2008; Begara-Morales,

Sánchez-Calvo, Luque, et al., 2014; Parani et al., 2004; Polverari et al., 2003) and other plant species (Besson-Bard et al., 2009). However, most of these studies have used sodium nitroprusside as an NO donor which must be handled with caution (Murgia, de Pinto, Delledonne, Soave, & De Gara, 2004), while GSNO has only been used to analyze NO-dependent transcriptomic responses in *Medicago truncatula* plants (Ferrarini et al., 2008). Recently, a broader study based on large-scale transcriptomic analyses, termed massively parallel sequencing or RNA-seq, has reported the analysis of genes modulated by GSNO. Arabidopsis plants were thus exposed through the roots to 1 mM GSNO, and GSNO-responsive genes were then identified in the main organs where 1945 GSNO-responsive genes were expressed differently in leaves and roots, with 114 of these corresponding exclusively to one of these organs (Begara-Morales, Sánchez-Calvo, Luque, et al., 2014).

4. FUNCTION OF SNOs UNDER ADVERSE ENVIRONMENTAL CONDITIONS

Plants are continually exposed to daily and seasonal environmental conditions such as temperature, light intensity, drought, humidity and mineral accessibility. However, they are also eventually exposed to adverse conditions such as salinity, heavy metal, mechanical damage and pathogen infection which can drastically change the physiological and metabolic response of plants to mitigate potential damage. The importance and functional significance of modulations in SNO content in a specific stress situation have received less attention when compared to the abundant literature on modulations in NO content (Lee, Wie, Fernandez, Feelisch, & Vierling, 2008; Leterrier et al., 2012).

It is frequently assumed that plants under specific adverse conditions respond to changes, usually an increase, in NO content or any NO-derived molecule. Thus, it has been suggested that an increase in tyrosine nitration could be a reliable marker of nitrooxidative stress (Corpas, del Río, & Barroso, 2007). An example of this phenomenon was provided by an analysis of olive leaves from plants exposed to salinity stress which showed a general increase in several parameters related to the NO metabolism such as L-arginine-dependent NO synthase activity, NO and SNO content and $ONOO^-$, with protein nitration levels being the final consequence of nitrooxidative stress (Valderrama et al., 2007). Similar behavioural results have been described in the leaves of pea plants exposed to low temperature and mechanical wounding (Corpas et al., 2008). Figure 2 model A depicts

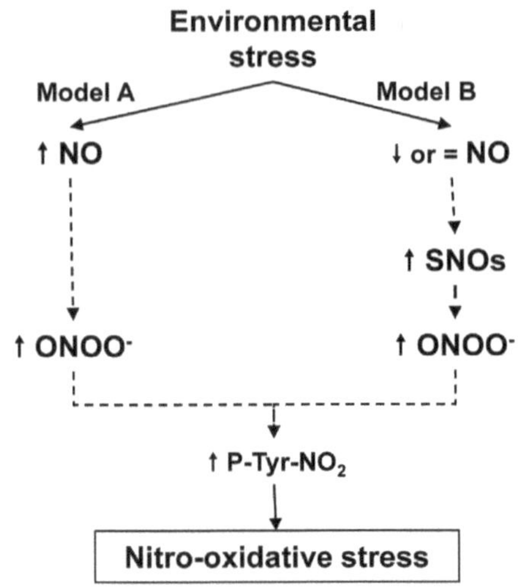

Figure 2 Sequence of events which can mediate nitrooxidative stress in higher plants, the increase in tyrosine nitration being regarded as a reliable marker of this kind of stress. **Model A**, a specific stress (i.e. salinity in olive plants, cold in pea leaves) provokes an increase in nitric oxide (NO) content, with a concomitant increase in peroxynitrite ($ONOO^-$) content which mediates protein nitration (P-Tyr-NO_2) (Corpas et al., 2008; Valderrama et al., 2007). **Model B**, although a specific stress (i.e. high temperature, mechanical wounding or biotic stress in sunflower seedlings) does not provoke a change in NO content, *S*-nitrosothiol (SNO) content increased drastically, with a concomitant rise in $ONOO^-$ and protein nitration (Chaki et al., 2009; Chaki, Valderrama, Fernández-Ocaña, Carreras, Gómez-Rodríguez, López-Jaramillo, et al., 2011; Chaki, Valderrama, Fernández-Ocaña, Carreras, Gómez-Rodríguez, Pedrajas, et al., 2011).

the sequence of events that provoke nitrooxidative stress directly mediated by a significant increase in NO content.

Nevertheless, there are several situations where the NO content appeared to be unaffected or even reduced. However, SNO content, as well as that of GSNO, is clearly upmodulated and an increase in $ONOO^-$ was observed, with tyrosine nitration being the final consequence of nitrooxidative stress. An example of this behaviour has been described in great detail in relation to sunflower seedlings exposed to high temperatures (38 °C for 4 h), where NO content was downregulated. However, a simultaneous accumulation of total SNOs including GSNO, with a concomitant increase in tyrosine nitration content, studied with the aid of high performance liquid chromatography in tandem with mass spectrometry (LC—MS/MS) and confocal laser scanning microscopy, has been observed (Chaki, Valderrama, Fernández-Ocaña, Carreras, Gómez-Rodríguez, López-Jaramillo, et al., 2011). Similar behaviour was observed in the compatible interaction between *A. thaliana*

Col-0 and *Pseudomonas syringae* DC3000 (*avrB*) where total content of SNOs increased after infection (Feechan et al., 2005). Furthermore, this increase in SNOs occurred in the absence of a specific increase in NO production, indicating that SNOs and NO may play distinct functional roles during the establishment of plant disease resistance. In fact, NO activity cannot be exclusively controlled at the NO synthesis level during this plant defence stage (Dahm, Moore, & Murphy, 2006; Feechan et al., 2005). In addition, an analogous response in sunflower hypocotyls after infection by the pathogen *Plasmopara halstedii* (Chaki et al., 2009) or mechanical injury has been reported (Chaki, Valderrama, Fernández-Ocaña, Carreras, Gómez-Rodríguez, Pedrajas, et al., 2011). Figure 2 model B depicts the sequence of events where NO content is low and the induced high content of SNOs appears to mediate the increase in $ONOO^-$, nitration and consequently nitrooxidative stress. To illustrate the importance of SNOs in these processes, Figure 3 shows experimental data where tyrosine nitration and peroxynitrite are significantly reduced in the

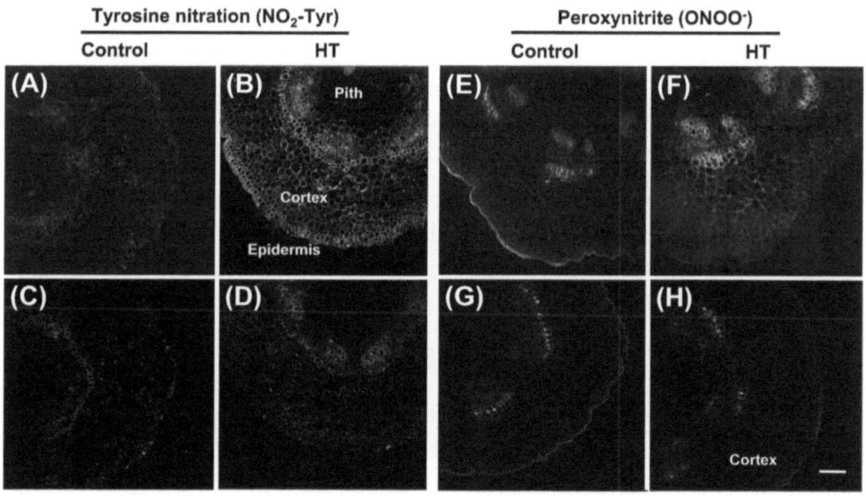

Figure 3 Confocal laser scanning microscope pictures showing the detection of protein 3-nitrotyrosine (A–D) and peroxynitrite (E–H) in cross sections of hypocotyls from sunflower seedlings subjected to high temperature and control. NO_2-Tyr was detected using a specific antibody and $ONOO^-$ was carried out using 3′-(*p*-aminophenyl) fluorescein as a fluorescent probe. To demonstrate the involvement of SNOs in the generation of protein nitration and peroxynitrite, the hypocotyl samples from sunflower seedlings exposed to high temperature (D and H) and control (C and G) were preincubated in a solution containing 1 mM ascorbate, 10 μM CuCl (molecules which can decompose SNOs) and 200 μM cPTIO (an NO scavenger). Bar = 200 mm. *Reproduced with permission of* Plant Cell and Environment *(2011), 34, 1803–1818.* (See colour plate)

presence of Cu^{1+} and ascorbate (as a reductant used as internal controls), as these compounds contribute to SNO decomposition (Chaki, Valderrama, Fernández-Ocaña, Carreras, Gómez-Rodríguez, López-Jaramillo, et al., 2011). In sunflower seedlings exposed to high temperature stress, increased production of tyrosine nitration (Figure 3, panels A and B) and $ONOO^-$ (Figure 3, panels E and F) was observed as compared to no-stress samples, evaluated with the aid of an antibody and specific fluorescent probe, respectively, and visualized by confocal laser scanning microscopy. When samples from control and those exposed to high temperature seedlings were preincubated in a solution containing 1 mM ascorbate, 10 μm CuCl (molecules that can break down SNOs) and 200 mM cPTIO (an NO scavenger), the content of both tyrosine nitration (Figure 3, panels C and D) and $ONOO^-$ (Figure 3, panels G and H) were significantly reduced, indicating that SNO decomposition mediates these two processes under high temperature stress conditions.

There are recent comparative analyses to identify and quantify protein targets of *S*-nitrosylation under adverse environmental conditions. For example, in Arabidopsis seedlings exposed at 4 °C for 4 h, 42 endogenously *S*-nitrosylated proteins were detected, out of which 11 were overnitrosylated following cold exposure (Puyaubert, Fares, Rézé, Peltier, & Baudouin, 2014). In the same way, the analysis of the *S*-nitrosoproteome of poplar (*Populus* × *canescens*) leaves after ozone fumigation revealed that 32 proteins were affected. Thus, the content of 9 *S*-nitrosylated proteins increased whereas the others 23 proteins were diminished (Vanzo et al., 2014).

5. CONCLUSIONS AND PERSPECTIVES

In higher plants, SNOs are now recognized as plant endogenous compounds that participate in signal transduction and stress responses. In this context, protein thiols are coming to be regarded as one of the major intracellular targets of NO. Thus, important plant proteins appear to undergo *S*-nitrosylation under biotic and abiotic stress conditions, and much work involving different experimental approaches has been carried out to identify these proteins. Great emphasis has been placed on identifying the specific residue(s) which is(are) affected and on evaluating the final effect on protein function. It is also important to point out that, as *S*-nitrosothiols in cell systems are more stable per se than NO, the importance of SNOs could be highlighted, especially in stress situations where NO content seems to be unaffected or even downregulated, with SNOs thus being responsible for the mechanism of response to these stress conditions. Improving the quantification and localization of these molecules (Diers, Keszler, & Hogg, 2014) in the different organs of higher plants and evaluating their importance under adverse environmental conditions present a technical challenge.

ACKNOWLEDGEMENT

Work in our laboratories is supported by an ERDF grant cofinanced by the Ministry of Economy and Competitiveness (project BIO2012-33904) and the Junta de Andalucía (groups BIO286 and BIO192) in Spain.

REFERENCES

Abat, J. K., & Deswal, R. (2009). Differential modulation of S-nitrosoproteome of *Brassica juncea* by low temperature: change in S-nitrosylation of Rubisco is responsible for the inactivation of its carboxylase activity. *Proteomics, 9*, 4368–4380.

Abat, J. K., Mattoo, A. K., & Deswal, R. (2008). S-nitrosylated proteins of a medicinal CAM plant *Kalanchoe pinnata*- ribulose-1,5-bisphosphate carboxylase/oxygenase activity targeted for inhibition. *The Federation of European Biochemical Societies Journal, 275*, 2862–2872.

Ahlfors, R., Brosche, M., Kollist, H., & Kangasjarvi, J. (2009). Nitric oxide modulates ozone-induced cell death, hormone biosynthesis and gene expression in *Arabidopsis thaliana*. *The Plant Journal, 58*, 1–12.

Airaki, M., Leterrier, M., Valderrama, R., Chaki, M., Begara-Morales, J. C., Barroso, J. B., et al. (2015). Spatial and temporal regulation of the metabolism of reactive oxygen and nitrogen species during the early development of pepper (*Capsicum annuum*) seedlings. *Annals of Botany, 116*, 679–693.

Airaki, M., Sánchez-Moreno, L., Leterrier, M., Barroso, J. B., Palma, J. M., & Corpas, F. J. (2011). Detection and quantification of S-nitrosoglutathione (GSNO) in pepper (*Capsicum annuum* L.) plant organs by LC-ES/MS. *Plant and Cell Physiology, 52*, 2006–2015.

Askew, S. C., Barnett, D. J., McAninly, J., & Williams, D. L. H. (1995). Catalysis by Cu^{2+} of nitric oxide release from S-nitrosothiols (RSNO). *Journal of the Chemical Society Perkin Transactions, 2*, 741–745.

Astier, J., Besson-Bard, A., Lamotte, O., Bertoldo, J., Bourque, S., Terenzi, H., et al. (2012). Nitric oxide inhibits the ATPase activity of the chaperone-like AAA^{+} ATPase CDC48, a target for S-nitrosylation in cryptogein signalling in tobacco cells. *Biochemical Journal, 447*, 249–260.

Astier, J., Kulik, A., Koen, E., Besson-Bard, A., Bourque, S., Jeandroz, S., et al. (2012). Protein S-nitrosylation: What's going on in plants? *Free Radical Biology & Medicine, 53*, 1101–1110.

Badri, D. V., Loyola-Vargas, V. M., Du, J., Stermitz, F. R., Broeckling, C. D., Iglesias-Andreu, L., et al. (2008). Transcriptome analysis of Arabidopsis roots treated with signaling compounds: a focus on signal transduction, metabolic regulation and secretion. *New Phytologist, 179*, 209–223.

Baudouin, E. (2011). The language of nitric oxide signalling. *Plant Biology (Stuttg), 13*, 233–242.

Begara-Morales, J. C., López-Jaramillo, F. J., Sánchez-Calvo, B., Carreras, A., Ortega-Muñoz, M., Santoyo-González, F., et al. (2013). Vinyl sulfone silica: application of an open preactivated support to the study of transnitrosylation of plant proteins by S-nitrosoglutathione. *BMC Plant Biology, 13*, 61.

Begara-Morales, J. C., Sánchez-Calvo, B., Chaki, M., Mata-Pérez, C., Valderrama, R., López-Jaramillo, J., et al. (2015). Differential molecular response of monodehydroascorbate reductase (MDAR) and glutathione reductase (GR) by nitration and *S*-nitrosylation. *Journal of Experimental Botany, 66*, 5983–5996.

Begara-Morales, J. C., Sánchez-Calvo, B., Chaki, M., Valderrama, R., Mata-Pérez, C., López-Jaramillo, J., et al. (2014). Dual regulation of cytosolic ascorbate peroxide (APX) by tyrosine nitration and *S*-nitrosylation. *Journal of Experimental Botany, 65*, 527–538.

Begara-Morales, J. C., Sánchez-Calvo, B., Luque, F., Leyva-Pérez, M. O., Leterrier, M., Corpas, F. J., et al. (2014). Differential transcriptomic analysis by RNA-seq of GSNO-responsive genes between Arabidopsis roots and leaves. *Plant and Cell Physiology, 55*, 1080–1095.

Belenghi, B., Romero-Puertas, M. C., Vercammen, D., Brackenier, A., Inzé, D., Delledonne, M., et al. (2007). Metacaspase activity of *Arabidopsis thaliana* is regulated by S-nitrosylation of a critical cysteine residue. *The Journal of Biological Chemistry, 282*, 1352–1358.

Besson-Bard, A., Astier, J., Rasul, S., Wawer, I., Dubreuil-Maurizi, C., Jeandroz, S., et al. (2009). Current view of nitric oxide-responsive genes in plants. *Plant Science, 177*, 302–309.

Besson-Bard, A., Pugin, A., & Wendehenne, D. (2008). New insights into nitric oxide signaling in plants. *Annual Review of Plant Biology, 59*, 21–39.

Broniowska, K. A., Diers, A. R., & Hogg, N. (2013). *S*-Nitrosoglutathione. *Biochimica et Biophysica Acta, 1830*, 3173–3181.

Broniowska, K. A., & Hogg, N. (2012). The chemical biology of *S*-nitrosothiols. *Antioxidants and Redox Signaling, 17*, 969–980.

Calcerrada, P., Peluffo, G., & Radi, R. (2011). Nitric oxide-derived oxidants with a focus on peroxynitrite: molecular targets, cellular responses and therapeutic implications. *Current Pharmaceutical Design, 17*, 3905–3932.

Camejo, D., Ortiz-Espín, A., Lázaro, J. J., Romero-Puertas, M. C., Lázaro-Payo, A., Sevilla, F., et al. (2015). Functional and structural changes in plant mitochondrial PrxII F caused by NO. *Journal of Proteomics, 119C*, 112–125.

Chaki, M., Álvarez de Morales, P., Ruiz, C., Begara-Morales, J. C., Barroso, J. B., Corpas, F. J., et al. (2015). Ripening of pepper (*Capsicum annuum*) fruit is characterized by an enhancement of protein tyrosine nitration. *Annals of Botany, 116*, 637–647.

Chaki, M., Fernández-Ocaña, A. M., Valderrama, R., Carreras, A., Esteban, F. J., Luque, F., et al. (2009). Involvement of reactive nitrogen and oxygen species (RNS and ROS) in sunflower-mildew interaction. *Plant and Cell Physiology, 50*, 265–279.

Chaki, M., Shekariesfahlan, A., Ageeva, A., Mengel, A., von Toerne, C., Durner, J., et al. (2015). Identification of nuclear target proteins for S-nitrosylation in pathogen-treated *Arabidopsis thaliana* cell cultures. *Plant Science, 238*, 115–126.

Chaki, M., Valderrama, R., Fernández-Ocaña, A. M., Carreras, A., Gómez-Rodríguez, M. V., López-Jaramillo, J., et al. (2011). High temperature triggers the metabolism of S-nitrosothiols in sunflower mediating a process of nitrosative stress which provokes the inhibition of ferredoxin-NADP reductase by tyrosine nitration. *Plant Cell and Environment, 34*, 1803–1818.

Chaki, M., Valderrama, R., Fernández-Ocaña, A. M., Carreras, A., Gómez-Rodríguez, M. V., Pedrajas, J. R., et al. (2011). Mechanical wounding induces a nitrosative stress by down-regulation of GSNO reductase and an increase in *S*-nitrosothiols in sunflower (*Helianthus annuus*) seedlings. *Journal of Experimental Botany, 62*, 1803–1813.

Corpas, F. J., Alché, J. D., & Barroso, J. B. (2013). Current overview of *S*-nitrosoglutathione (GSNO) in higher plants. *Frontiers in Plant Science, 4*, 126.

Corpas, F. J., & Barroso, J. B. (2015). Nitric oxide from a "green" perspective. *Nitric Oxide, 45*, 15–19.

Corpas, F. J., Chaki, M., Fernández-Ocaña, A., Valderrama, R., Palma, J. M., Carreras, A., et al. (2008). Metabolism of reactive nitrogen species in pea plants under abiotic stress conditions. *Plant and Cell Physiology, 49*, 1711–1722.

Corpas, F. J., Leterrier, M., Valderrama, R., Airaki, M., Chaki, M., Palma, J. M., et al. (2011). Nitric oxide imbalance provokes a nitrosative response in plants under abiotic stress. *Plant Science, 181*, 604–611.

Corpas, F. J., Palma, J. M., del Río, L. A., & Barroso, J. B. (2009). Evidence supporting the existence of L-arginine-dependent nitric oxide synthase activity in plants. *New Phytologist, 184*, 9–14.

Corpas, F. J., del Río, L. A., & Barroso, J. B. (2007). Need of biomarkers of nitrosative stress in plants. *Trends in Plant Science, 12*, 436–438.

Couturier, J., Chibani, K., Jacquot, J. P., & Rouhier, N. (2013). Cysteine-based redox regulation and signaling in plants. *Frontiers in Plant Science, 4*, 105.

Dahm, C. C., Moore, K., & Murphy, M. P. (2006). Persistent *S*-nitrosation of complex I and other mitochondrial membrane proteins by *S*-nitrosothiols but not nitric oxide or peroxynitrite: implications for the interaction of nitric oxide with mitochondria. *The Journal of Biological Chemistry, 281*, 10056–10065.

Dicks, A. P., Beloso, P. H., & Williams, D. L. H. (1997). Decomposition of S-nitrosothiols: the effects of added thiols. *Journal of the Chemical Society Perkin Transactions, 2*, 1429–1434.

Diers, A. R., Keszler, A., & Hogg, N. (2014). Detection of *S*-nitrosothiols. *Biochimica et Biophysica Acta, 1840*, 892–900.

Elviri, L., Speroni, F., Careri, M., Mangia, A., di Toppi, L. S., & Zottini, M. (2010). Identification of *in vivo* nitrosylated phytochelatins in *Arabidopsis thaliana* cells by liquid chromatography-direct electrospray-linear ion trap-mass spectrometry. *Journal of Chromatography A, 1217*, 4120–4126.

Fares, A., Nespoulous, C., Rossignol, M., & Peltier, J. B. (2014). Simultaneous identification and quantification of nitrosylation sites by combination of biotin switch and ICAT labeling. *Methods in Molecular Biology, 1072*, 609–620.

Fares, A., Rossignol, M., & Peltier, J. B. (2011). Proteomics investigation of endogenous S-nitrosylation in Arabidopsis. *Biochemical and Biophysical Research Communications, 416*, 331–336.

Feechan, A., Kwon, E., Yun, B. W., Wang, Y., Pallas, J. A., & Loake, G. J. (2005). A central role for S-nitrosothiols in plant disease resistance. *Proceedings of the National Academy of Sciences of the United States of America, 102*, 8054–8059.

Feng, J., Wang, C., Chen, Q., Chen, H., Ren, B., Li, X., et al. (2013). *S*-nitrosylation of phosphotransfer proteins represses cytokinin signaling. *Nature Communications, 4*, 1529.

Ferrarini, A., De Stefano, M., Baudouin, E., Pucciariello, C., Polverari, A., Puppo, A., et al. (2008). Expression of *Medicago truncatula* genes responsive to nitric oxide in pathogenic and symbiotic conditions. *Molecular Plant–Microbe Interactions, 21*, 781–790.

Foster, M. W., McMahon, T. J., & Stamler, J. S. (2003). S-nitrosylation in health and disease. *Trends in Molecular Medicine, 9*, 160–168.

Foster, M. W., Pawloski, J. R., Singel, D. J., & Stamler, J. S. (2005). Role of circulating *S*-nitrosothiols in control of blood pressure. *Hypertension, 45*, 15–17.

Fröhlich, A., & Durner, J. (2011). The hunt for plant nitric oxide synthase (NOS): is one really needed? *Plant Science, 181*, 401–404.

Gaston, B., Fry, E., Sears, S., Heroman, W. M., Ignarro, L., & Stamler, J. S. (1998). Umbilical arterial S-nitrosothiols in stressed newborns: role in perinatal circulatory transition. *Biochemical and Biophysical Research Communications, 253*, 899–901.

Hess, D. T., Matsumoto, A., Kim, S. O., Marshall, H. E., & Stamler, J. S. (2005). Protein S-nitrosylation: purview and parameters. *Nature Reviews Molecular Cell Biology, 6*, 150–166.
Hogg, N. (2002). The biochemistry and physiology of *S*-nitrosothiols. *Annual Review of Pharmacology & Toxicology, 42*, 585–600.
Holtgrefe, S., Gohlke, J., Starmann, J., Druce, S., Klocke, S., Altmann, B., et al. (2008). Regulation of plant cytosolic glyceraldehyde 3-phosphate dehydrogenase isoforms by thiol modifications. *Physiologia Plantarum, 133*, 211–228.
Holzmeister, C., Fröhlich, A., Sarioglu, H., Bauer, N., Durner, J., & Lindermayr, C. (2011). Proteomic analysis of defense response of wildtype *Arabidopsis thaliana* and plants with impaired NO- homeostasis. *Proteomics, 11*, 1664–1683.
Hu, J., Huang, X., Chen, L., Sun, X., Lu, C., Zhang, L., et al. (2015). Site-specific nitrosoproteomic identification of endogenously *S*-nitrosylated proteins in Arabidopsis. *Plant Physiology, 167*, 1731–1746.
Kato, H., Takemoto, D., & Kawakita, K. (2013). Proteomic analysis of S-nitrosylated proteins in potato plant. *Physiologia Plantarum, 148*, 371–386.
Kevil, C. G., & Patel, R. P. (2010). S-Nitrosothiol biology and therapeutic potential in metabolic disease. *Current Opinion in Investigational Drugs, 11*, 1127–1134.
Lamotte, O., Bertoldo, J. B., Besson-Bard, A., Rosnoblet, C., Aimé, S., Hichami, S., et al. (2015). Protein S-nitrosylation: specificity and identification strategies in plants. *Frontiers in Chemistry, 2*, 114.
Laver, J. R., McLean, S., Bowman, L. A., Harrison, L. J., Read, R. C., & Poole, R. K. (2013). Nitrosothiols in bacterial pathogens and pathogenesis. *Antioxidant & Redox Signaling, 18*, 309–322.
Lee, U., Wie, C., Fernandez, B. O., Feelisch, M., & Vierling, E. (2008). Modulation of nitrosative stress by *S*-nitrosoglutathione reductase is critical for thermotolerance and plant growth in Arabidopsis. *The Plant Cell, 20*, 786–802.
Leterrier, M., Airaki, M., Palma, J. M., Chaki, M., Barroso, J. B., & Corpas, F. J. (2012). Arsenic triggers the nitric oxide (NO) and *S*-nitrosoglutathione (GSNO) metabolism in Arabidopsis. *Environmental Pollution, 166C*, 136–143.
Leterrier, M., Chaki, M., Airaki, M., Valderrama, R., Palma, J. M., Barroso, J. B., et al. (2011). Function of *S*-nitrosoglutathione reductase (GSNOR) in plant development and under biotic/abiotic stress. *Plant Signaling and Behavior, 6*, 789–793.
Lindermayr, C., Saalbach, G., Bahnweg, G., & Durner, J. (2006). Differential inhibition of Arabidopsis methionine adenosyltransferases by protein *S*-nitrosylation. *The Journal of Biological Chemistry, 281*, 4285–4291.
Lindermayr, C., Saalbach, G., & Durner, J. (2005). Proteomic identification of S-nitrosylated proteins in Arabidopsis. *Plant Physiology, 137*, 921–930.
Lindermayr, C., Sell, S., Müller, B., Leister, D., & Durner, J. (2010). Redox regulation of the NPR1-TGA1 system of *Arabidopsis thaliana* by nitric oxide. *The Plant Cell, 22*, 2894–2907.
Malik, S. I., Hussain, A., Yun, B. W., Spoel, S. H., & Loake, G. J. (2011). GSNOR-mediated de-nitrosylation in the plant defence response. *Plant Science, 181*, 540–544.
Murgia, I., de Pinto, M. C., Delledonne, M., Soave, C., & De Gara, L. (2004). Comparative effects of various nitric oxide donors on ferritin regulation, programmed cell death, and cell redox state in plant cells. *Journal of Plant Physiology, 161*, 777–783.
Noctor, G., Queval, G., Mhamdi, A., Chaouch, S., & Foyer, C. H. (2011). Glutathione. *Arabidopsis Book, 9*, e0142.
Parani, M., Rudrabhatla, S., Myers, R., Weirich, H., Smith, B., Leaman, D. W., et al. (2004). Microarray analysis of nitric oxide responsive transcripts in Arabidopsis. *Plant Biotechnology Journal, 2*, 359–366.
Pfeiffer, S., Mayer, B., & Hemmens, B. (1999). Nitric oxide: chemical puzzles posed by a biological messenger. *Angewandte Chemie International Edition, 38*, 1714–1731.

Polverari, A., Molesini, B., Pezzotti, M., Buonaurio, R., Marte, M., & Delledonne, M. (2003). Nitric oxide-mediated transcriptional changes in *Arabidopsis thaliana*. *Molecular Plant—Microbe Interactions, 16*, 1094—1105.

Puyaubert, J., Fares, A., Rézé, N., Peltier, J. B., & Baudouin, E. (2014). Identification of endogenously S-nitrosylated proteins in Arabidopsis plantlets: effect of cold stress on cysteine nitrosylation level. *Plant Science, 215—216*, 150—156.

del Río, L. A., Corpas, F. J., & Barroso, J. B. (2004). Nitric oxide and nitric oxide synthase activity in plants. *Phytochemistry, 65*, 783—792.

Romero-Puertas, M. C., Laxa, M., Mattè, A., Zaninotto, F., Finkemeier, I., Jones, A. M., et al. (2007). *S*-nitrosylation of peroxiredoxin II E promotes peroxynitrite-mediated tyrosine nitration. *The Plant Cell, 19*, 4120—4130.

Sánchez-Calvo, B., Barroso, J. B., & Corpas, F. J. (2013). Hypothesis: nitro-fatty acids play a role in plant metabolism. *Plant Science, 199—200*, 1—6.

Scheler, C., Durner, J., & Astier, J. (2013). Nitric oxide and reactive oxygen species in plant biotic interactions. *Current Opinion in Plant Biology, 16*, 534—539.

Serpa, V., Vernal, J., Lamattina, L., Grotewold, E., Cassia, R., & Terenzi, H. (2007). Inhibition of AtMYB2 DNA-binding by nitric oxide involves cysteine S-nitrosylation. *Biochemical and Biophysical Research Communications, 361*, 1048—1053.

Shapiro, A. D. (2005). Nitric oxide signaling in plants. *Vitamins and Hormones, 72*, 339—398.

Smith, B. C., & Marletta, M. A. (2012). Mechanisms of S-nitrosothiol formation and selectivity in nitric oxide signaling. *Current Opinion in Chemical Biology, 16*, 498—506.

Souza, J. M., Peluffo, G., & Radi, R. (2008). Protein tyrosine nitration—functional alteration or just a biomarker? *Free Radical Biology & Medicine, 45*, 357—366.

Tada, Y., Spoel, S. H., Pajerowska-Mukhtar, K., Mou, Z., Song, J., Wang, C., et al. (2008). Plant immunity requires conformational changes of NPR1 via *S*-nitrosylation and thioredoxins. *Science, 321*, 952—956.

Tanou, G., Filippou, P., Belghazi, M., Job, D., Diamantidis, G., Fotopoulos, V., et al. (2012). Oxidative and nitrosative-based signaling and associated post-translational modifications orchestrate the acclimation of citrus plants to salinity stress. *The Plant Journal, 72*, 585—599.

Tanou, G., Job, C., Rajjou, L., Arc, E., Belghazi, M., Diamantidis, G., et al. (2009). Proteomics reveals the overlapping roles of hydrogen peroxide and nitric oxide in the acclimation of citrus plants to salinity. *The Plant Journal, 60*, 795—804.

Tavares, C. P., Vernal, J., Delena, R. A., Lamattina, L., Cassia, R., & Terenzi, H. (2014). *S*-nitrosylation influences the structure and DNA binding activity of AtMYB30 transcription factor from *Arabidopsis thaliana*. *Biochimica et Biophysica Acta, 1844*, 810—817.

Terrile, M. C., París, R., Calderón-Villalobos, L. I., Iglesias, M. J., Lamattina, L., Estelle, M., et al. (2012). Nitric oxide influences auxin signaling through *S*-nitrosylation of the Arabidopsis TRANSPORT INHIBITOR RESPONSE 1 auxin receptor. *The Plant Journal, 70*, 492—500.

Trostchansky, A., Bonilla, L., González-Perilli, L., & Rubbo, H. (2013). Nitro-fatty acids: formation, redox signaling, and therapeutic potential. *Antioxidants & Redox Signaling, 19*, 1257—1265.

Valderrama, R., Corpas, F. J., Carreras, A., Fernández-Ocaña, A., Chaki, M., Luque, F., et al. (2007). Nitrosative stress in plants. *FEBS Letters, 581*, 453—461.

Vanin, A. F., Malenkova, I. V., & Serezhenkov, V. A. (1997). Iron catalyzes both decomposition and synthesis of *S*-nitrosothiols: optical and electron paramagnetic resonance studies. *Nitric Oxide, 1*, 191—203.

Vanzo, E., Ghirardo, A., Merl-Pham, J., Lindermayr, C., Heller, W., Hauck, S. M., et al. (2014). *S*-nitroso-proteome in poplar leaves in response to acute ozone stress. *PLoS One, 9*(9), e106886.

Wang, P., Du, Y., Hou, Y., Zhao, Y., Hsu, C., Yuan, F., et al. (2015). Nitric oxide negatively regulates abscisic acid signaling in guard cells by *S*-nitrosylation of OST1. *Proceedings of the National Academy of Sciences of the United States of America, 112*, 613–618.
Wojtaszek, P. (2000). Nitric oxide in plants: to NO or not to NO. *Phytochemistry, 54*, 1–4.
Yang, H., Mu, J., Chen, L., Feng, J., Hu, J., Li, L., et al. (2015). *S*-Nitrosylation positively regulates ascorbate peroxidase activity during plant stress responses. *Plant Physiology, 167*, 1604–1615.
Yu, M., Yun, B. W., Spoel, S. H., & Loake, G. J. (2012). A sleigh ride through the SNO: regulation of plant immune function by protein *S*-nitrosylation. *Current Opinion in Plant Biology, 15*, 424–430.
Yun, B. W., Feechan, A., Yin, M., Saidi, N. B., Le Bihan, T., Yu, M., et al. (2011). *S*-nitrosylation of NADPH oxidase regulates cell death in plant immunity. *Nature, 478*, 264–268.
Zaffagnini, M., Morisse, S., Bedhomme, M., Marchand, C. H., Festa, M., Rouhier, N., et al. (2013). Mechanisms of nitrosylation and denitrosylation of cytoplasmic glyceraldehyde-3-phosphate dehydrogenase from *Arabidopsis thaliana*. *The Journal of Biological Chemistry, 288*, 22777–22789.
Zhang, Y., & Hogg, N. (2005). S-Nitrosothiols: cellular formation and transport. *Free Radical Biology & Medicine, 38*, 831–838.

CHAPTER SIX

Costs and Benefits of Nitric Oxide Generation in Plants Exposed to Cadmium

Magdalena Arasimowicz-Jelonek*[,1], Jolanta Floryszak-Wieczorek[§], Karolina Izbiańska*

*Department of Plant Ecophysiology, Faculty of Biology, Adam Mickiewicz University, Poznań, Poland
[§]Department of Plant Physiology, Poznań University of Life Sciences, Poznań, Poland
[1]Corresponding author: E-mail: marasimowicz@wp.pl

Contents

Abstract

The functional role of nitric oxide (NO) produced endogenously in plants affected by cadmium (Cd) has been thoroughly studied over the past decade. However, the fundamental question of the fate of the NO message during the heavy metal (HM) stress is still open. So far, the dual role of NO in Cd toxicity has been demonstrated in various model plants. On the one hand, NO may buffer cell homoeostasis and constitutes a key element in the expression of gene sets responsible for stress tolerance. Conversely, NO contributes to Cd toxicity by promoting cellular redox misbalance, Cd uptake and

Advances in Botanical Research, Volume 77
ISSN 0065-2296
http://dx.doi.org/10.1016/bs.abr.2015.10.009

metal-induced reduction of root growth. Trying to resolve how NO switches plant responses to cope with HM stress, the current review is focused mainly on potential costs and benefits arising from NO accumulation during Cd exposure.

1. INTRODUCTION

Cadmium (Cd) is a nonessential heavy metal (HM) that has deleterious effects on all living organisms. In plants the metal's toxicity is often associated with severe disturbances of crucial metabolic processes such as water and mineral uptake and transport, nitrogen metabolism, photosynthesis and respiration. Due to its high solubility in water, the metal is rapidly taken up by plant roots, where the most evident phytotoxic symptoms are observed (DalCorso, Manara, & Furini, 2013; di Toppi & Gabbrielli, 1999). Cadmium stress causes various anatomical and structural changes in plant tissues resulting in growth reduction and organ abnormalities. On the metabolic level, signs of early response to Cd include the accumulation of reactive oxygen species (ROS), nitric oxide (NO) generation and disturbance in the antioxidant cell capacity. In consequence, these changes promote oxidative stress conditions at least partially responsible for Cd toxicity (Garnier et al., 2006; Rodríguez-Serrano et al., 2009; Sandalio, Dalurzo, Gomez, Romero-Puertas, & Del Rio, 2001).

Nitric oxide was identified as a crucial signaling molecule effective in triggering plant responses against a broad range of stress factors (e.g. Leitner, Vandelle, Gaupels, Bellin, & Delledonne, 2009). In general, this gaseous free radical is synthesized by plants as a local NO burst (the NO-hot spots type), stimulating a further sequence of defence events as soon as the first minutes after the stress stimuli have been identified. However, timing and intensity of NO generation during Cd stress vary greatly (e.g. Arasimowicz-Jelonek et al., 2012; Barroso et al., 2006; Bartha, Kolbert, & Erdei, 2005; Besson-Bard et al., 2009; De Michele et al., 2009). Cd-induced NO production in vivo observed in various plant species seems to be strictly dependent on the used metal form and concentration, duration of stress treatment, model plant and its developmental phase as well as on analyzed plant tissues or organs (Arasimowicz-Jelonek, Floryszak-Wieczorek, & Gwóźdź, 2011; Xiong, Fu, Tao, & Zhu, 2010). It appears that short-term Cd stress enhances NO synthesis in plant roots already within the first several hours of HM exposure (Arasimowicz-Jelonek et al., 2012; Besson-Bard et al., 2009; Mahmood, Gupta, & Kaiser, 2009). In contrast, a prolonged metal

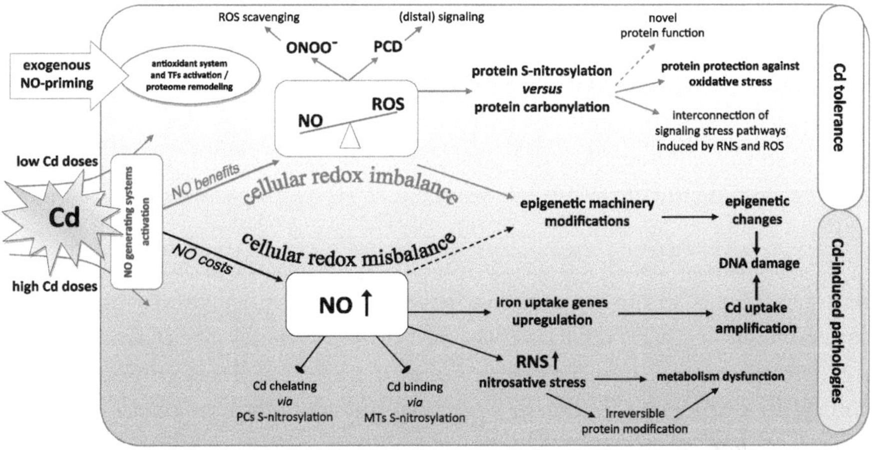

Figure 1 A potential plant's costs and benefits associated with NO generation during Cd stress. Abbreviations: MTs, metallothioneins; NO, nitric oxide; $ONOO^-$, peroxynitrite anion; PCD, programmed cell death; PCs, phytochelatins; RNS, reactive nitrogen species; ROS, reactive oxygen species; TFs, transcription factors.

treatment visibly diminishes NO content in roots (Rodriguez-Serrano et al., 2006; Xu, Wang, et al., 2010). Besides differences in timing of NO synthesis in plant cells challenged with Cd, the functional role of this molecule during HM stress is ambiguous or even opposite.

Since plants activate defence/tolerance mechanisms to protect themselves from a broad range of environmental stresses, it is likely that in certain circumstances NO generation comes at a price and the associated autotoxicity costs may occur. To understand functional NO heterogeneity in plants challenged by Cd, reported in literature, we discuss in the present review about potential costs and benefits associated with Cd-induced NO generation (Figure 1). The role of NO and ROS in the programmed cell death (PCD) triggered by Cd is further discussed chapter "Nitric Oxide and Reactive Oxygen Species in PCD Signaling".

2. NO COSTS IN CADMIUM STRESS: FROM SENSING TO AMPLIFYING CD-INDUCED PATHOLOGY

After sensing the HM the plant activates various molecular responses, from the specific gene transcription to the Cd-associated complex of regulatory networks to counteract the stress stimuli. As it was mentioned before, the success or collapse of the defence strategy of plants exposed to Cd depends on several factors concerning the severity and duration of metal stress and plant resistance to Cd exposure (Arasimowicz-Jelonek et al., 2011). It has been well documented that under prolonged stress conditions, NO contributes to cadmium toxicity in plants by promoting cellular redox misbalance. Cadmium induced oxidative stress in wheat seedlings at concentrations in soil between 3.3 and 10 mg/kg (Lin et al., 2007). A similar toxic and critical Cd concentration between 5 and 10 mg/kg was found for maize plants (Cunha & Nascimento, 2009). At these Cd concentrations, up to a 30% reduction of the dry biomass was noted and maize plants accumulated between 27 and 35 mg/kg of Cd in leaves for soils without and with liming, respectively. At the cellular level the prolonged storage of high (millimolar) concentrations of Cd indirectly induces a huge ROS generation by the impaired antioxidant machinery, which was not able to cope with the enhanced oxidative stress (De Michele et al., 2009; Valko, Morris, & Cronin, 2005). Under severe Cd stress endogenously produced NO supports metal toxicity by favouring Cd instead of Ca uptake, by dysregulation of iron homoeostasis and by oxidative/nitrosative stress promoting cell metabolic dysfunction (Besson-Bard et al., 2009). Published reports have provided data that high doses of Cd (100–200 μM $CdCl_2$) rapidly elevated endogenous NO content in the cell (Bartha et al., 2005; Besson-Bard et al., 2009; Groppa, Rosales, Iannone, & Benavides, 2008), while a longer period of plant exposure to severe Cd stress decreased NO production, probably due to the inactivation of the enzymes involved in NO synthesis (Gill, Hasanuzzaman, Nahar, Macovei, & Tuteja, 2013; Rodriguez-Serrano et al., 2009).

2.1 NO Antagonizes Cd Chelation by S-Nitrosylation or Release of Bound Cd from Metal-Binding Ligands

Plant molecular response to Cd stress is characterized by the synthesis of specific metal-chelating compounds and signaling molecules engaged in the metal uptake, transport and immobilization (Dalcorso et al., 2010). As a thiol-affectionate metal, free Cd mainly targets reduced glutathione (GSH), the most important sulphur-containing antioxidant and redox buffer in plants, playing an essential role in plant metabolism and stress tolerance (Foyer, Trebst, & Noctor, 2006). Generally, HMs induce a decrease in GSH storage, and a subsequent increase in the level of the oxidized GSH form. A severe cellular damage observed in *Arabidopsis thaliana* plants exposed to Cd was reflected in a higher oxidized glutathione (GSSG)/GSH ratio in the root tissue (Cuypers et al., 2011). Similarly, under Cd excess the GSH content was reported to decrease in shoots of barley and lettuce (Akhter, 2012). Additionally, phytochelatins (PCs) rapidly synthetized from GSH upon Cd exposure, amplified drainage of its pool (De Michele et al., 2009; Ortega-Villasante, Rellán-Álvarez, Del Campo, Carpena-Ruiz, & Hernández, 2005). In turn, when analyzing the function of NO in cadmium-challenged plants, it should be noticed that under physiological conditions NO reacts with GSH to form S-nitrosoglutathione (GSNO), having a significant role as a mobile reservoir of NO bioactivity (Barroso et al., 2006). Stress-associated huge generation of NO might modify the total pool of S-nitrosothiols (SNOs) and convert NO-exported bioactivity into altered metabolism via protein modification (Astier et al., 2011). Under cadmium stress with high doses of exogenous NO in *Boehmeria nivea* leaves, and when S-nitrosoglutathione reductase (GSNOR) activity was pharmacologically inhibited, an increase in SNO level was observed, leading to reduced GSH contents (Wang et al., 2015). Moreover, when pea plants were grown at a toxic Cd concentration (50 μM) both GSH and GSNO contents were drastically reduced, while GSNOR activity and its transcript expression were downregulated as well (Barroso et al., 2006). The process of denitrosylation is also a system precisely regulated via GSH/GSNO reductase and thioredoxin; however, it represents a less-described aspect of NO signaling in stress. Recently Vanzo et al. (2014) showed that the increase of the phenylalanine ammonia–lyase activity in response to acute ozone exposure was partially regulated by denitrosylation, which might activate the phenylpropanoid pathway within minutes after stress exposure. A decrease in *S*-nitrosylation of catalase (CAT), the protein involved in H_2O_2 detoxification, was found by Romero-Puertas, Rodríguez-Serrano, and Sandalio (2013) under cadmium treatment.

The plant S-nitrosoproteome has intensively been explored for the last 10 years and so far more than 200 protein *S*-nitrosylation targets have been proposed (e.g. Abramowski, Arasimowicz-Jelonek, Izbiańska, Billert, & Floryszak-Wieczorek, 2015; Kato, Takemoto, & Kawakita, 2013; Lindermayr, Saalbach, & Durner, 2005). *S*-nitrosylation and denitrosylation together generate the S-nitrosoproteome of a cell, so special emphasis must be given to perform a functional analysis of NO-modified proteins under Cd stress. The first important step was done with the detection and characterization of in vivo *S*-nitrosylation of PCs expressed in cadmium-stressed *A. thaliana* cells by both LC-electrospray (ESI)-MS and MS/MS analysis (De Michele et al., 2009; Elviri et al., 2010). Presented data indicate that S-nitrosylated PCs are probably less effective in intracellular Cd sequestration and detoxification, implying that excess of NO might create a negative regulatory loop and promote Cd cytotoxicity (Arasimowicz-Jelonek et al., 2011). Metallothioneins (MTs), which are small cysteine-rich proteins that chelate metals forming Cd-detoxified complexes, have been observed in some plant species (Ernst, Krauss, Verkleij, & Wesenberg, 2008; Macovei, Balestrazzi, Confalonieri, Faè, & Carbonera, 2011; Shim et al., 2009). Similarly to mammalian MTs, when subjected to NO plant MTs could trigger a release of Cd from chelating compounds and thereby play an important role in augmenting deleterious effects of free Cd (Khatai, Goessler, Lorencova, & Zangger, 2004; Misra et al., 1996).

2.2 NO-Mediated Nitrosative Stress Enhances Harmful Effects of Cd

Metabolic costs of Cd-induced NO at high concentrations may be related with the formation of peroxynitrite ($ONOO^-$) and other reactive nitrogen species (RNS), which have most widely been attributed to irreversible protein modification via tyrosine nitration and inhibition of some antioxidative enzyme activities, leading to the accumulation of ROS (Arasimowicz-Jelonek et al., 2011). Corpas and Barroso (2014) found that under 150 μM $CdCl_2$ stress, the generation of NO and of the superoxide anion ($O_2^{\bullet -}$) increased $ONOO^-$ level significantly. The latter was cell localized in peroxisomes and in the cytosol of *Arabidopsis* roots and contributed to nitro-oxidative stress conditions. Unfortunately, the knowledge of the connection between RNS-induced protein nitration and its implications to plant metabolism under cadmium stress is quite limited. Recently, the rise in the protein pool undergoing tyrosine nitration was demonstrated in response to arsenic (Leterrier et al., 2012).

2.3 Cd-Triggered NO Production Affects Divalent Ions Status

Several studies revealed that Cd enters the cells by the same uptake systems as those used by cations such as Fe, Ca and Zn. It has been well documented that Cd overabundance could compete with those divalent ions for the transporters, promoting a limitation in their uptake and leading to a nutrient deficiency (Rodrigues-Serrano et al., 2009). In turn, the mechanism related to the Cd-NO dialogue has been extensively studied with regard to NO sensing in iron deficiency (Graziano & Lamattina, 2007; Ramirez, Graziano, & Lamattina, 2008). It has been established that an enhanced NO generation in roots of tomato seedlings under iron deficiency was required for the expression and activity of iron uptake components, e.g. FRO1, IRT1 and FER. Then it was shown in *A. thaliana* roots that NO upregulates the expression of genes, including *IRT1* under iron deprivation caused by Cd (Besson-Bard et al., 2009; Besson-Bard & Wendehenne, 2009). Since the observed NO overproduction appeared to be responsible for the upstream uptake of Cd instead of Fe by IRT1 and additionally for Cd-induced perturbation in Ca accumulation, therefore it might be concluded that NO contributes to Cd toxicity and promotes its harmful effects in plants. The beneficial effect of exogenous gibberellic acid on reducing NO and alleviating Cd toxicity in wild *Arabidopsis* plants in contrast to the *IRT1* knockout mutant *irt1* confirmed the involvement of this transporter in Cd uptake (Zhu et al., 2012). Moreover, these effects were partially reversed in *Arabidopsis* roots supplemented with GSNO. Recently, a study on *Arabidopsis* HY1 (AtHO1) mutants revealed new data on upregulation of intracellular haeme oxygenase 1 (*HO1*) that might decrease NO production, increase Cd exclusion and generally improve iron homoeostasis in *Arabidopsis* root tissues (Han et al., 2014).

2.4 The Link of Cd-NO as the Causative Factor in Various Pathophysiological Abnormalities

Prolonged exposure of plant tissue to Cd excess disrupts physiological signaling processes via affecting cell surface receptors, second messengers, gene transcription and regulation, which account for Cd cytotoxicity. Overproduced NO can provoke harmful effects in plant cells as well, because its negative impact depends mainly on the NO concentration. Therefore, common Cd-NO implication may also amplify cell response to the other pathological effects, such as direct damage of enzymes and transporters, thiol oxidation or ROS/RNS production. *Arabidopsis* cell suspension cultures subjected to severe Cd stress (100 and 150 μM $CdCl_2$) revealed an NO-dependent expression of the marker *senescence-associated gene 12* (*SAG12*), preceding the internucleosomal fragmentation of DNA when the cells started to die (De Michele et al., 2009). The effects of premature senescence as the symptom of chronic exposure to sublethal amounts of Cd were earlier observed in plants (McCarthy et al., 2001; Rodriguez-Serano et al., 2006; Sandalio et al., 2001). Studies on Cd-induced ROS and DNA damage in plants should also include oxidation of RNA and the consequences of abnormal cell functioning under Cd stress, e.g. premature termination of translation or degradation of translated proteins (Chmielowska-Bąk, Izbiańska, & Deckert, 2015). Published reports have provided data that RNA is more susceptible to oxidation than DNA and its oxidation precedes cell death (Liu et al., 2012; Shan, Chang, &

Lin, 2007). So far, the link between Cd-triggered epigenetic alterations has mainly been explored in mammals (Fragou, Fragou, Kouidou, Njau, & Kovatsi, 2011). Interestingly, it was documented that it is not ROS, but rather Cd that directly interferes with the epigenome resulting in hypomethylation or hypermethylation of DNA (Huang, Zhang, Qi, Chen, & Ji, 2008). Cadmium provoked hypermethylation also in garden cress *Lepidium sativum*; however, increasing intensity of Cd stress resulted in the reduced level of DNA methylation (Yanez Barrientos, Wrobel, Lopez Torres, Gutiérrez Corona, & Wrobel, 2013). In turn, in the red alga *Gracilaria dura* Cd exposure caused a decrease in methylated DNA (Kumar, Bijo, Baghel, Reddy, & Jha, 2012).

3. BENEFITS OF NO GENERATION: FROM NO PRIMING TO CD TOLERANCE

Multiple studies have indicated that Cd exposure disturbs the redox balance inducing oxidative stress, which has been recognized as a central mechanism of Cd-induced pathologies at the cellular level. Cadmium is not a Fenton metal, therefore it is not capable of directly inducing the production of ROS (Salin, 1988). However, indirectly Cd provokes oxidative stress by various mechanisms including a displacement of redox-active metals, depletion of redox scavengers, inhibition of antioxidant enzymes and inhibition of the electron transport chain resulting in mitochondrial damage (Nair, DeGheselle, Smeets, Van Kerkhove, & Cuypers, 2013).

Synthesis of ROS is most frequently parallel with that of NO in plants challenged by Cd. Seeking the potential benefits of NO accumulation under Cd stress, it may be stated that the signaling molecule is a key component in plant cells balancing the intensity of Cd-induced oxidative stress. The cellular redox balance, resulting from the complex chemistry between NO and ROS, may therefore modify a positive stress imbalance or negative stress misbalance.

3.1 NO Priming for Cd Tolerance

Priming is widely accepted as the plant capacity to mobilize faster and more potent defence responses to a subsequent biotic or abiotic stress (Conrath, 2011). NO involvement in the priming phenomenon was highlighted by our team earlier. We showed that a precise control of synthesized NO and NO-dependent reversible modifications play an important role in integrating and coordinating defence responses during plant priming via systemic acquired resistance (SAR) inducers (Floryszak-Wieczorek et al., 2012; Janus et al., 2013). When analyzing experimental designs using NO donors for plant pretreatment, it may be observed that NO application itself could impose stress to the plants, therefore acting as the priming stimulus or a novel priming-induced compound (Groβ, Durner, & Gaupels, 2013).

As summarized by Groβ et al. (2013), plant treatment with exogenous NO always improved tolerance to abiotic stresses by favouring a decrease in the level of H_2O_2 and limiting the propagation of lipid oxidation. Alternatively, the application of NO donors upregulated the cellular antioxidant machinery under stress conditions at both gene expression and enzyme activity levels. What is more, a majority of published data have shown that the beneficial effects of exogenous NO were observed after the timing of donor decomposition. This suggests that NO released from the donor did not have a direct influence on stress-induced ROS levels, but rather facilitated more potent responses involved in the induction of signaling events controlling the cellular redox status, similarly as in plant responses to various priming agents (Groβ et al., 2013). In this sense the NO-mediated priming effect by acclimating citrus plants before a salt stress event was earlier described by Tanou et al. (2009).

Exogenously applied NO also effectively alleviated Cd toxicity. For example, pretreatment with the NO donor sodium nitroprusside (SNP) reduced the toxic effects of the HM in yellow lupine, rice, sunflower, wheat and barley (Chen et al., 2010; Kopyra & Gwóźdź, 2003; Laspina, Groppa, Tomaro, & Benavides, 2005; Singh, Batish, Kaur, Arora, & Kohli, 2008; Xiong, An, Lu, & Zhu, 2009). Increased activities of superoxide dismutase (SOD), CAT, ascorbate peroxidase (APX), guaiacol peroxidase (GPX) and glutathione reductase (GR), as well as an elevated pool of nonenzymatic antioxidants, i.e. ascorbic acid and GSH, were commonly recorded in Cd-stressed plants pretreated with NO (He, Ren, Chen, & Chen, 2014; Wang et al., 2013). Various enzymatic antioxidants were also upregulated at the transcript level (Chen et al., 2010). Interestingly, NO augmentation in different forms and doses triggered efficient mechanisms against the deleterious effects of HM stress, not only due to the attenuation of metal-induced oxidative stress (Hsu & Kao, 2005; Kopyra & Gwóźdź, 2003; Singh et al., 2008; Xu, Wang, et al., 2010). NO priming for Cd tolerance resulted also in reduced metal accumulation (He et al., 2014; Singh & Shah, 2014; Xiong et al., 2009) and in limited Cd translocation from roots to shoots as observed in rice and ryegrass seedlings, respectively (Wang et al., 2013). The mechanisms by which exogenous NO enhances cadmium tolerance could involve increasing pectin and hemicellulose contents in root cell walls favouring Cd deposition and, in consequence, decreasing Cd accumulation in the aerial parts of plants (Xiong et al., 2009).

Finally, as a good plant inducer the commonly used SNP was found to favour plant vigour. Application of SNP increased Cd-affected uptake of macro- and micronutrients, reduced the symptoms of the metal toxicity and promoted plant growth (e.g. Groppa et al., 2008; Panda, Nath, Chanu, Sharma, & Panda, 2011; Xiong et al., 2009; Xu, Wang, et al., 2010). The mitigation of the Cd-induced negative effect on root growth inhibition could be attributed to NO participation in maintaining the auxin equilibrium by reducing auxin degradation via indoleacetic acid (IAA) oxidase activity as found in roots of *Medicago truncatula* (Xu, Wang, et al., 2010).

Other benefits of exogenous NO application prior to exposure to Cd stress, as well as benefits resulting from endogenously produced NO, could involve changes in the expression of gene-coding proteins engaged in modulating the redox status at the molecular level during Cd stress. For example, the use of an NO modulator (L-NAME, a mammalian nitric oxide synthase (NOS) inhibitors) in *Arabidopsis* plants cotreated with 30 μM $CdCl_2$ revealed a set of completely suppressed genes in comparison to Cd treatment alone. This included, for instance, genes encoding a copper chaperone showing similarity to the yeast Antioxidant1 protein-protecting cells against ROS, a putative peroxidase (POX) and a germin-like protein (Besson-Bard, Pugin, & Wendehenne, 2008). In turn, in roots of soybean seedlings the NO scavenger 2-phenyl-4,4,5,5-tetramethylimidazoline-1-oxyl 3-oxide (PTIO) can regulate cellular response via induction of Cd-dependent signaling-associated genes, including *ACS*, *MAPKK2*, *DOF1* and *MYBZ2* (Chmielowska-Bąk & Deckert, 2013).

Taken together, there is evidence that plants alter their metabolism in response to a prior NO experience, being in consequence more tolerant to subsequent Cd exposure, often accompanied by endogenous NO overproduction. Molassiotis, Tanou, and Diamantidis (2010) compared NO plant preexposure to an early phase of NO generation followed by a later phase of NO formation during plant acclimation to osmotic stress, which could have a decisive role in shaping stress tolerance.

3.2 NO, PCD and Distal Signaling

The process of PCD plays an important role in various biological events, leading to controlled cellular death during the adaptation to changing environmental conditions, organ remodelling throughout development and senescence. Importantly, one of the concepts assumes that PCD is required also for an enhanced effectiveness of protective responses in neighbouring cells or in order to generate a mobile distal signal (Overmyer, Brosché, & Kangasjärvi, 2003). Thus, it is possible that through stress-induced PCD plant roots as organs sensing Cd may generate a mobile distal signal facilitating enhanced tolerance responses of other upper parts of the plant, not directly exposed to the stress stimuli, e.g. leaves. The hypothesis was supported in Cd-challenged plants, where PCD symptoms instigated by Cd pollution were engaged in root-to-leaves signaling. The terminal deoxynucleotidyl transferase-mediated dUTP nick-end labeling (TUNEL) positive reaction of lupine root cells was correlated with an overaccumulation of the main components of the plant defence machinery in leaves, including NO, H_2O_2 and salicylic acid (Arasimowicz-Jelonek et al., 2012). The observed Cd-induced PCD in lupine roots was preceded by a relatively early burst of NO. Both RNS and ROS were found to be implicated in PCD induction in cell suspensions (De Michele et al., 2009; Ma et al., 2010) and the whole plant model systems exposed to Cd stress (Arasimowicz-Jelonek et al., 2012; Ye, Li, & Xing, 2013). Most probably NO, next to ROS, participates in the controlling of the threshold for triggering PCD in plants. In *Arabidopsis* cells, both H_2O_2 and NO were necessary to trigger PCD, whereas in lupine roots $O_2^{\bullet -}$, rather than H_2O_2, functions as the molecule that synergizes with NO to switch the PCD program in response to Cd stress (Arasimowicz-Jelonek et al., 2012; De Michele et al., 2009). Additionally, Ye et al. (2013) documented that the mechanism of NO function in Cd induced PCD in *Arabidopsis* involved MPK6-mediated caspase-3-like protease activation. Chapter "Nitric Oxide and Reactive Oxygen Species in PCD Signaling" provides further details on the role of NO and ROS in Cd-induced PCD.

3.3 $ONOO^-$ Formation: NO Signal Resetting or Nitro-oxidative Stress

The maintenance of cell homoeostasis in plants suffering from Cd-mediated oxidative stress could be ascribed also to other NO-derived compounds. It is known that in the biological milieu NO and $O_2^{\bullet -}$ may interact to form $ONOO^-$. The presence of $ONOO^-$ has been demonstrated in vivo in plant cells (Arasimowicz-Jelonek et al., 2011; Corpas & Barroso, 2014; Corpas, Hayashi, Mano, Nishimura, & Barroso, 2009; Saito, Yamamoto-Katou, Yoshioka, Doke, & Kawakita, 2006). Using the *Arabidopsis* model Corpas and Barroso (2014) documented that peroxisomes are endogenous sources of $ONOO^-$ in root and guard cells, where RNS is overproduced under HM stress. Due to its chemical nature, $ONOO^-$ is an important biological oxidant and nitrated compound, contributing to oxidative and nitrosative stress in living cells (Arasimowicz-Jelonek & Floryszak-Wieczorek, 2011). However, it should be noted that the tyrosine nitration phenomenon, assumed as a reliable marker of nitro-oxidative stress, may be provoked by other cellular nitrating agents (Eiserich et al., 1998). Peroxynitrite formation could therefore provide an important regulatory loop for NO bioactivity under Cd stress, since the $ONOO^-$-mediated tyrosine nitration phenomenon can be assumed as a regulatory mechanism for protein activity. As revealed by a pioneer study (Delledonne, Zeier, Marocco, & Lamb, 2001), application of the $ONOO^-$ donor SIN-1 to soybean cell suspensions failed to reduce its viability at a concentration of up to 5 mM. Also in lupine roots, exogenous $ONOO^-$ did not promote cell death and even augmented cell viability during a short-term (24 h) 89 μM Cd exposure (Arasimowicz-Jelonek et al., 2012). This observation, suggesting that $ONOO^-$ formation might modulate cell viability by resetting the boosted NO signal and/or $O_2^{\bullet-}$ overproduction, was reported in plant cells upon Cd challenge. To combat the excess of $ONOO^-$, plant cells could then switch on effective scavenging mechanisms including enzymatic $ONOO^-$ detoxification by thiol-dependent POXs (Romero-Puertas et al., 2007; Romero-Puertas, Perazzolli, Zago, & Delledonne, 2004; Sakamoto et al., 2003).

3.4 *S*-Nitrosylation versus Carbonylation—Sophisticated Mechanisms of Cd Tolerance

The reversible covalent binding of an NO moiety to the thiol side chain of protein cysteine leads to the formation of SNOs. The NO-coded message may be stored in this posttranslational modification (PTM) of proteins and controlled by the thioredoxin system and by *trans*-nitrosylation reactions with low-molecular mass thiols, governed mainly by GSNO reductase (Chaki & Lindermayr, 2014). *S*-nitrosylation was found to be involved in the regulation of oxidative signaling under Cd stress. This PTM affected the activity of the two main H_2O_2-removing and producing enzymes in pea peroxisomes, i.e. CAT and glycolate oxidase regulating the H_2O_2 level during stress. Additionally, NO itself can affect ROS synthesis through the *S*-nitrosylation of the NADPH oxidase AtRBOHD (Yun et al., 2011).

Cysteine residues may also be targeted by ROS, resulting in oxidative protein carbonylation. Since selected cysteine residues are sensitive to both ROS and NO, they could combine two pathways that coordinate an early NO- and ROS-dependent signaling (Yu, Lamattina, Spoel, & Loake, 2014). What is more, ROS and NO may compete for same cysteine residues, which could be reactive hubs and act as cellular redox sensors capable of monitoring and coordinating the cell redox balance (Chung, Wang, Venkatraman, Murray, & Van Eyk, 2013; see also chapter "Alone NO Longer: Interactions of Nitric Oxide with Reactive Oxygen Species and Hydrogen Sulfide"). The competition between protein carbonylation and *S*-nitrosylation was earlier observed during acclimation of citrus plants to salinity stress. The authors suggested that *S*-nitrosylation provokes conformational changes in specific proteins to temporarily lock the protein structure in a particular state, under which they are no longer sensitive to irreversible carbonylation induced by ROS (Lounifi et al., 2013; Tanou et al., 2009, 2010). The NO-mediated protection of critical protein thiols from oxidation was also shown under seed desiccation. The results by Bai et al. (2011) indicated that NO reinforces seed desiccation tolerance of the recalcitrant plant *Antiaris toxicaria* by regulating antioxidant enzyme activities to limit H_2O_2 accumulation. During desiccation tolerance, NO activated the ascorbate–GSH cycle via protein *S*-nitrosylation and alleviated the inactivation of the ascorbate–GSH cycle by repressing desiccation-induced protein carbonylation (Bai et al., 2011).

ROS-mediated PTMs of proteins were extensively studied under HM stress (Gonçalves et al., 2009; Pena, Pasquini, Tomaro, & Gallego, 2006, 2007; Romero-Puertas, Palma, Gómez, Del Río, & Sandalio, 2002). Cadmium increased protein carbonylation in plants, altering the redox cell status mainly by modifying the antioxidant system and amplifying ROS overproduction at the cellular level (Romero-Puertas et al., 2004). Thus, an excess of protein carbonylation causes an irreversible oxidative process and contributes to the inhibition or impairment of multiple enzymes, affecting their biological functions. A typical example is the drastically decreased photosynthesis resulting from the carbonylation of RuBisCO subunits in pea plants exposed to Cd stress (Romero-Puertas et al., 2002; Sandalio et al., 2001). Interestingly, there is evidence that a tight link between oxidation and nitrosylation exists also in plant cells challenged by Cd. As proposed by Wang et al. (2015), artificially induced *S*-nitrosylation could reverse Cd-induced enzyme protein carbonylation in leaves of *B. nivea*. Moreover, exogenous NO was able to decrease the level of oxidized proteins in soybean seedlings exposed to the metal (Kopyra & Gwóźdź, 2003). In contrast, the diminished formation of endogenous NO (Barroso et al., 2006; Rodriguez-Serrano et al., 2009) was found to coincide with the unchanged pattern of SNOs in pea plants treated with Cd (Ortega-Galisteo et al., 2012) and an elevated protein oxidation level (Romero-Puertas et al., 2002).

4. IS THERE ANY UNIVERSALITY OF NO RESPONSE DURING HM STRESS?

It is well documented that Cd stress modulates NO generation in plants starting from the first several hours of Cd exposure (Arasimowicz-Jelonek et al., 2012; Besson-Bard et al., 2009; Mahmood et al., 2009; Pérez-Chaca et al., 2014). Although Cd is the most studied metal, a growing body of evidence suggests that NO may play an essential role in plant responses against other HMs. The production of NO has been demonstrated in vivo in several plant species exposed to various HMs (Table 1). For example, in roots of the metal accumulator plant *Brassica juncea* and of the crop plant *Pisum sativum*, the NO signal was recorded in response to 100 μM Cd, copper (Cu) or zinc (Zn). However, different NO levels have been obtained with the different metal ions. In the same model plants, NO production in response to Cu was seven to eight times higher in the first 3 h as compared to Cd treatment, which induced only a slow rise of NO (Bartha et al., 2005). A similar tendency was observed in cells of white poplar (*Populus alba* L.) exposed to Cd and Cu. The amount of NO detected 30 min following Cd exposure was approximately 30% lower than in the case of Cu treatment (Ballestrazi et al., 2009). More recently, a strong NO overproduction triggered by Zn was found in *Vicia faba* roots. The Zn-induced NO generation was significantly reduced by mammalian NOS inhibitors, i.e. L-NAME and PBITU, suggesting that an NOS-like enzyme is engaged in NO synthesis (Zou, Zheng, Yuan, Yuan, & Wang, 2012). NOS-like activity was also induced by Cd in barley roots and the dynamic distribution of endogenous NO coincided with the induced NADPH-diaphorase activity, known as a marker for mammalian NOS activity (Valentovicová, Halusková, Huttová, Mistrík, & Tamás, 2010). NOS-like dependent NO synthesis was observed also in Cd-treated *A. thaliana* (Besson-Bard et al., 2009) and pea plants (Rodríguez-Serrano et al., 2009). In turn, application of both NOS and nitrate reductase (NR) inhibitors on tomato seedlings exposed to Cu stress revealed that the Cu-induced NO accumulation is related to both NOS-like and NR activities (Wang, Yang, et al., 2010). In a recent study by Yu et al. (2012), NR was found to be the main source of NO in Pb-exposed *Pogonatherum crinitum* root cells. Interestingly, the application of 300 μM $FeSO_4$ to *Arabidopsis* cell suspension cultures was shown to lead to a rapid NO production in plastids, but the used pharmacological and genetic approaches indicated that this burst involved neither NOS-like nor NR activities (Arnaud et al., 2006).

Table 1 The Effects of Various Toxic Metals on NO Generation in Plants

Heavy Metal/Metal	Species (Organs)	Concentration and Duration of Heavy Metal/Metal Exposure	Changes in NO Level	References
Cd	*Glycine max* (suspension cultures)	4 and 7 μM Cd^{2+} 72 h	Increase	Kopyra, Stachoń-Wilk, and Gwóźdź (2006)
	Arabidopsis thaliana (suspension cultures)	150 μM Cd^{2+} 48 h	Increase	De Michele et al. (2009)
	Populus alba (suspension cultures)	150 μM Cd^{2+} 30 min	Increase	Balestrazzi et al. (2009)
	Nicotiana tabacum (suspension cultures)	150 μM Cd^{2+} 12 h	Increase	Ma et al. (2010)
	Triticum aestivum (roots)	100 μM Cd^{2+} 5 d	Increase	Groppa et al. (2008)
	T. aestivum (roots)	1–10 μM Cd^{2+} 3 h and 4 weeks	Increase	Mahmood et al. (2009)
	A. thaliana (roots)	200 μM Cd^{2+} 7 h	Increase	Besson-Bard et al. (2009)
	A. thaliana (leaves)	50 μM Cd^{2+} 96 h	Increase	Besson-Bard et al. (2009)
	Hordeum vulgare (root tips)	1 mM μM Cd^{2+} 24 h	Increase	Valentovicová et al. (2010)
	Lupinus luteus (roots)	89 μM Cd^{2+} 12 and 24 h	Increase	Arasimowicz-Jelonek et al. (2012)
	Pisum sativum (roots)	50 μM Cd^{2+} 14 d	Decrease	Rodríguez-Serrano et al. (2006)
	P. sativum (leaves)	50 μM Cd^{2+} 14 d	Decrease	Rodríguez-Serrano et al. (2009)
	Oryza sativa (roots)	100 μM Cd^{2+} 24 h	Decrease	Xiong et al. (2009)
	Medicago truncatula (roots)	50 μM Cd^{2+} 48 h	Decrease	Xu, Wang, et al. (2010)

(Continued)

Table 1 The Effects of Various Toxic Metals on NO Generation in Plants—cont'd

Heavy Metal/Metal	Species (Organs)	Concentration and Duration of Heavy Metal/Metal Exposure	Changes in NO Level	References
Fe	*A. thaliana* (suspension cultures)	300 μM Fe^{2+} 0–30 min	Increase	Arnaud et al. (2006)
Cu	*Brassica juncea, P. sativum L* (roots)	100 μM Cu^{2+} 7 days	Increase	Bartha et al. (2005)
	Panax ginseng (roots)	50 μM Cu^{2+} 24 h	Increase	Tewari, Hahn, and Paek (2008)
	P. alba (suspension cultures	150 μM Cu^{2+} 0–30 min	Increase	Ballestrazi et al. (2009)
	Lycopersicon esculentum (leaves/roots)	1 μM Cu^{2+} 24 h, 6 and 12 d	Increase	Wang, Yang, et al. (2010)
	Vicia faba (roots)	1 mM Cu^{2+} 4 h	Increase	Zou et al. (2012)
	L. esculentum (leaves)	1, 2.5 and 5 mM Cu^{2+} 24 h	Increase	Chakraborty, Chandra, and Acharya (2015)
Zn	*B. juncea, P. sativum L* (roots)	100 μM Zn^{2+} 7 days	Increase	Bartha et al. (2005)
	Solanum nigrum (roots)	200 and 400 μM Zn^{2+} 0–10 d	Initial increase (up to 2 days) then decrease	Xu, Yin, et al. (2010)
As	*Festuca arundinacea* (leaves)	25 μMAs 4 and 8 d	Increase	Jin et al. (2010)
Pb	*Pogonatherum crinitum* (root cells)	100 μM Pb^{2+} 0–40 min	Increase	Yu et al. (2012)
Al	*A. thaliana* (roots)	90 μM Al^{3+} 1 h	Decrease	Illéš et al. (2006)
	Hibiscus moscheutos (roots)	100 μM Al^{3+} 20 min	Decrease	Tian et al. (2007)

In contrast to these observations, several authors reported a negative effect of HMs stress on NO production. For example, NO levels were strongly reduced by Cd after both short- and long-term metal treatment (Rodriguez-Serrano et al., 2006, 2009; Xiong et al., 2009). In the hyperaccumulator plant *Solanum nigrum* exposed to Zn, an enhanced NO production was followed by a marked decrease in the signal generation (Xu, Wang, et al., 2010). In turn, Illes et al. (2006) found in *Arabidopsis* roots that a 1-h treatment with toxic aluminium (Al) substantially reduces NO accumulation via the inhibition of an NOS-like activity. Tian, Sun, Zhao, and Zhang (2007) also discovered that treatment with Al for 20 min induces a rapid decrease of NO in roots of *Hibiscus moscheutos* seedlings. Thus, it can be concluded that NO synthesis in response to different HMs could be achieved by different sources acting separately or jointly to deal with the stress.

Peroxynitrite generation and tyrosine residue nitration seem to be another common elements of NO metabolism during HM stress. Short-term Cd exposure (24 h) of yellow lupine plants resulted in $ONOO^-$ formation in the differentiation and elongation zone of roots (Arasimowicz-Jelonek et al., 2011). An early increase of $ONOO^-$-dependent fluorescence was also found in pea roots treated with $CuSO_4$. Interestingly, Cd supplied as 100 μM $CdCl_2$ for 24 and 48 h induced the generation of NO without $ONOO^-$ formation in pea root cells (Lechotai et al., 2011). More recently, Feigl et al. (2015) documented a strong NO production concomitant with $ONOO^-$ formation in *B. juncea* roots at 7 days after 150 μM $ZnSO_4$ treatment. The Zn-triggered changes were accompanied by protein tyrosine nitration in both control and Zn-treated plants.

Although the functional role of endogenous NO during plant responses to different types of HMs seems to be puzzling, several lines of evidence highlighted the participation of NO in HM-induced PCD events. Nitric oxide implication in PCD events was observed not only in response to Cd as discussed above (Arasimowicz-Jelonek et al., 2012; De Michele et al., 2009; Ma et al., 2010; Ye et al., 2013), but also during Zn and Cu exposure (Balestrazzi et al., 2009; Xu, Yin, Li, & Liu, 2010). Xu, Yin, et al. (2010) presented data indicating that the interplay between NO and ROS promoted Zn-induced PCD in *S. nigrum* root tips and subsequently modulated the root system architecture to adapt to Zn toxicity. Scavenging of the endogenous NO or blocking of NO synthesis by the NOS inhibitor L-NAME significantly reduced the number of observed TUNEL-positive nuclei, indicating that the signal is required for Zn-induced PCD in primary root tips (Xu, Yin, et al., 2010). Also Balestrazzi et al. (2009) noted that white poplar cell cultures exposed to Cd, Cu and Zn showed the morphological hallmarks of both PCD and necrosis, which were associated with the increase of NO production.

As we mentioned earlier, a majority of published data have shown that application of exogenous NO in different forms and doses facilitates efficient mechanisms against abiotic stresses. Similarly, Zhang, Han, Chen, Jin, and Cui (2009) noted that an exogenous application of NO alleviates the toxic effect of Cu excess through the activation of antioxidative enzymes, decreased accumulation of H_2O_2 and an adjustment of H^+-ATPase and H^+-PPase in tomato plants. Moreover, the addition of NO to Cu-stressed ryegrass plants restricted the accumulation of the metal in leaves (Dong et al., 2014). Exogenous NO was able to mitigate also nickel (Ni)-induced oxidative stress in *B. juncea* (Kazemi, Khavari-Nejad, Fahimi, Saadatmand, & Nejad-Sattari, 2010), tomato (Kazemi, 2012) and wheat (Wang, Zhang, et al., 2010). The application of SNP enhanced the activities of GPX, APX, SOD, CAT, POXs, GR, and GST and reduced the translocation of Ni from roots to shoots. Importantly, NO inhibited also the translocation of Cd and lead (Pb) from roots to the shoots of perennial ryegrass plants, thus alleviating HM toxicity in this plant (Bai et al., 2015; Wang et al., 2013). These observations are consistent with the postulated role of exogenous NO as a priming molecule upregulating the antioxidant machinery and diminishing cellular damages caused by HM stress.

5. CONCLUSIONS

Cadmium is neither desirable nor necessary for plants and is a significant environmental problem. Therefore, scientists are exploring various ways to learn how plants gained adaptation to cope with this toxic metal. A growing body of conflicting evidence suggests that NO combined with Cd can regulate cellular responses to alleviate or potentiate metal toxicity. Most recently, a pharmacological approach performed by Shi, Ye, and Chan (2014) demonstrated an interaction between NO and another gaseous signal molecule, hydrogen sulfide (H_2S), during Cd stress, which might be essential for the fate of the plant stress response to the HM. So far, the initial peak of endogenous NO in response to short-term Cd stress could play a signaling function in iron homoeostasis, PCD and root growth, whereas in long-term Cd exposure, NO is associated with an induced senescence process stemming from an excess of ROS and ethylene (Romero-Puertas et al., 2013). Despite the great progress made in this field, partly described in this paper, there are still numerous issues waiting to be discovered and applied in phytoremediation.

REFERENCES

Abramowski, D., Arasimowicz-Jelonek, M., Izbiańska, K., Billert, H., & Floryszak-Wieczorek, J. (2015). Nitric oxide modulates redox-mediated defense in potato challenged with *Phytophthora infestans*. *European Journal of Plant Pathology, 143*, 237–260.

Akhter, F. (2012). *Cadmium accumulation and distribution in lettuce and barley*. Electronic Thesis and Dissertation Repository Paper 756. Ontario, Canada: The School of Graduate and Postdoctoral Studies Western University London.

Arasimowicz-Jelonek, M., & Floryszak-Wieczorek, J. (2011). Understanding the fate of peroxynitrite in plant cells – from physiology to pathophysiology. *Phytochemistry, 72*, 681–688.

Arasimowicz-Jelonek, M., Floryszak-Wieczorek, J., Deckert, J., Rucińska-Sobkowiak, R., Gzyl, J., Pawlak-Sprada, S., et al. (2012). Nitric oxide implication in cadmium-induced programmed cell death in roots and signaling response of yellow lupine plants. *Plant Physiology and Biochemistry, 58*, 124–134.

Arasimowicz-Jelonek, M., Floryszak-Wieczorek, J., & Gwóźdź, E. A. (2011). The message of nitric oxide in cadmium challenged plants. *Plant Science, 181*, 612–620.

Arnaud, N., Murgia, I., Boucherez, J., Briat, J.-F., Cellier, F., & Gaymard, F. (2006). An iron-induced nitric oxide burst precedes ubiquitin-dependent protein degradation for *Arabidopsis AtFer1* Ferritin gene expression. *The Journal of Biological Chemistry, 281*, 23579–23588.

Astier, J., Rasul, S., Koen, E., Manzoor, H., Besson-Bard, A., Lamotte, O., et al. (2011). *S*-nitrosylation: an emerging post-translational protein modification in plants. *Plant Science, 181*, 527–533.

Bai, X., Yang, L., Tian, M., Chen, J., Shi, J., Yang, Y., et al. (2011). Nitric oxide enhances desiccation tolerance of recalcitrant *Antiaris toxicaria* seeds *via* protein *S*-nitrosylation and carbonylation. *PLoS One, 6*. http://dx.doi.org/10.1371/journal.pone.0020714.

Bai, X. Y., Dong, Y. J., Wang, Q. H., Xu, L. L., Kong, J., & Liu, S. (2015). Effects of lead and nitric oxide on photosynthesis, antioxidative ability, and mineral element content of perennial ryegrass. *Biologia Plantarum, 59*, 163–170.

Ballestrazi, A., Macovei, A., Testoni, C., Raimondi, E., Doná, M., & Carbonera, D. (2009). Nitric oxide biosynthesis in White Poplar (*Populus alba* L.) suspension cultures challenged with heavy metals. *Plant Stress, 3*, 1–6.

Barroso, J. B., Corpas, F. J., Carreras, A., Rodriguez-Serrano, M., Esteban, F. J., Fernandez-Ocana, A., et al. (2006). Localization of S-nitrosoglutathione and expression of S-nitrosoglutathione reductase in pea plants under cadmium stress. *Journal of Experimental Botany, 57*, 1785–1793.

Bartha, B., Kolbert, Z., & Erdei, L. (2005). Nitric oxide production induced by heavy metals in *Brassica juncea* L. Czern. and *Pisum sativum* L. *Acta Biologica Szegediensis, 49*, 9–12.

Besson-Bard, A., Gravot, A., Richaud, P., Auroy, P., Duc, C., Gaymard, F., et al. (2009). Nitric oxide contributes to cadmium toxicity in *Arabidopsis* by promoting cadmium accumulation in roots and by up-regulating genes related to iron uptake. *Plant Physiology, 149*, 1302–1315.

Besson-Bard, A., Pugin, A., & Wendehenne, D. (2008). New insights into nitric oxide signaling in plants. *Annual Review of Plant Biology, 59*, 21–39.

Besson-Bard, A., & Wendehenne, D. (2009). NO contributes to cadmium toxicity in *Arabidopsis thaliana* by mediating an iron deprivation response. *Plant Signaling and Behavior, 4*, 252–254.

Chaki, M., & Lindermayr, C. (2014). S-Nitrosoglutathione reductase: key regulator of plant development and stress response. In M. N. Khan, M. Mobin, F. Mohammad, & F. J. Corpas (Eds.), *Nitric oxide in plants: Metabolism and role in stress physiology* (pp. 139–151). Springer International Publishing.

Chakraborty, N., Chandra, S., & Acharya, K. (2015). Sublethal heavy metal stress stimulates innate immunity in tomato. *The Scientific World Journal*. http://dx.doi.org/10.1155/2015/208649.

Chen, F., Wang, F., Sun, H., Cai, Y., Mao, W., Zhang, G., et al. (2010). Genotype-dependent effect of exogenous nitric oxide on Cd-induced changes in antioxidative metabolism, ultrastructure, and photosynthetic performance in barley seedlings (*Hordeum vulgare*). *Journal of Plant Growth Regulation, 29*, 394–408.

Chmielowska-Bąk, J., & Deckert, J. (2013). Nitric oxide mediates Cd-dependent induction of signaling-associated genes. *Plant Signaling & Behavior, 8*, e26664.

Chmielowska-Bąk, J., Izbiańska, K., & Deckert, J. (2015). Products of lipid, protein and RNA oxidation as signals and regulators of gene expression in plants. *Frontiers in Plant Science, 6*, 405.

Chung, H. S., Wang, S. B., Venkatraman, V., Murray, C. I., & Van Eyk, J. E. (2013). Cysteine oxidative posttranslational modifications emerging regulation in the cardiovascular system. *Circulation Research, 112*, 367–381.

Conrath, U. (2011). Molecular aspects of defence priming. *Trends in Plant Science, 16*, 524–531.

Corpas, F. J., & Barroso, J. B. (2014). Peroxynitrite ($ONOO^-$) is endogenously produced in *Arabidopsis* peroxisomes and is overproduced under cadmium stress. *Annals of Botany, 113*, 87–96.

Corpas, F. J., Hayashi, M., Mano, S., Nishimura, M., & Barroso, J. B. (2009). Peroxisomes are required for in vivo nitric oxide accumulation in the cytosol following salinity stress of *Arabidopsis* plants. *Plant Physiology, 151*, 2083–2094.

Cunha, K. P. V., & Nascimento, C. W. A. (2009). Silicon effects on metal tolerance and structural changes in maize (*Zea mays* L.) grown on a cadmium and zinc enriched soil. *Water, Air & Soil Pollution, 197*, 323–330.

Cuypers, A., Smeets, K., Ruytinx, J., Opdenakker, K., Keunen, E., Remans, T., et al. (2011). The cellular redox state as a modulator in cadmium and copper responses in *Arabidopsis thaliana* seedlings. *Journal of Plant Physiology, 168*, 309–316.

DalCorso, G., Manara, A., & Furini, A. (2013). An overview of heavy metal challenge in plants: from roots to shoots. *Metallomics, 5*, 1117–1132.

De Michele, R., Vurro, E., Rigo, C., Costa, A., Elviri, L., Di Valentin, M., et al. (2009). Nitric oxide is involved in cadmium-induced programmed cell death in *Arabidopsis* suspension cultures. *Plant Physiology, 150*, 217–228.

Delledonne, M., Zeier, J., Marocco, A., & Lamb, C. (2001). Signal interactions between nitric oxide and reactive oxygen intermediates in the plant hypersensitive disease resistance response. *Proceeding of the National Academy of Sciences of United States of America, 98*, 13454–13459.

Dong, Y., Xu, L., Wang, Q., Fan, Z., Kong, J., & Bai, X. (2014). Effects of exogenous nitric oxide on photosynthesis, antioxidative ability, and mineral element contents of perennial ryegrass under copper stress. *Journal of Plant Interaction, 9*, 402–411.

Eiserich, J. P., Hristova, M., Cross, C. E., Jones, A. D., Freeman, B. A., Halliwell, B., et al. (1998). Formation of nitric oxide-derived inflammatory oxidants by myeloperoxidase in neutrophils. *Nature, 391*, 393–397.

Elviri, L., Speroni, F., Careri, M., Mangia, A., Di Toppi, L. S., & Zottini, M. (2010). Identification of in vivo nitrosylated phytochelatins in *Arabidopsis thaliana* cells by liquid chromatography-direct electrospray-linear ion trap-mass spectrometry. *Journal of Chromatography A, 1217*, 4120–4126.

Ernst, W. H. O., Krauss, G. J., Verkleij, J. A. C., & Wesenberg, D. (2008). Interaction of heavy metals with the sulphur metabolism in angiosperms from an ecological point of view. *Plant, Cell & Environment, 31*, 123–143.

Feigl, G., Lehotai, N., Molnár, Á., Ördög, A., Rodríguez-Ruiz, M., Palma, J. M., et al. (2015). Zinc induces distinct changes in the metabolism of reactive oxygen and nitrogen species (ROS and RNS) in the roots of two *Brassica* species with different sensitivity to zinc stress. *Annals of Botany, 116*, 613–625.

Floryszak-Wieczorek, J., Arasimowicz-Jelonek, M., Milczarek, G., Janus, Ł., Pawlak-Sprada, S., Abramowski, D., et al. (2012). Nitric oxide-mediated stress imprint in potato as an effect of exposure to a priming agent. *Molecular Plant-Microbe Interactions, 25*, 1469–1477.

Foyer, C. H., Trebst, A., & Noctor, G. (2006). Protective and signalling functions of ascorbate, glutathione and tocopherol in chloroplasts. In B. Demmig-Adams, & W. W. Adams (Eds.), *Advances in photosynthesis and respiration: Photoprotection, photoinhibition, gene regulation, and environment* (pp. 241–268). Dordrecht, The Netherlands: Springer Science.

Fragou, D., Fragou, A., Kouidou, S., Njau, S., & Kovatsi, L. (2011). Epigenetic mechanisms in metal toxicity. *Toxicology Mechanisms and Methods, 21*, 343–352.

Garnier, L., Simon-Plas, F., Thuleau, P., Agnel, J. P., Blein, J. P., Ranjeva, R., et al. (2006). Cadmium affects tobacco cells by a series of three waves of reactive oxygen species that contribute to cytotoxicity. *Plant, Cell & Environment, 29*, 1956–1969.

Gill, S. S., Hasanuzzaman, M., Nahar, K., Macovei, A., & Tuteja, N. (2013). Importance of nitric oxide in cadmium stress tolerance in crop plants. *Plant Physiology and Biochemistry, 63*, 254–261.

Gonçalves, J. F., Antes, F. G., Maldaner, J., Pereira, L. B., Tabaldi, L. A., Rauber, R., et al. (2009). Cadmium and mineral nutrient accumulation in potato plantlets grown under cadmium stress in two different experimental culture conditions. *Plant Physiology and Biochemistry, 47*, 814–821.

Graziano, M., & Lamattina, L. (2007). Nitric oxide accumulation is required for molecular and physiological responses to iron deficiency in tomato roots. *The Plant Journal, 52*, 949–960.

Groß, F., Durner, J., & Gaupels, F. (2013). Nitric oxide, antioxidants and prooxidants in plant defence responses. *Frontiers in Plant Science, 29*. http://dx.doi.org/10.3389/fpls.2013.00419.

Groppa, M. D., Rosales, E. P., Iannone, M. F., & Benavides, M. P. (2008). Nitric oxide, polyamines and Cd-induced phytotoxicity in wheat roots. *Phytochemistry, 69*, 2609–2615.

Han, B., Yang, Z., Xie, Y., Nie, L., Cui, J., & Shen, W. (2014). Arabidopsis HY1 confers cadmium tolerance by decreasing nitric oxide production and improving iron homeostasis. *Molecular Plant, 7*, 388–403.

He, J., Ren, Y., Chen, X., & Chen, H. (2014). Protective roles of nitric oxide on seed germination and seedling growth of rice (*Oryza sativa* L.) under cadmium stress. *Ecotoxicology and Environmental Safety, 108*, 114–119.

Hsu, Y. T., & Kao, C. H. (2005). Abscisic acid accumulation and cadmium tolerance in rice seedlings. *Physiologia Plantarum, 124*, 71–80.

Huang, D., Zhang, Y., Qi, Y., Chen, C., & Ji, W. (2008). Global DNA hypomethylation, rather than reactive oxygen species (ROS), a potential facilitator of cadmium-stimulated K562 cell proliferation. *Toxicology Letters, 179*, 43–47.

Illéš, P., Schlicht, M., Pavlovkin, J., Lichtscheidl, I., Baluška, F., & Ovecka, M. (2006). Aluminum toxicity in plants: internalization of aluminum into cells of the transition zone in *Arabidopsis* root apices relates to changes in plasma membrane potential, endosomal behavior and nitric oxide production. *Journal of Experimental Botany, 57*, 4201–4213.

Janus, Ł., Milczarek, G., Arasimowicz-Jelonek, M., Abramowski, D., Billert, H., & Floryszak-Wieczorek, J. (2013). Normoergic NO-dependent changes, triggered by a SAR inducer in potato, create more potent defense responses to *Phytophthora infestans*. *Plant Science, 211*, 23–34.

Jin, J.-W., Xu, Y.-F., & Huang, Y.-F. (2010). Protective effect of nitric oxide against arsenic-induced oxidative damage in tall fescue leaves. *African Journal of Biotechnology, 9*, 1619–1627.

Kato, H., Takemoto, D., & Kawakita, K. (2013). Proteomic analysis of S-nitrosylated proteins in potato plant. *Physiologia Plantarum, 148*, 371–386.

Kazemi, N. (2012). Effect of exogenous nitric oxide on alleviating nickel-induced oxidative stress in leaves of tomato plants. *International Journal of Agricultural Sciences, 2*, 799–809.

Kazemi, N., Khavari-Nejad, R. A., Fahimi, H., Saadatmand, S., & Nejad-Sattari, T. (2010). Effects of exogenous salicylic acid and nitric oxide on lipid peroxidation and antioxidant enzyme activities in leaves of *Brassica napus* L. under nickel stress. *Scientia Horticulturae, 126*, 402–407.

Khatai, L., Goessler, W., Lorencova, H., & Zangger, K. (2004). Modulation of nitric oxide-mediated metal release from metallothionein by the redox state of glutathione in vitro. *European Journal of Biochemistry, 271*, 2408–2416.

Kopyra, M., & Gwóźdź, E. A. (2003). Nitric oxide stimulates seed germination and counteracts the inhibitory effect of heavy metals and salinity on root growth of *Lupinus luteus*. *Plant Physiology and Biochemistry, 441*, 1011–1017.

Kopyra, M., Stachoń-Wilk, M., & Gwóźdź, E. A. (2006). Effects of exogenous nitric oxide on the antioxidant capacity of cadmium-treated soybean cell suspension. *Acta Physiologiae Plantarum, 28*, 525–536.

Kumar, M., Bijo, A. J., Baghel, R. S., Reddy, C. R. K., & Jha, B. (2012). Selenium and spermine alleviate cadmium induced toxicity in the red seaweed *Gracilaria dura* by regulating antioxidants and DNA methylation. *Plant Physiology and Biochemistry, 51*, 129–138.

Laspina, N. V., Groppa, M. D., Tomaro, M. L., & Benavides, M. P. (2005). Nitric oxide protects sunflower leaves against Cd-induced oxidative stress. *Plant Science, 169*, 323–330.

Lehotai, N., Pető, A., Bajkán, S., Erdei, L., Tari, I., & Kolbert, Z. (2011). *In vivo* and *in situ* visualization of early physiological events induced by heavy metals in pea root meristem. *Acta Physiologiae Plantarum, 33*, 2199–2207.

Leitner, M., Vandelle, E., Gaupels, F., Bellin, D., & Delledonne, M. (2009). NO signals in the haze. Nitric oxide signalling in plant defence. *Current Opinion in Plant Biology, 12*, 451–458.

Leterrier, M., Airaki, M., Palma, J. M., Chaki, M., Barroso, J. B., & Corpas, F. J. (2012). Arsenic triggers the nitric oxide (NO) and S-nitrosoglutathione (GSNO) metabolism in *Arabidopsis*. *Environmental Pollution, 166*, 136–143.

Lin, R., Wang, X., Luo, Y., Du, W., Guo, H., & Yin, D. (2007). Effects of soil cadmium on growth, oxidative stress and antioxidant system in wheat seedlings (*Triticum aestivum* L.). *Chemosphere, 69*, 89–98.

Lindermayr, C., Saalbach, G., & Durner, J. (2005). Proteomic identification of S-nitrosylated proteins in *Arabidopsis*. *Plant Physiology, 137*, 921–930.

Liu, M., Gong, X., Alluri, R. K., Wu, J., Sablo, T., & Li, Z. (2012). Characterization of RNA damage under oxidative stress in *Escherichia coli*. *Biological Chemistry, 393*, 123–132.

Lounifi, I., Arc, E., Molassiotis, A., Job, D., Rajjou, L., & Tanou, G. (2013). Interplay between protein carbonylation and nitrosylation in plants. *Proteomics, 13*, 568–578.

Ma, W., Xu, W., Xu, H., Chen, Y., He, Z., & Ma, M. (2010). Nitric oxide modulates cadmium influx during cadmium-induced programmed cell death in tobacco BY-2 cells. *Planta, 232*, 325–335.

Macovei, A., Balestrazzi, A., Confalonieri, M., Faè, M., & Carbonera, D. (2011). New insights on the barrel medicMtOGG1 and MtFPG functions in relation to oxidative stress response in planta and during seed imbibition. *Plant Physiology and Biochemistry, 49*, 1040–1050.

Mahmood, T., Gupta, K. J., & Kaiser, W. M. (2009). Cadmium stress stimulates nitric oxide production by wheat roots. *Pakistan Journal of Botany, 41*, 1285–1290.

McCarthy, I., Romero-Puertas, M. C., Palma, J. M., Sandalio, L. M., Corpas, F. J., Gomez, M., et al. (2001). Cadmium induces senescence symptoms in leaf peroxisomes of pea plants. *Plant Cell & Environment, 24*, 1065–1073.

Misra, R. R., Hochadel, J. F., Smith, G. T., Cook, J. C., Waalkes, M. P., & Wink, D. A. (1996). Evidence that nitric oxide enhances cadmium toxicity by displacing the metal from metallothionein. *Chemical Research in Toxicology, 9*, 326–332.

Molassiotis, A., Tanou, G., & Diamantidis, G. (2010). NO says more than 'YES' to salt tolerance. Salt priming and systemic nitric oxide signaling in plants. *Plant Signaling & Behavior, 5*, 209–212.

Nair, A. R., DeGheselle, O., Smeets, K., Van Kerkhove, E., & Cuypers, A. (2013). Cadmium-induced pathologies: where is the oxidative balance lost (or not)? *International Journal of Molecular Sciences, 14*, 6116–6143.

Ortega-Galisteo, A. P., Rodríguez-Serrano, M., Pazmiño, D. M., Gupta, D. K., Sandalio, L. M., & Romero-Puertas, M. C. (2012). S-nitrosylated proteins in pea (*Pisum sativum* L.) leaf peroxisomes: changes under abiotic stress. *Journal of Experimental Botany, 63*, 2089–2103.

Ortega-Villasante, C., Rellán-Álvarez, R., Del Campo, F. F., Carpena-Ruiz, R. O., & Hernández, L. E. (2005). Cellular damage induced by cadmium and mercury in *Medicago sativa*. *Journal of Experimental Botany, 56*, 2239–2251.

Overmyer, K., Brosché, M., & Kangasjärvi, J. (2003). Reactive oxygen species and hormonal control of cell death. *Trends in Plant Science, 8*, 335–342.

Panda, P., Nath, S., Chanu, T. T., Sharma, G. D., & Panda, S. K. (2011). Cadmium stress induced oxidative stress and role of nitric oxide in rice (*Oryza sativa* L.). *Acta Physiologiae Plantarum, 33*, 1737–1747.

Pena, L. B., Pasquini, L. A., Tomaro, M. L., & Gallego, S. M. (2006). Proteolytic system in sunflower (*Helianthus annuus* L.) leaves under cadmium stress. *Plant Science, 171*, 531–537.

Pena, L. B., Pasquini, L. A., Tomaro, M. L., & Gallego, S. M. (2007). 20S proteasome and accumulation of oxidized and ubiquitinated proteins in maize leaves subjected to cadmium stress. *Phytochemistry, 68*, 1139–1146.

Pérez-Chaca, M. V., Rodríguez-Serrano, M., Molina, A. S., Pedranzani, H. E., Zirulnik, F., Sandalio, L. M., et al. (2014). Cadmium induces two waves of reactive oxygen species in *Glycine max* (L.) roots. *Plant, Cell & Environment, 37*, 1672–1687.

Ramírez, L., Graziano, M., & Lamattina, L. (2008). Decoding plants adaptive responses to iron deficiency: is nitric oxide a central player? *Plant Signaling and Behavior, 3*, 795–797.

Rodríguez-Serrano, M., Romero-Puertas, M. C., Pazmino, D. M., Testillano, P. S., Risueno, M. C., Del Río, L. A., et al. (2009). Cellular response of pea plants to cadmium toxicity: cross talk between reactive oxygen species, nitric oxide, and calcium. *Plant Physiology, 150*, 229–243.

Rodríguez-Serrano, M., Romero-Puertas, M. C., Zabalza, A., Corpas, F. J., Gomez, M., Del Río, L. A., et al. (2006). Cadmium effect on oxidative metabolism of pea (*Pisum sativum* L.) roots. Imaging of reactive oxygen species and nitric oxide accumulation in vivo. *Plant, Cell & Environment, 29*, 1532–1544.

Romero-Puertas, M. C., Laxa, M., Mattè, A., Zaninotto, F., Finkemeier, I., Jones, A. M., et al. (2007). *S*-nitrosylation of peroxiredoxin II E promotes peroxynitrite-mediated tyrosine nitration. *Plant Cell, 19*, 4120–4130.

Romero-Puertas, M. C., Palma, J. M., Gómez, L. A., Del Río, L. A., & Sandalio, L. M. (2002). Cadmium causes oxidative modification of proteins in plants. *Plant, Cell & Environment, 25*, 677–686.

Romero-Puertas, M. C., Perazzolli, M., Zago, E. D., & Delledonne, M. (2004). Nitric oxide signalling functions in plant-pathogen interactions. *Cellular Microbiology, 6*, 795–803.
Romero-Puertas, M. C., Rodríguez-Serrano, M., & Sandalio, L. M. (2013). Protein *S*-nitrosylation in plants under abiotic stress: an overview. *Frontiers in Plant Science, 4*. http://dx.doi.org/10.3389/fpls.2013.00373.
Saito, S., Yamamoto-Katou, A., Yoshioka, H., Doke, N., & Kawakita, K. (2006). Peroxynitrite generation and tyrosine nitration in defense responses in tobacco BY-2 cells. *Plant and Cell Physiology, 47*, 689–697.
Sakamoto, A., Tsukamoto, S., Yamamoto, H., Ueda-Hashimoto, M., Takahashi, M., Suzuki, H., & Morikawa, H. (2003). Functional complementation in yeast reveals a protective role of chloroplast 2-Cys peroxiredoxin against reactive nitrogen species. *Plant Journal, 33*, 841–851.
Salin, M. L. (1988). Toxic oxygen species and protective systems of the chloroplast. *Physiologia Plantarum, 72*, 681–689.
Sandalio, L. M., Dalurzo, H. C., Gomez, M., Romero-Puertas, M. C., & Del Rio, L. A. (2001). Cadmium-induced changes in the growth and oxidative metabolism of pea plants. *Journal of Experimental Botany, 52*, 2115–2126.
Shan, X., Chang, Y., & Lin, C. L. (2007). Messenger RNA oxidation is an early event preceding cell death and causes reduced protein expression. *The FASEB Journal, 21*, 2753–2764.
Shi, H., Ye, T., & Chan, Z. (2014). Nitric oxide-activated hydrogen sulfide is essential for cadmium stress response in bermudagrass (*Cynodon dactylon* (L). Pers.). *Plant Physiology and Biochemistry, 74*, 99–107.
Shim, D., Hwang, J. U., Lee, J., Lee, S., Choi, Y., An, G., et al. (2009). Orthologues of the class A4 heat shock transcription factor HsfA4a confer cadmium tolerance in wheat and rice. *Plant Cell, 21*, 4031–4043.
Singh, H. P., Batish, D. R., Kaur, G., Arora, K., & Kohli, R. K. (2008). Nitric oxide (as sodium nitroprusside) supplementation ameliorates Cd toxicity in hydroponically grown wheat roots. *Environmental and Experimental Botany, 63*, 158–167.
Singh, P., & Shah, K. (2014). Evidences for reduced metal-uptake and membrane injury upon application of nitric oxide donor in cadmium stressed rice seedlings. *Plant Physiology and Biochemistry, 83*, 180–184.
Tanou, G., Job, C., Belghazi, M., Molassiotis, A., Diamantidis, G., & Job, D. (2010). Proteomic signatures uncover hydrogen peroxide and nitric oxide cross-talk signaling network in citrus plants. *Journal of Proteome Research, 9*, 5994–6006.
Tanou, G., Job, C., Rajjou, L., Arc, E., Belghazi, M., Diamantidis, G., et al. (2009). Proteomics reveals the overlapping roles of hydrogen peroxide and nitric oxide in the acclimation of citrus plants to salinity. *The Plant Journal, 60*, 795–804.
Tewari, R. K., Hahn, E.-J., & Paek, K.-Y. (2008). Modulation of copper toxicity-induced oxidative damage by nitric oxide supply in the adventitious roots of *Panax ginseng*. *Plant Cell Reports, 27*, 171–181.
Tian, Q. Y., Sun, D. H., Zhao, M. G., & Zhang, W. H. (2007). Inhibition of nitric oxide synthase (NOS) underlies aluminum-induced inhibition of root elongation in *Hibiscus moscheutos*. *New Phytologist, 174*, 322–331.
di Toppi, L. S., & Gabbrielli, R. (1999). Response to cadmium in higher plants. *Environmental and Experimental Botany, 41*, 105–130.
Valentovicová, K., Halusková, L., Huttová, J., Mistrík, I., & Tamás, L. (2010). Effect of cadmium on diaphorase activity and nitric oxide production in barley root tips. *Journal of Plant Physiology, 167*, 10–14.
Valko, M., Morris, H., & Cronin, M. T. D. (2005). Metals, toxicity and oxidative stress. *Current Medicinal Chemistry, 12*, 1161–1208.

Vanzo, E., Ghirardo, A., Merl-Pham, J., Lindermayr, C., Heller, W., Hauck, S. M., et al. (2014). S-Nitroso-proteome in poplar leaves in response to acute ozone stress. *PLoS One, 9*. http://dx.doi.org/10.1371/journal.pone.0106886.

Wang, D., Liu, Y., Tan, X., Liu, H., Zeng, G., Hu, X., et al. (2015). Effect of exogenous nitric oxide on antioxidative system and *S*-nitrosylation in leaves of *Boehmeria nivea* (L.) Gaud under cadmium stress. *Environmental Science and Pollution Research, 22*, 3489–3497.

Wang, L., Yang, L., Yanga, F., Li, X., Song, Y., Wang, X., et al. (2010). Involvements of H_2O_2 and metallothionein in NO-mediated tomato tolerance to copper toxicity. *Journal of Plant Physiology, 167*, 1298–1306.

Wang, Q. H., Liang, X., Dong, Y. J., Xu, L. L., Zhang, X. W., Hou, J., et al. (2013). Effects of exogenous nitric oxide on cadmium toxicity, element contents and antioxidative system in perennial ryegrass. *Plant and Growth Regulation, 69*, 11–20.

Wang, S. H., Zhang, H., Jiang, S. J., Zhang, L., He, Q. Y., & He, H. Q. (2010). Effects of the nitric oxide donor sodium nitroprusside on antioxidant enzymes in wheat seedling roots under nickel stress. *Russian Journal of Plant Physiology, 57*, 833–839.

Xiong, J., An, L., Lu, H., & Zhu, C. (2009). Exogenous nitric oxide enhances cadmium tolerance of rice by increasing pectin and hemicellulose contents in root cell wall. *Planta, 230*, 755–765.

Xiong, J., Fu, G., Tao, L., & Zhu, C. (2010). Roles of nitric oxide in alleviating heavy metal toxicity in plants. *Archives of Biochemistry and Biophysics, 497*, 13–20.

Xu, J., Wang, W., Yin, H., Liu, X., Sun, H., & Mi, Q. (2010). Exogenous nitric oxide improves antioxidative capacity and reduces auxin degradation in roots of *Medicago truncatula* seedlings under cadmium stress. *Plant and Soil, 326*, 321–330.

Xu, J., Yin, H., Li, Y., & Liu, X. (2010). Nitric oxide is associated with long-term zinc tolerance in *Solanum nigrum*. *Plant Physiology, 154*, 1319–1334.

Yanez Barrientos, E., Wrobel, K., Lopez Torres, A., Gutiérrez Corona, F., & Wrobel, K. (2013). Application of reversed-phase high-performance liquid chromatography with fluorimetric detection for simultaneous assessment of global DNA and total RNA methylation in *Lepidium sativum*: effect of plant exposure to Cd(II) and Se(IV). *Analytical and Bioanalytical Chemistry, 405*, 2397–2404.

Ye, Y., Li, Z., & Xing, D. (2013). Nitric oxide promotes MPK6-mediated caspase-3-like activation in cadmium-induced *Arabidopsis thaliana* programmed cell death. *Plant, Cell & Environment, 36*, 1–15.

Yu, M., Lamattina, L., Spoel, S. H., & Loake, G. J. (2014). Nitric oxide function in plant biology: a redox cue in deconvolution. *New Phytologist, 202*, 1142–1156.

Yu, Q., Sun, L., Jin, H., Chen, Q., Chen, Z., & Xu, M. (2012). Lead-induced nitric oxide generation plays a critical role in lead uptake by *Pogonatherum crinitum* root cells. *Plant and Cell Physiology, 53*, 1728–1736.

Yun, B.-W., Feechan, A., Yin, M., Saidi, N. B. B., LeBihan, T., Yu, M., et al. (2011). *S*-nitrosylation of NADPH oxidase regulates cell death in plant immunity. *Nature, 478*, 264–268.

Zhang, Y., Han, X., Chen, X., Jin, H., & Cui, X. (2009). Exogenous nitric oxide on antioxidative system and ATPase activities from tomato seedlings under copper stress. *Scientia Horticulturae, 123*, 217–223.

Zhu, X. F., Jiang, T., Wang, Z. W., Lei, G. J., Shi, Y. Z., Li, G. X., et al. (2012). Gibberellic acid alleviates cadmium toxicity by reducing nitric oxide accumulation and expression of IRT1 in *Arabidopsis thaliana*. *Journal of Hazardous Materials, 239–240*, 302–307.

Zou, T., Zheng, L. P., Yuan, H. Y., Yuan, Y. F., & Wang, J. W. (2012). The nitric oxide production and NADPH-diaphorase activity in root tips of *Vicia faba* L. under copper toxicity. *Plant Omics Journal, 5*, 115–121.

CHAPTER SEVEN

Role of NO-dependent Posttranslational Modifications in Switching Metabolic Pathways

María C. Romero-Puertas[1], Luisa M. Sandalio
Department of Biochemistry and Cellular and Molecular Biology of Plants, Estación Experimental del Zaidín (CSIC), Granada, Spain
[1]Corresponding author: E-mail: maria.romero@eez.csic.es

Contents

Abstract

Cellular metabolism is organized in a complex network of multiple steps called metabolic pathways to complete daily procedures and to maintain a metabolite balance in a coordinated way. As environmental conditions both inside and outside the cell are continuously changing, these pathways need to be highly regulated. Enzymes usually have different regulatory mechanisms at both the transcriptional and posttranscriptional level. A key feature of the signaling molecule nitric oxide (NO) is the direct alteration of proteins through posttranslational modifications (PTMs) such as *S*-nitrosylation and nitration. Certain proteins have been described as putative targets of NO-dependent PTMs and as comprising a broad range of categories, with the cellular metabolism being one of the most important. The functional effect of these NO-dependent PTMs on enzymes, the metabolic pathway flux and crosstalk between the different PTMs is

Advances in Botanical Research, Volume 77
ISSN 0065-2296
http://dx.doi.org/10.1016/bs.abr.2015.10.005

still unclear. In this review, we will discuss the impact of NO-dependent PTMs on metabolic enzymes and their possible physiological effects on the regulation of the main pathways affected by these modifications.

1. INTRODUCTION

Nitric oxide (NO) is a free radical that can gain or lose an electron to reach energetically more favourable structures, namely the nitrosonium cation (NO^+) or the nitroxyl radical (NO^-; Neill et al., 2008). Since the first study of NO generation in plants (Klepper, 1979) and studies showing the function of this molecule (Delledonne, Xia, Dixon, & Lamb, 1998; Durner, Wendehenne, & Klessig, 1998), considerable data have confirmed its role as a signaling molecule involved in physiological processes such as seed dormancy and germination, plant growth and development and also in response to numerous biotic and abiotic stresses (Beligni & Lamattina, 2000; Mur et al., 2013; Neill et al., 2008). NO is now recognized as an ubiquitous intra- and intercellular signaling molecule involved in the regulation of an impressive range of diverse cellular functions in phylogenetically distant species (Besson-Bard, Pugin, & Wendehenne, 2008; Chen, Vandelle, Bellin, & Delledonne, 2014; León, Castillo, Coego, Lozano-Juste, & Mir, 2014; Sanz et al., 2015; Yu, Lamattina, Spoel, & Loake, 2014). It is also important to note that NO, whose properties depend on the rate and location of production, has both cytotoxic and cyto-protecting/stimulating features (Beligni & Lamattina, 2001; Neill, Desikan, Clarke, Hurst, & Hancock, 2002).

While NO is a reactive molecule capable of rapidly diffusing through biological membranes, the ways in which NO functions are still largely unknown. It has been shown that NO may regulate different processes by inducing gene transcription or by activating secondary messengers (Besson-Bard et al., 2008; Gaupels, Kuruthukulangarakoola, & Durner, 2011; Palmieri et al., 2008). Changes in plant gene expression in response to gaseous NO or NO donors have been studied using medium- and large-scale transcriptional studies including cDNA-AFLP, microarrays and RNAseq analyses. Multiple NO target genes are involved in signal transduction, defence, cell death, transport, basic metabolism, reactive oxygen species (ROS) production and even degradation processes (Besson-Bard et al., 2009).

In addition, a key feature of NO's mode of action is the regulation of different biological processes by directly modifying proteins (Martínez-Ruiz,

Cadenas, & Lamas, 2011). NO is able to regulate proteins through covalent posttranslational modifications (PTMs) by connecting the metal centres of the proteins and by affecting their tyrosine (Tyr) and cysteine (Cys) residues (nitration and *S*-nitrosylation (SNO), respectively), thus changing their location, aggregation state and activity (Martínez-Ruiz et al., 2011; Souza, Peluffo, & Radi, 2008). The principal functional effects of NO may be SNO, the covalent binding of an NO group to a Cys residue (Stamler, Lamas, & Fang, 2001). Over the last 10 years, putative SNO targets in plants in response to NO donors under physiological and stress conditions have been described using small-, medium- and large-scale proteomic analyses (Astier et al., 2011; Romero-Puertas, Rodríguez-Serrano, & Sandalio, 2013). Currently, more than 1000 proteins involved in a wide range of biological processes in *Arabidopsis thaliana* and other plant species are known to be targets of SNO (Hu et al., 2015; Kovacs & Lindermayr, 2013), although the functional effect of this modification has been analyzed in only around 2% of these proteins (Astier et al., 2012; Kovacs & Lindermayr, 2013; Romero-Puertas et al., 2013). Though studied to a lesser degree, Tyr nitration is another NO-dependent PTM in which 3-nitrotyrosine is formed after a nitro group ($-NO_2$) is added to the *ortho* position of Tyr residues (N-Tyr; Vandelle & Delledonne, 2011). Nitration is a highly specific effect of NO signaling which includes changes in enzymatic activity, proteolytic degradation and protein phosphorylation (Abello, Kerstjens, Postma, & Bischoff, 2009). In this review, we will focus on the current state of knowledge regarding metabolic pathway regulation by NO-dependent PTMs. We will also elucidate the function of NO as a signaling mechanism in plants under physiological and stress conditions.

2. NO IN PLANTS: PRODUCTION AND TURNOVER

2.1 NO Production

Most NO produced in animal systems is due to a well-characterized family of enzymes called nitric oxide synthase (NOS; Bruckdorfer, 2005). NO biosynthesis in plants, however, is not such a simple process, as no genes homologous to mammalian NOS have been identified in higher plants up to now; only a plant NOS has been identified from the green alga *Ostreococcus tauri* which was found to be 45% similar to human NOS (Foresi et al., 2010). The oxidative (arginine or hydroxylamine-dependent) and reductive (nitrate-dependent) pathways for NO synthesis have been described in plants (Fröhlich & Durner, 2011; Gupta, Fernie, Kaiser, & van Dongen, 2011; Mur et al., 2013, Figure 1). Various reductive pathways for NO biosynthesis have been studied, including nitrate reductase (NR), the best known pathway for NO production in plants (Rockel, Strube, Rockel, Wildt, & Kaiser, 2002) which is capable of reducing nitrite to NO, depending on nitrite accumulation and pH levels; it has been suggested that the plasma membrane nitrite:NO reductase produces NO in roots, probably in the apoplast (Stöhr, Strube, Marx, Ullrich, & Rockel, 2001); xanthine oxidoreductase may produce NO under specific conditions such as anaerobic or phosphate deficiency (Godber et al., 2000; Wang et al., 2010) and mitochondrial nitrite reduction via cytochrome c oxidase/reductase (complexes III–V), probably under anoxic conditions (Stoimenova, Igamberdiev, Gupta, & Hill, 2007). In oxidative pathways, the existence of arginine-dependent NOS-like (NOS_l) activities has been described in several studies using NOS inhibitors and by measuring NO after the incubation of plant extracts with arginine (Moreau, Lindermayr, Durner, & Klessig, 2010; Yamasaki & Cohen, 2006). It has also been suggested that arginine-dependent NO can

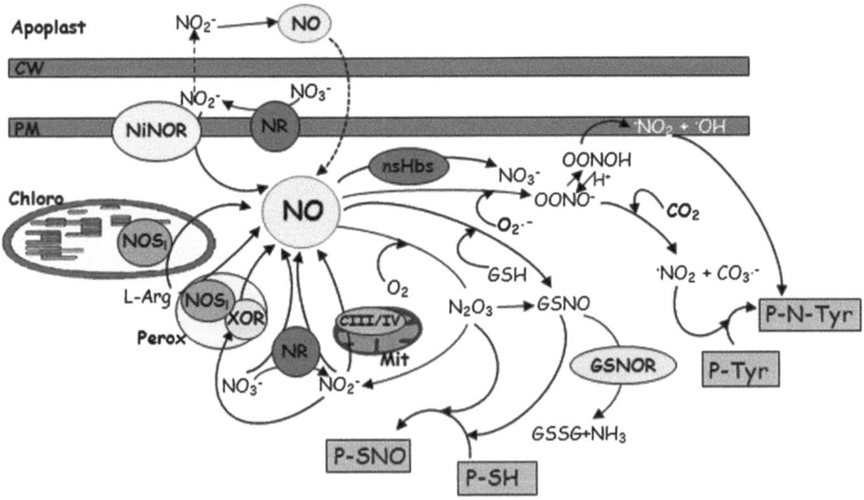

Figure 1 Overview of NO production, metabolism and scavenging in plants. The figure shows a diagram of the main sources described for NO production, including both pathways, oxidative (arginine- or hydroxylamine-dependent) and reductive (nitrate-dependent) and the main scavengers for NO including haemoglobins, oxygen, GSH and superoxide ion. Abbreviations: GSH, glutathione; GSNO, nitrosoglutathione, GSNOR, GSNO reductase; nsHbs, non-symbiotic haemoglobins; NOS_l, activity that resemble NO production as catalyzed by the animal enzyme NOS; NiNOR, plasma membrane-bound NiNOR; NR, nitrate reductase; XOR, xanthine oxidoreductase. (See colour plate)

be produced by polyamines in *Arabidopsis* seedlings (Tun et al., 2006). Plants could also produce NO as part of a ROS-dependent mechanism through incubation with exogenous hydroxylamines (Rümer, Kapuganti, & Kaiser, 2009), although the significance of this pathway is still unclear as the presence of hydroxylamines in plants has not yet been confirmed.

Most NO production described takes place at specific localizations under highly specific conditions, although the problem of determining the source(s) responsible for NO production as well as its effects and location in precise situations still needs to be addressed.

2.2 NO Metabolism and Scavenging

Given that the level of NO is essential in order to define its function and impact, NO needs to be highly regulated, removed or metabolized (Figure 1). However, our knowledge of the NO catabolism is still limited, possibly due to its inherent properties; NO has a half-life of only a few seconds and, once produced, quickly reacts with O_2 to form nitrogen dioxide (NO_2) which degrades to nitrite (a precursor to NO) and nitrate in aqueous solutions (Figure 1; Neill et al., 2008). Also, in the presence of superoxide anion ($O_2^{\bullet-}$), NO's reaction with this free radical is approximately three times faster than that of $O_2^{\bullet-}$ with its scavenger superoxide dismutase (Ischiropoulos & Al-Mehdi, 1995). Peroxynitrite ($ONOO^-$) is therefore expected to be produced in the presence of both these radicals. Tyr nitration, an $ONOO^-$ production track, incorporates NO_2 into the phenolic ring of Tyr residues to produce a 3-nitrotyrosine residue as previously mentioned. Nitration may alter the function of proteins involved in signal transduction and can be regarded as an NO-dependent signaling mechanism (see below).

NO can react reversibly with thiol groups in the Cys residues of proteins generating SNO, a key feature of NO signaling (Lindermayr & Durner, 2009) or can also react with glutathione (GSH) producing nitrosoglutathione (GSNO), considered to be a reservoir of NO (Liu et al., 2001; Sakamoto, Ueda, & Morikawa, 2002). As GSH concentrations are approximately 2–3 mM in plant cells, GSNO production may have a considerable impact on the level of NO (Neill et al., 2008). There is no data on the mechanism or reaction associated with GSNO production in plants, although it has been reported to be metabolized by GSNO reductase (GSNOR). GSNOR, which controls the level of NO and nitrosothiols, has been shown to be a key enzyme in most NO-regulated processes such as root development, pathogen defense and nitrogen assimilation (Feechan et al., 2005;

Frungillo, Skelly, Loake, Spoel, & Salgado, 2014; Rustérucci, Espunya, Díaz, Chabannes, & Martínez, 2007).

In addition to reacting with thiol groups, NO can also react with transition metals; a clear example of this is the soluble enzyme guanylate cyclase (sGC) in animals, which is stimulated by the binding of NO to its haem domain, producing an increase in the level of cGMP, a secondary messenger which, in turn, activates a number of targets related, for instance, to Ca^{2+} signaling. Although an NO-sensitive sGC homologue has not yet been identified in plants, there appears to be NO- and cGMP-dependent signaling pathways similar to those encountered in animals (Durner et al., 1998; Neill et al., 2008). On the other hand, non-symbiotic haemoglobins (nsHbs) from maize, barley and *Arabidopsis* have been shown to react with NO which they can move to a solution producing nitrate (Gupta, Hebelstrup, Mur, & Igamberdiev, 2011; Perazzolli, Romero-Puertas, & Delledonne, 2006). The AHb1 protein, which protects against an increase in NO produced under hypoxia but not against an NO burst in response to *Pseudomonas*, has been found to play a physiological role in *Arabidopsis*, suggesting that the scavenging of NO by Hb1 may depend on the source and level of the NO produced (Romero-Puertas, Perazzolli, Zago, & Delledonne, 2004). The expression of Hb1 is induced under low oxygen stress in different plant species when an increase in NO due to NR is assumed to occur (Gupta, Fernie, et al., 2011). Additionally, nsHbs appear to play a role in the establishment of symbiosis as overexpression of LjnHb1 or AfnHb1 in *Lotus japonicus* induces an increase in the number of nodules with a concomitant decrease in NO production (Hichri et al., 2015; Shimoda et al., 2009).

3. NO-DEPENDENT PTM REGULATION IN PLANTS

3.1 *S*-Nitrosylation

SNO appears to be a short-term NO-dependent mechanism where the reactive nitrogen species (RNS) N_2O_3 may be the best candidate for becoming an SNO agent (Figure 1; Martínez-Ruiz & Lamas, 2009). It has been suggested that chemically favourable conditions for SNO would require a high level of NO production to avoid the predominance of nitration and oxidation (Espey, Thomas, Miranda, & Wink, 2002), suggesting that local concentrations of different RNSs may be essential for NO-dependent PTMs to occur. The specificity of SNO is also governed by the reactivity of some Cys residues possibly supported by an acid–base mechanism (Lamotte et al., 2015; Seth & Stamler, 2011). However, an EXC sequence, where E is a glutamate residue, X indicates any amino acid residue and C represents a Cys residue, has recently been described as a putative consensus sequence present in more than 140 SNO peptides identified in plants (Hu et al., 2015). A further nine consensus sequences for SNO were identified in the same study; while an acidic amino acid residue was found to be present in all predicted SNO motifs, basic amino acid residues did not appear to be essential for the SNO motifs (Hu et al., 2015). Under physiological conditions, the SNO group is relatively labile, with regulation being carried out by reversible SNO. Although little is known about the mechanisms regulating this modification, especially in plants, more data on several denitrosylation pathways is emerging. Thus, nitrosothiols react with other reduced thiols (mainly GSH) through transnitrosylation, giving rise to the glutathionylation of the protein or to GSNO (Martínez-Ruiz et al., 2011). GSNOR activity, involved in maintaining GSNO at low levels, has been shown to be caused by the alcohol dehydrogenase in mammals and plants (Feechan et al., 2005; Jensen, Belka, & Du Bois, 1998; Liu et al., 2001), with the cycle being completed by glutathione reductase which reduces GSSG to GSH in an NADPH-dependent manner. A direct denitrosylation pathway, involving the enzyme thioredoxin (Trx) combined with Trx reductase, has been described, firstly in mammals and more recently in plants (Benhar, Forrester, Hess, & Stamler, 2008; Kneeshaw, Gelineau, Tada, Loake, & Spoel, 2014). Interestingly, *Arabidopsis* Trx h5 is able to discriminate between different substrates, suggesting specificity to protein-SNO signaling in plant defenses (Kneeshaw et al., 2014).

Putative *S*-nitrosylated targets have been extensively described in plants since the initial proteomic studies carried out on *Arabidopsis* under NO donor conditions and in response to biotic and abiotic stress (Abat, Mattoo, & Deswal, 2008; Lindermayr, Saalbach, & Durner, 2005; Romero-Puertas et al., 2008). Although over 1000 targets involved in major cellular processes have now been identified, only about 20 proteins have been characterized in precise detail (Astier et al., 2012; Kovacs & Lindermayr, 2013; Romero-Puertas et al., 2013; Yu, Yun, Spoel, & Loake, 2012).

3.2 Nitration

Following $ONOO^-$ detection, several studies have been carried out to identify related protein targets, and Tyr nitration was then defined (Pacher, Beckman, & Liaudet, 2007). Although $ONOO^-$ does not appear to be a direct nitrating agent, the simultaneous presence of both oxidants and

RNS (particularly $ONOO^-$) in all the pathways identified is essential (Souza et al., 2008). As with SNO, the increase in nitration has been shown to be selective under different conditions and for certain proteins in both animals and plants, although the factors determining this selectivity are still unclear (Abello et al., 2009; Ischiropoulos, 2003; Vandelle & Delledonne, 2011). However, nitration appears to occur on a subcellular selectivity basis, as the SNO, which differs according to the cell compartment involved and is highly pH-dependent (Abello et al., 2009). $ONOO^-$ reacts with CO_2, giving rise to powerful nitrating reagents, and catalysis caused by interaction with metal centers may also stimulate the nitration process (Abello et al., 2009; Szabó, Ischiropoulos, & Radi, 2007). Tyr nitration is a thermodynamically stable modification and has, for a long time, been considered to be irreversible. Nevertheless, in animal tissue, some studies have shown that N-Tyr levels fall in different tissues in a time-, concentration- and temperature-dependent manner, suggesting the presence of denitrase activity. Two different denitrase mechanisms, one dependent on reducing agents and another not requiring such agents (Abello et al., 2009; Kuo, Kanadia, Shanbhag, & Toro, 1999), have been detected; other nonenzymatic denitrating reactions have also been found (Akaike, Fujii, Sawa, & Ihara, 2010).

Around 130 proteins susceptible to nitration, involved in biological processes such as photosynthesis, glycolysis and nitrate assimilation, have been found in plants using proteomic analyses (Cecconi et al., 2009; Lozano-Juste, Colom-Moreno, & León, 2011; Mounira Chaki et al., 2009), although the functional significance of these modifications and most nitrated sites have been described only for a few proteins (Alvarez et al., 2011; Begara-Morales et al., 2015, 2014; Chaki et al., 2013; Galetskiy et al., 2011; Lozano-Juste et al., 2011).

4. METABOLIC PATHWAYS AFFECTED BY NO-DEPENDENT PTMs

Many metabolic enzymes and their regulators undergo PTM, resulting in changes, including degradation/stabilization and mobilization, in their active state, which optimize metabolic flux (Friso & van Wijk, 2015). About 1000 plant proteins have been identified as putative targets of SNO up to now, including 926 proteins identified in a recent study of *A. thaliana* (Hu et al., 2015), some of them described in previous studies (Lamotte et al., 2015; Romero-Puertas et al., 2013). Classification of these proteins has shown that almost 30% of the total are part of the plant metabolism

(Hu et al., 2015; Sehrawat & Deswal, 2014a). Thus, gene ontology (GO) classification of SNO targets showed that the major GO category relates to the metabolism and that the main organelle is associated with the chloroplast/plastid (Hu et al., 2015; Sehrawat & Deswal, 2014a). On the other hand, although the nitration targets studied in plants account for around 10% of the SNO targets described, a large proportion of the proteins identified have metabolic functions (Lozano-Juste et al., 2011). These results suggest that regulation of metabolic pathways may depend on NO-dependent PTMs of different enzymes under physiological and pathophysiological conditions. A concise overview of the two main pathways that may be regulated by NO in plants is provided below.

4.1 The Photorespiratory Pathway

For the purposes of CO_2 fixation, all oxygenic phototrophs use the ribulose 1,5-bisphosphate carboxylase/oxygenase (RubisCO), the most abundant protein on earth, which has two competing substrates, CO_2 and O_2, and whose carboxylation to oxygenation depends, to a great extent, on the CO_2/O_2/temperature ratio (Foyer, Bloom, Queval, & Noctor, 2009). Oxygenation of RubisCO produces 2-P-glycolate which is rescued in the photorespiratory pathway (also known as the C_2 cycle), releasing CO_2 and NH_3 and consuming ATP and NADPH (Figure 2). Photorespiration is a complex pathway involving 16 enzymes and several translocators that are distributed in three cell organelles: the chloroplast, peroxisomes and mitochondria (Figure 2). Initially, photorespiration was considered to be a wasteful, counterproductive process for CO_2 fixation under daylight conditions, leading to a decline in plant efficiency (Reumann & Weber, 2006). However, this pathway affects several plant processes such as photosystem II, plant bioenergy, the carbon and nitrogen metabolism, respiration and redox homoeostasis, especially under stress conditions (Foyer et al., 2009; Maurino & Peterhansel, 2010). Photorespiration is also the main source of H_2O_2 in photosynthetic cells and influences many signaling pathways, particularly those governing plant growth, hormonal responses to the environment and defense responses including programmed cell death (Foyer & Noctor, 2009; Hagemann et al., 2013; Maurino & Peterhansel, 2010), thus explaining why the regulation of this cycle is probably extremely important for the plant's metabolism. As many photorespiratory enzymes have been shown to be targets of phosphorylation, the photorespiratory cycle appears to be partly regulated by this PTM (Hodges, Jossier, Boex-Fontvieille, & Tcherkez, 2013). Recently, several photorespiration pathway enzymes

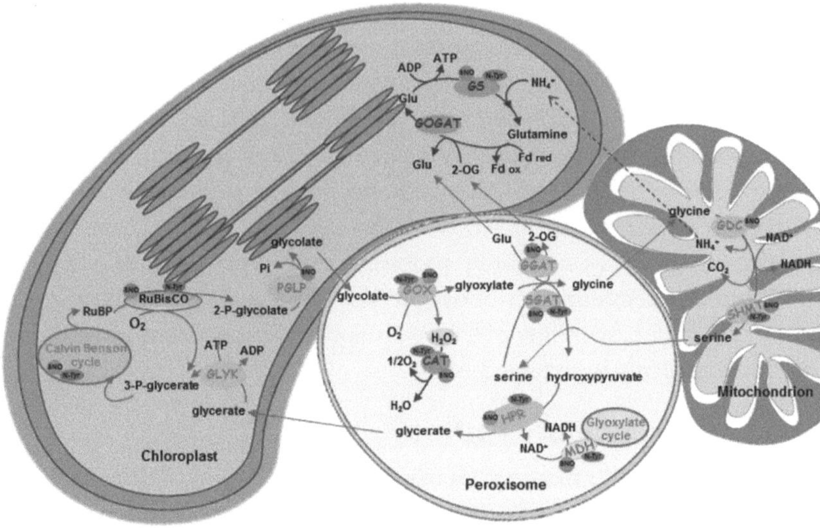

Figure 2 Targets of NO-dependent PTMs in the photorespiratory pathway and some related enzymes. Most of the enzymes involved in photorespiration pathway, involving the three organelles, chloroplasts, peroxisomes and mitochondrions, are targets for S-nitrosylation (SNO) and tyrosine nitration (N-Tyr). Abbreviations: CAT, catalase; GDC, glycine decarboxylase complex; GGAT, glutamate glyoxylate aminotransferase; GLYK, glycerate kinase; GOGAT, ferredoxin-dependent glutamate synthase; GOX, glycolate oxidase; GS, glutamine synthetase; HPR, hydroxypyruvate reductase; MDH, malate dehydrogenase; 2OG, 2-oxoglutarate; PGLP, phosphoglycolate phosphatase; RuBP, ribulose-1,5-bisphosphate; RuBisCO, RuBP carboxylase/oxygenase; SGAT, serine glyoxylate aminotransferase; SHMT, serine hydroxymethyltransferase. (See colour plate)

have been identified as targets of NO-dependent PTMs (Figure 2), suggesting that NO may also regulate this photorespiration pathway. All enzymes involved in the peroxisomal photorespiration stages: hydroxypyruvate reductase (HPR), glycolate oxidase (GOX), serine–glyoxylate aminotransferase (SGAT) and aminotransferase 1 have been found to be targets of SNO in peroxisomes isolated from pea leaves (Ortega-Galisteo et al., 2012). One of these proteins, glutamate:glyoxylate aminotransferase, has been described as differentially *S*-nitrosylated in *Arabidopsis* plantlets in response to cold stress (Puyaubert, Fares, Rézé, Peltier, & Baudouin, 2014). In addition, the key peroxisomal enzyme in this pathway, GOX, which releases H_2O_2 after oxidizing glycolate, has also been identified as a target for SNO in *Kalanchoe pinnata* plants (Abat et al., 2008) and in citrus leaves together with HPR (Tanou et al., 2009). Furthermore, GOX activity is inhibited by NO donors and the pattern of GOX-SNO changes under stress conditions, suggesting

NO-dependent regulation of photorespiration-dependent H_2O_2 production under both physiological and pathophysiological conditions (Ortega-Galisteo et al., 2012). It should be noted that catalase (CAT), one of the main enzymes involved in degrading H_2O_2 produced in peroxisomes, is also regulated by SNO and nitration (Chaki et al., 2015; Lozano-Juste et al., 2011; Ortega-Galisteo et al., 2012). Interestingly, GOX showed significantly higher SNO levels in *atgsnor* mutants (Holzmeister et al., 2011), which may need to inhibit GOX-dependent H_2O_2 production in order to avoid higher nitrosative stress levels caused by decreased GSNOR activity. Two of these peroxisomal enzymes, GOX and SGAT, were found to have been nitrated in *Arabidopsis* seedlings (Lozano-Juste et al., 2011).

Two enzymes involved in the mitochondrion steps, Ser hydroxymethyltransferase (SHMT) and Gly decarboxylase (GDC), which convert Gly to Ser, have been described as targets of SNO (Camejo et al., 2013; Holzmeister et al., 2011; Palmieri, Lindermayr, Bauwe, Steinhauser, & Durner, 2010). Notably, their patterns of SNO change under stress conditions such as salinity and infection with *Pseudomonas*. SHMT has also been identified as a target of nitration in *Arabidopsis* seedlings under normal conditions (Lozano-Juste et al., 2011). Finally, phosphoglycolate phosphatase, one of two chloroplastic photorespiration enzymes, is also regarded as a target of SNO in *Arabidopsis* plants (Hu et al., 2015).

Malate dehydrogenase (MDH), involved in fatty acid β-oxidation in germinating seeds (Pracharoenwattana, Cornah, & Smith, 2007) and associated with photorespiration during which NADH is supplied to HPR (Reumann & Weber, 2006), has also been identified as a putative *S*-nitrosylated peroxisomal protein (Ortega-Galisteo et al., 2012). Peroxisomal MDH activity is inhibited by NO donors in a concentration-dependent manner in pea leaves (Ortega-Galisteo et al., 2012); it would be interesting to determine whether this PTM affects the NAD/NADH ratio in the peroxisomes and sugar bioavailability. The mitochondrial MDH isoform is also regarded as a candidate for SNO in *Arabidopsis* leaves undergoing hypersensitive responses (Romero-Puertas et al., 2008). The chloroplastic and glyoxisomal MDH isoenzymes have also been identified as targets of nitration in *Arabidopsis* (Lozano-Juste et al., 2011).

4.2 The Calvin–Benson Cycle

The Calvin–Benson cycle (CBC), also known as the C_3 carbon fixation pathway, first discovered in the 1950s (Bassham, Benson, & Calvin, 1950), causes CO_2 fixation and its conversion into carbohydrate, which constitutes the basis of the food chain (Buchanan, Gruissem, & Rusell, 2000). The CBC is composed of three phases (carboxylation, reduction and regeneration), requiring 13 enzymes (all located in the stroma), energy (three ATP molecules in each cycle) and reduction power (two NADPH molecules in each cycle) (Buchanan et al., 2000). As many enzymes are commonly found in the CBC and glycolytic pathways (for carbohydrate degradation purposes), specific and tight regulation is necessary to ensure that, under light conditions, the synthetic apparatus is active and the degradative apparatus is inactive in order to avoid vain cycling and to ensure optimal functioning (Buchanan et al., 2000). Stromal pH and Mg^{2+} concentration are known to be key regulators of CBC enzymes such as RubisCo, fructose-1,6-bisphosphatase and phosphoribulokinase. Additionally, redox modifications of several CBC enzymes mediated by the ferredoxin—thioredoxin system provide a mechanism for light-dependent activation and deactivation, thus ensuring that carbohydrate synthesis will occur under light conditions (Foyer & Noctor, 2005; Michelet et al., 2013; Schürmann & Buchanan, 2008). Like redox-dependent proteins, all CBC enzymes have been identified as possible targets of SNO, while several target Cys residues of these proteins have also been described (Hu et al., 2015; Michelet et al., 2013). However, the effect of this PTM on the protein activity of most of these enzymes and its relationship to other redox modifications have not been studied. Triosephosphate isomerase is partially inhibited by SNO in *Chlamydomonas* (Zaffagnini et al., 2014); chloroplastic GAPDH may also be inhibited by this PTM as it exhibits strong similarity to cytosolic GAPDH which is totally inhibited by *S*-nitrosylation (Zaffagnini et al., 2013); total GAPDH activity in *Arabidopsis* extracts is reduced by up to 10% after GSNO incubation (Lindermayr et al., 2005). Additionally, GAPDH shows a transitory increase in its SNO levels in BY-2 tobacco cells in response to salt, although this process did not influence its interaction with the osmotic stress-activated protein kinase NtOSAK nor the activity of this latter (Wawer et al., 2010).

The best characterized effect of SNO has been on the RuBisCo CBC enzyme, recurrently identified as a target of SNO due to its abundance. In fact, a study after RuBisCo removal has been made in order to facilitate detection of less abundant proteins (Sehrawat, Abat, & Deswal, 2013). SNO of both the small and large subunits of RuBisCO has shown differential patterns under cold stress and inhibition of its activity by cold-mediated SNO (Abat & Deswal, 2009). Inhibition by the SNO of the CBC enzymes suggests an NO-dependent negative regulation of the cycle, possibly in order to save energy or the reduction power necessary to handle stressful conditions. It should be noted that at least half of the CBC enzymes, including RuBisCO, have also been identified as targets of nitration (Chaki et al., 2009; Lozano-Juste et al., 2011), although little data concerning the effect of this NO-dependent PTM on these enzymes has been presented. Nevertheless, treatment of *Arabidopsis* seedlings with SIN-1, a peroxynitrite donor, has been shown to partially inhibit total GAPDH activity (Lozano-Juste et al., 2011), which is similar to the data on GAPDH from yeast and mammals (Buchczyk, Briviba, Hartl, & Sies, 2000; Palamalai & Miyagi, 2010).

4.3 Nitrogen, Sulphur and Other Metabolisms

Sulphur and nitrogen metabolisms are also important targets of NO-dependent PTMs in plants (Abat et al., 2008; Sehrawat & Deswal, 2014b). Glutamine synthetase (GS), which catalyzes glutamine biosynthesis from glutamate via ATP and ammonium and plays a key role in the nitrogen metabolism, has been identified as a target of SNO, whose effect on the protein has not, however, been studied (Abat et al., 2008). Nevertheless, the nitration of the protein negatively regulates its activity in the root nodules of *M. truncatula* which affects the root nodule metabolism (Melo, Silva, Ribeiro, Seabra, & Carvalho, 2011). The nitration of GS has also been described in *Arabidopsis* seedlings (Lozano-Juste et al., 2011) and plants undergoing hypersensitive responses (Cecconi et al., 2009). In *K. pinnata* and *Arabidopsis* seedlings, methionine synthase (MS), which catalyzes the first step of methionine biosynthesis and is linked to both the carbon and sulphur metabolism, has been identified as a target of SNO and nitration (Abat et al., 2008; Lozano-Juste et al., 2011). Although MS activity has been reported to be regulated by NO in mammals and plant tissues (Brouwer, Chamulitrat, Ferruzzi, Sauls, & Weinberg, 1996; Danishpajooh et al., 2001; Nicolaou, Kenyon, Gibbons, Ast, & Gibbons, 1996), the specific effects of both these NO-dependent PTMs have not yet been described. MS is not the only protein involved in methionine biosynthesis to be described as a target of NO-dependent PTMs; two other enzymes, *S*-adenosylmethionine synthetases 1 and 2 and *S*-adenosylhomocysteinase 1, have been found to be targets of SNO and nitration, suggesting that NO may regulate the biosynthesis of this amino acid as well as ethylene in plants (Chaki et al., 2009; Lindermayr, Saalbach, Bahnweg, & Durner, 2006; Lozano-Juste et al., 2011). Additionally, the last step in the sulphur assimilation process is carried out by *O*-acetylserine(thiol)lyase, whose cytosolic isoenzyme has been shown to be inhibited by the nitration of the Tyr residue 302 (Alvarez et al., 2011). This activity is essential in order to control the biosynthesis of the key molecules Cys, Met and GSH, especially under stress conditions.

Glucosinolates, which are sulphur- and nitrogen-containing metabolites in the Brassicaceae family and which have been associated with nutritional status, growth, stress responses and plant—pathogen interactions, may also be regulated by NO-dependent PTMs through the SNO of myrosinase which hydrolyzes these metabolites (Sehrawat & Deswal, 2014a). Other enzymes involved in the catabolism of glucosinolates, such as the myrosinase-associated protein, the epithiospecifier protein in *Brassica juncea* (Sehrawat et al., 2013; Sehrawat & Deswal, 2014a) and the myrosinase-binding protein in *Arabidopsis* (Lindermayr et al., 2005), have been described as targets of SNO, thus confirming the role played by NO in the regulation of glucosinolate levels.

On the other hand, almost 30% of the proteins associated with the chlorophyll metabolism have been identified as possible targets of SNO (Hu et al., 2015). These proteins are involved in both the biosynthesis and degradation of chlorophyll, suggesting that NO plays a key role in chlorophyll homoeostasis. Although no physiological role for SNO has been proven in these proteins, chlorophyll content has been found to be lower in *gsnor* mutants than in WT plants (Hu et al., 2015; Lee, Wie, Fernandez, Feelisch, & Vierling, 2008), suggesting that there is an inverse relation between NO levels and chlorophyll content through the SNO of enzymes involved in the chlorophyll metabolism.

5. CONCLUSIONS AND FUTURE RESEARCH

There is a large set of plant proteins susceptible to direct modification and regulation by NO through posttranscriptional modifications, a high percentage of which belong to metabolic pathways that enable plants to function as a complex system and to complete their life cycle under normal conditions. These metabolic pathway networks need to be subjected to finely tuned regulation, especially when plants have to deal with changes in environmental conditions. Thus, NO-dependent PTMs are emerging as key mechanisms in the regulation of metabolic pathways, which are deemed to be as important as other modifications such as phosphorylation. However, the way in which PTMs affect metabolic enzymes or pathways in vivo is still not clear, partly due to technical difficulties, such as the measurement of the input and output of reactions, and partly due to flux through the enzyme or pathways. NO-dependent PTMs are known to be specific and their localization is a key factor in the development of the modification and its effects, which is necessary for the metabolic pathway flux. The major challenge is to study the crosstalk between multiple PTMs, especially between NO-dependent PTMs which probably depend on the specific RNS formed, microenvironmental conditions such as pH and metabolites concentration and their specific impact on the enzyme. Therefore, unravelling the PTM puzzle, involving a predictive and thoughtful study of the combinatorial functions of PTMs, is needed to convert the growing collection of data into an understanding of plant metabolic regulation. The SNO of proteins, kinases and phosphatases has now been shown to affect an extensive range of phosphorylation-/dephosphorylation-dependent signal transduction pathways in animals, though not yet in plants (Hess & Stamler, 2012). On the other hand, protein phosphorylation interferes with SNO levels and consequently protein activity (Chen, Mathias, Falero-Perez, & Kim, 2015), thus demonstrating a functional crosstalks between these two PTMs. A similar crosstalk has been shown in relation to Tyr nitration and phosphorylation (Greenacre & Ischiropoulos, 2001; Schopfer, Baker, & Freeman, 2003). Further research into vegetal tissue is needed in order to clarify the existence of crosstalk between different NO-dependent PTMs and phosphorylation.

Additionally, crosstalk among SNO and alternative redox modifications may be anticipated, as SNO could prevent further oxidation of protein thiols, catalyse disulphide formation, enhance glutathionylation and even regulate oxidases and reductases (Hess & Stamler, 2012). Analysis of interplays and crosstalks between other redox modifications and SNO, especially in metabolic pathways such as CBCs which are tightly regulated by these redox modifications, should provide us with a basis to determine the physiological roles of different modifications in cellular metabolisms and their capacity to respond to environmental cues.

ACKNOWLEDGEMENTS

This work was supported by ERDF-cofinanced grant BIO2012-36742 within the framework of the Spanish Research and Development Plan and by the EX12-BIO-296 and BIO-337 grants from the Junta de Andalucía in Spain. The authors would like to thank Ms. Laura Terrón for the organelle illustration and Michael O'Shea for the English editing.

REFERENCES

Abat, J. K., & Deswal, R. (2009). Differential modulation of S-nitrosoproteome of *Brassica juncea* by low temperature: change in S-nitrosylation of Rubisco is responsible for the inactivation of its carboxylase activity. *Proteomics, 9*(18), 4368–4380.

Abat, J. K., Mattoo, A. K., & Deswal, R. (2008). S-nitrosylated proteins of a medicinal CAM plant *Kalanchoe pinnata* – ribulose-1,5-bisphosphate carboxylase/oxygenase activity targeted for inhibition. *FEBS Journal, 275*(11), 2862–2872.

Abello, N., Kerstjens, H. A. M., Postma, D. S., & Bischoff, R. (2009). Protein tyrosine nitration: selectivity, physicochemical and biological consequences, denitration, and proteomics methods for the identification of tyrosine-nitrated proteins. *Journal of Proteome Research, 8*(7), 3222–3238.

Akaike, T., Fujii, S., Sawa, T., & Ihara, H. (2010). Cell signaling mediated by nitrated cyclic guanine nucleotide. *Nitric Oxide: Biology and Chemistry/Official Journal of the Nitric Oxide Society, 23*(3), 166–174.

Alvarez, C., Lozano-Juste, J., Romero, L. C., García, I., Gotor, C., & León, J. (2011). Inhibition of *Arabidopsis* O-acetylserine(thiol)lyase A1 by tyrosine nitration. *The Journal of Biological Chemistry, 286*(1), 578–586.

Astier, J., Kulik, A., Koen, E., Besson-Bard, A., Bourque, S., Jeandroz, S., et al. (2012). Protein S-nitrosylation: What's going on in plants? *Free Radical Biology and Medicine, 53*(5), 1101–1110.

Astier, J., Rasul, S., Koen, E., Manzoor, H., Besson-Bard, A., Lamotte, O., et al. (2011). S-nitrosylation: an emerging post-translational protein modification in plants. *Plant Science, 181*(5), 527–533.

Bassham, J. A., Benson, A. A., & Calvin, M. (1950). The path of carbon in photosynthesis. *The Journal of Biological Chemistry, 185*(2), 781–787 (Retrieved).

Begara-Morales, J. C., Sanchez-Calvo, B., Chaki, M., Mata-Perez, C., Valderrama, R., Padilla, M. N., et al. (2015). Differential molecular response of monodehydroascorbate reductase and glutathione reductase by nitration and S-nitrosylation. *Journal of Experimental Botany, 66*(19), 5983–5996.

Begara-Morales, J. C., Sánchez-Calvo, B., Chaki, M., Valderrama, R., Mata-Pérez, C., López-Jaramillo, J., et al. (2014). Dual regulation of cytosolic ascorbate peroxidase (APX) by tyrosine nitration and S-nitrosylation. *Journal of Experimental Botany, 65*(2), 527–538.

Beligni, M. V., & Lamattina, L. (2000). Nitric oxide stimulates seed germination and de-etiolation, and inhibits hypocotyl elongation, three light-inducible responses in plants. *Planta, 210*(2), 215–221.

Beligni, M. V., & Lamattina, L. (2001). Nitric oxide: a non-traditional regulator of plant growth. *Trends in Plant Science, 6*(11), 508–509.

Benhar, M., Forrester, M. T., Hess, D. T., & Stamler, J. S. (2008). Regulated protein denitrosylation by cytosolic and mitochondrial thioredoxins. *Science (New York, NY), 320*(5879), 1050–1054.

Besson-Bard, A., Astier, J., Rasul, S., Wawer, I., Dubreuil-Maurizi, C., Jeandroz, S., et al. (2009). Current view of nitric oxide-responsive genes in plants. *Plant Science, 177*(4), 302–309.

Besson-Bard, A., Pugin, A., & Wendehenne, D. (2008). New insights into nitric oxide signaling in plants. *Annual Review of Plant Biology, 59*, 21–39.

Brouwer, M., Chamulitrat, W., Ferruzzi, G., Sauls, D. L., & Weinberg, J. B. (1996). Nitric oxide interactions with cobalamins: biochemical and functional consequences. *Blood, 88*(5), 1857–1864.

Bruckdorfer, R. (2005). The basics about nitric oxide. *Molecular Aspects of Medicine, 26*(1–2), 3–31.

Buchanan, B. B., Gruissem, W., & Rusell, L. J. (2000). *Biochemistry and molecular biology of plants*. Rockville, MD: American Society of Plant Physiologists.

Buchczyk, D. P., Briviba, K., Hartl, F. U., & Sies, H. (2000). Responses to peroxynitrite in yeast: glyceraldehyde-3-phosphate dehydrogenase (GAPDH) as a sensitive intracellular target for nitration and enhancement of chaperone expression and ubiquitination. *Biological Chemistry, 381*(2), 121–126.

Camejo, D., Romero-Puertas, M. D. C., Rodríguez-Serrano, M., Sandalio, L. M., Lázaro, J. J., Jiménez, A., et al. (2013). Salinity-induced changes in S-nitrosylation of pea mitochondrial proteins. *Journal of Proteomics, 79*, 87–99.

Cecconi, D., Orzetti, S., Vandelle, E., Rinalducci, S., Zolla, L., & Delledonne, M. (2009). Protein nitration during defense response in *Arabidopsis thaliana*. *Electrophoresis, 30*(14), 2460–2468.

Chaki, M., Alvarez de Morales, P., Ruiz, C., Begara-Morales, J. C., Barroso, J. B., Corpas, F. J., et al. (2015). Ripening of pepper (*Capsicum annuum*) fruit is characterized by an enhancement of protein tyrosine nitration. *Annals of Botany*, 1–11.

Chaki, M., Carreras, A., López-Jaramillo, J., Begara-Morales, J. C., Sánchez-Calvo, B., Valderrama, R., et al. (2013). Tyrosine nitration provokes inhibition of sunflower carbonic anhydrase (β-CA) activity under high temperature stress. *Nitric Oxide: Biology and Chemistry/Official Journal of the Nitric Oxide Society, 29*, 30–33.

Chaki, M., Valderrama, R., Fernández-Ocaña, A. M., Carreras, A., López-Jaramillo, J., Luque, F., et al. (2009). Protein targets of tyrosine nitration in sunflower (*Helianthus annuus* L.) hypocotyls. *Journal of Experimental Botany, 60*(15), 4221–4234.

Chen, J., Vandelle, E., Bellin, D., & Delledonne, M. (2014). Detection and function of nitric oxide during the hypersensitive response in *Arabidopsis thaliana*: Where there's a will there's a way. *Nitric Oxide: Biology and Chemistry/Official Journal of the Nitric Oxide Society, 43*(July), 81–88.

Chen, Y., Mathias, L., Falero-Perez, J. M., & Kim, S. F. (2015). PKA-mediated phosphorylation of Dexras1 suppresses iron trafficking by inhibiting S-nitrosylation. *FEBS Letters*, 1–8.

Danishpajooh, I. O., Gudi, T., Chen, Y., Kharitonov, V. G., Sharma, V. S., & Boss, G. R. (2001). Nitric oxide inhibits methionine synthase activity in vivo and disrupts carbon flow through the folate pathway. *The Journal of Biological Chemistry, 276*(29), 27296–27303.

Delledonne, M., Xia, Y., Dixon, R. A., & Lamb, C. (1998). Nitric oxide functions as a signal in plant disease resistance. *Nature, 394*(6693), 585–588.

Durner, J., Wendehenne, D., & Klessig, D. F. (1998). Defense gene induction in tobacco by nitric oxide, cyclic GMP, and cyclic ADP-ribose. *Proceedings of the National Academy of Sciences of the United States of America, 95*(17), 10328–10333.

Espey, M. G., Thomas, D. D., Miranda, K. M., & Wink, D. A. (2002). Focusing of nitric oxide mediated nitrosation and oxidative nitrosylation as a consequence of reaction with superoxide. *Proceedings of the National Academy of Sciences of the United States of America, 99*(17), 11127–11132.

Feechan, A., Kwon, E., Yuri, B., Wang, Y., Pallas, J., & Loake, G. (2005). A central role for S-nitrosothiols in plant disease resistance. *Proceedings of the National Academy of Sciences of the United States of America, 102*(22), 8054–8059.

Foresi, N., Correa-Aragunde, N., Parisi, G., Caló, G., Salerno, G., & Lamattina, L. (2010). Characterization of a nitric oxide synthase from the plant kingdom: NO generation from the green alga *Ostreococcus tauri* is light irradiance and growth phase dependent. *The Plant Cell, 22*(11), 3816–3830.

Foyer, C. H., Bloom, A. J., Queval, G., & Noctor, G. (2009). Photorespiratory metabolism: genes, mutants, energetics, and redox signaling. *Annual Review of Plant Biology, 60*, 455–484.

Foyer, C. H., & Noctor, G. (2005). Redox homeostasis and antioxidant signaling: a metabolic interface between stress perception and physiological responses. *The Plant Cell, 17*(7), 1866–1875.
Foyer, C. H., & Noctor, G. (2009). Redox regulation in photosynthetic organisms: signaling, acclimation, and practical implications. *Antioxidants & Redox Signaling, 11*(4), 861–905.
Friso, G., & van Wijk, K. J. (2015). Update: Post-translational protein modifications in plant metabolism. *Plant Physiology*. pii:pp.01378.2015.
Fröhlich, A., & Durner, J. (2011). The hunt for plant nitric oxide synthase (NOS): is one really needed? *Plant Science, 181*(4), 401–404.
Frungillo, L., Skelly, M. J., Loake, G. J., Spoel, S. H., & Salgado, I. (2014). S-nitrosothiols regulate nitric oxide production and storage in plants through the nitrogen assimilation pathway. *Nature Communications, 5*, 1–10.
Galetskiy, D., Lohscheider, J. N., Kononikhin, A. S., Popov, I. A., Nikolaev, E. N., & Adamska, I. (2011). Phosphorylation and nitration levels of photosynthetic proteins are conversely regulated by light stress. *Plant Molecular Biology*, 77(4–5), 461–473.
Gaupels, F., Kuruthukulangarakoola, G. T., & Durner, J. (2011). Upstream and downstream signals of nitric oxide in pathogen defence. *Current Opinion in Plant Biology, 14*(6), 707–714.
Godber, B. L. J., Doel, J. J., Sapkota, G. P., Blake, D. R., Stevens, C. R., Eisenthal, R., et al. (2000). Reduction of nitrite to nitric oxide catalyzed by xanthine oxidoreductase. *The Journal of Biological Chemistry, 275*(11), 7757–7763.
Greenacre, S. A., & Ischiropoulos, H. (2001). Tyrosine nitration: localisation, quantification, consequences for protein function and signal transduction. *Free Radical Research, 34*(6), 541–581.
Gupta, K. J., Fernie, A. R., Kaiser, W. M., & van Dongen, J. T. (2011). On the origins of nitric oxide. *Trends in Plant Science, 16*(3), 160–168.
Gupta, K. J., Hebelstrup, K. H., Mur, L. A. J., & Igamberdiev, A. U. (2011). Plant hemoglobins: Important players at the crossroads between oxygen and nitric oxide. *FEBS Letters, 585*(24), 3843–3849.
Hagemann, M., Fernie, A. R., Espie, G. S., Kern, R., Eisenhut, M., Reumann, S., et al. (2013). Evolution of the biochemistry of the photorespiratory C2 cycle. *Plant Biology, 15*(4), 639–647.
Hess, D. T., & Stamler, J. S. (2012). Regulation by S-nitrosylation of protein post-translational modification. *The Journal of Biological Chemistry, 287*(7), 4411–4418.
Hichri, I., Boscari, A., Castella, C., Rovere, M., Puppo, A., & Brouquisse, R. (2015). Nitric oxide: a multifaceted regulator of the nitrogen-fixing symbiosis. *Journal of Experimental Botany, 66*(10), 2877–2887.
Hodges, M., Jossier, M., Boex-Fontvieille, E., & Tcherkez, G. (2013). Protein phosphorylation and photorespiration. *Plant Biology, 15*(4), 694–706.
Holzmeister, C., Fröhlich, A., Sarioglu, H., Bauer, N., Durner, J., & Lindermayr, C. (2011). Proteomic analysis of defense response of wildtype *Arabidopsis thaliana* and plants with impaired NO- homeostasis. *Proteomics, 11*(9), 1664–1683.
Hu, J., Huang, X., Chen, L., Sun, X., Lu, C., Zhang, L., et al. (2015). Site-specific nitrosoproteomic identification of endogenously S -Nitrosylated proteins in Arabidopsis. *Plant Physiology, 167*(4), 1731–1746.
Ischiropoulos, H. (2003). Biological selectivity and functional aspects of protein tyrosine nitration. *Biochemical and Biophysical Research Communications, 305*(3), 776–783.
Ischiropoulos, H., & Al-Mehdi, A. B. (1995). Peroxynitrite-mediated oxidative protein modifications. *FEBS Letters, 364*(3), 279–282.
Jensen, D. E., Belka, G. K., & Du Bois, G. C. (1998). S-Nitrosoglutathione is a substrate for rat alcohol dehydrogenase class III isoenzyme. *The Biochemical Journal, 331*(Pt 2), 659–668.

Klepper, L. A. (1979). Effects of certain herbicides and their combinations on nitrate and nitrite reduction. *Plant Physiology, 64*(2), 273–275.

Kneeshaw, S., Gelineau, S., Tada, Y., Loake, G. J., & Spoel, S. H. (2014). Selective protein denitrosylation activity of thioredoxin-h5 modulates plant immunity. *Molecular Cell, 56*(1), 153–162.

Kovacs, I., & Lindermayr, C. (May 2013). Nitric oxide-based protein modification: formation and site-specificity of protein S-nitrosylation. *Frontiers in Plant Science, 4*, 1–10.

Kuo, W. N., Kanadia, R. N., Shanbhag, V. P., & Toro, R. (1999). Denitration of peroxynitrite-treated proteins by "protein nitratases" from rat brain and heart. *Molecular and Cellular Biochemistry, 201*(1–2), 11–16.

Lamotte, O., Bertoldo, J. B., Besson-Bard, A., Rosnoblet, C., Aimé, S., Hichami, S., et al. (January 2015). Protein S-nitrosylation: specificity and identification strategies in plants. *Frontiers in Chemistry, 2*, 1–10.

Lee, U., Wie, C., Fernandez, B. O., Feelisch, M., & Vierling, E. (2008). Modulation of nitrosative stress by S-nitrosoglutathione reductase is critical for thermotolerance and plant growth in Arabidopsis. *The Plant Cell, 20*(3), 786–802.

León, J., Castillo, M. C., Coego, A., Lozano-Juste, J., & Mir, R. (2014). Diverse functional interactions between nitric oxide and abscisic acid in plant development and responses to stress. *Journal of Experimental Botany, 65*(4), 907–921.

Lindermayr, C., & Durner, J. (2009). S-Nitrosylation in plants: pattern and function. *Journal of Proteomics, 73*(1), 1–9.

Lindermayr, C., Saalbach, G., Bahnweg, G., & Durner, J. (2006). Differential inhibition of *arabidopsis* methionine adenosyltransferases by protein S-nitrosylation. *The Journal of Biological Chemistry, 281*(7), 4285–4291.

Lindermayr, C., Saalbach, G., & Durner, J. (March 2005). Proteomic identification of S-nitrosylated proteins. *Plant Physiology, 137*, 921–930.

Liu, L., Hausladen, A., Zeng, M., Que, L., Heitman, J., & Stamler, J. S. (2001). A metabolic enzyme for S-nitrosothiol conserved from bacteria to humans. *Nature, 410*(6827), 490–494.

Lozano-Juste, J., Colom-Moreno, R., & León, J. (2011). In vivo protein tyrosine nitration in *Arabidopsis thaliana*. *Journal of Experimental Botany, 62*(10), 3501–3517.

Martínez-Ruiz, A., Cadenas, S., & Lamas, S. (2011). Nitric oxide signaling: classical, less classical, and nonclassical mechanisms. *Free Radical Biology and Medicine, 51*(1), 17–29.

Martínez-Ruiz, A., & Lamas, S. (2009). Two decades of new concepts in nitric oxide signaling: from the discovery of a gas messenger to the mediation of nonenzymatic posttranslational modifications. *IUBMB Life, 61*(2), 91–98.

Maurino, V. G., & Peterhansel, C. (2010). Photorespiration: current status and approaches for metabolic engineering. *Current Opinion in Plant Biology, 13*(3), 249–256.

Melo, P. M., Silva, L. S., Ribeiro, I., Seabra, A. R., & Carvalho, H. G. (2011). Glutamine synthetase is a molecular target of nitric oxide in root nodules of *Medicago truncatula* and is regulated by tyrosine nitration. *Plant Physiology, 157*(3), 1505–1517.

Michelet, L., Zaffagnini, M., Morisse, S., Sparla, F., Pérez-Pérez, M. E., Francia, F., et al. (November 2013). Redox regulation of the Calvin-Benson cycle: something old, something new. *Frontiers in Plant Science, 4*, 470.

Moreau, M., Lindermayr, C., Durner, J., & Klessig, D. F. (2010). NO synthesis and signaling in plants – Where do we stand? *Physiologia Plantarum, 138*(4), 372–383.

Mur, L. A. J., Mandon, J., Persijn, S., Cristescu, S. M., Moshkov, I. E., Novikova, G. V., et al. (2013). Nitric oxide in plants: an assessment of the current state of knowledge. *AoB PLANTS, 5*, 1–17.

Neill, S., Bright, J., Desikan, R., Hancock, J., Harrison, J., & Wilson, I. (2008). Nitric oxide evolution and perception. *Journal of Experimental Botany, 59*(1), 25–35.

Neill, S. J., Desikan, R., Clarke, A., Hurst, R. D., & Hancock, J. T. (2002). Hydrogen peroxide and nitric oxide as signalling molecules in plants. *Journal of Experimental Botany, 53*(372), 1237–1247.

Nicolaou, A., Kenyon, S. H., Gibbons, J. M., Ast, T., & Gibbons, W. A. (1996). In vitro inactivation of mammalian methionine synthase by nitric oxide. *European Journal of Clinical Investigation, 26*(2), 167–170.

Ortega-Galisteo, A. P., Rodríguez-Serrano, M., Pazmiño, D. M., Gupta, D. K., Sandalio, L. M., & Romero-Puertas, M. C. (2012). S-Nitrosylated proteins in pea (*Pisum sativum* L.) leaf peroxisomes: changes under abiotic stress. *Journal of Experimental Botany, 63*(5), 2089–2103.

Pacher, P., Beckman, J. S., & Liaudet, L. (2007). Nitric oxide and peroxynitrite in health and disease. *Physiological Reviews, 87*(1), 315–424.

Palamalai, V., & Miyagi, M. (2010). Mechanism of glyceraldehyde-3-phosphate dehydrogenase inactivation by tyrosine nitration. *Protein Science: A Publication of the Protein Society, 19*(2), 255–262.

Palmieri, M. C., Lindermayr, C., Bauwe, H., Steinhauser, C., & Durner, J. (2010). Regulation of plant glycine decarboxylase by s-nitrosylation and glutathionylation. *Plant Physiology, 152*(3), 1514–1528.

Palmieri, M. C., Sell, S., Huang, X., Scherf, M., Werner, T., Durner, J., et al. (2008). Nitric oxide-responsive genes and promoters in *Arabidopsis thaliana*: a bioinformatics approach. *Journal of Experimental Botany, 59*(2), 177–186.

Perazzolli, M., Romero-Puertas, M. C., & Delledonne, M. (2006). Modulation of nitric oxide bioactivity by plant haemoglobins. *Journal of Experimental Botany, 57*(3), 479–488.

Pracharoenwattana, I., Cornah, J. E., & Smith, S. M. (2007). *Arabidopsis* peroxisomal malate dehydrogenase functions in beta-oxidation but not in the glyoxylate cycle. *The Plant Journal: For Cell and Molecular Biology, 50*(3), 381–390.

Puyaubert, J., Fares, A., Rézé, N., Peltier, J.-B., & Baudouin, E. (2014). Identification of endogenously S-nitrosylated proteins in *Arabidopsis* plantlets: effect of cold stress on cysteine nitrosylation level. *Plant Science, 215–216*, 150–156.

Reumann, S., & Weber, A. P. M. (2006). Plant peroxisomes respire in the light: some gaps of the photorespiratory C2 cycle have become filled-Others remain. *Biochimica et Biophysica Acta - Molecular Cell Research, 1763*(12), 1496–1510.

Rockel, P., Strube, F., Rockel, A., Wildt, J., & Kaiser, W. M. (2002). Regulation of nitric oxide (NO) production by plant nitrate reductase in vivo and in vitro. *Journal of Experimental Botany, 53*(366), 103–110.

Romero-Puertas, M. C., Campostrini, N., Mattè, A., Righetti, P. G., Perazzolli, M., Zolla, L., et al. (2008). Proteomic analysis of S-nitrosylated proteins in *Arabidopsis thaliana* undergoing hypersensitive response. *Proteomics, 8*(7), 1459–1469.

Romero-Puertas, M. C., Perazzolli, M., Zago, E. D., & Delledonne, M. (2004). Nitric oxide signalling functions in plant-pathogen interactions. *Cellular Microbiology, 6*(9), 795–803.

Romero-Puertas, M. C., Rodríguez-Serrano, M., & Sandalio, L. M. (September 2013). Protein S-nitrosylation in plants under abiotic stress: an overview. *Frontiers in Plant Science, 4*, 373.

Rümer, S., Kapuganti, J. G., & Kaiser, W. M. (2009). Oxidation of hydroxylamines to NO by plant cells. *Plant Signaling & Behavior, 4*(9), 853–855.

Rustérucci, C., Espunya, M. C., Díaz, M., Chabannes, M., & Martínez, M. C. (2007). S-nitrosoglutathione reductase affords protection against pathogens in *Arabidopsis*, both locally and systemically. *Plant Physiology, 143*(3), 1282–1292.

Sakamoto, A., Ueda, M., & Morikawa, H. (2002). *Arabidopsis* glutathione-dependent formaldehyde dehydrogenase is an S-nitrosoglutathione reductase. *FEBS Letters, 515*(1–3), 20–24.

Sanz, L., Albertos, P., Mateos, I., Sanchez-Vicente, I., Lechon, T., Fernandez-Marcos, M., et al. (2015). Nitric oxide (NO) and phytohormones crosstalk during early plant development. *Journal of Experimental Botany, 66*(10), 2857–2868.

Schopfer, F. J., Baker, P. R. S., & Freeman, B. A. (2003). NO-dependent protein nitration: a cell signaling event or an oxidative inflammatory response? *Trends in Biochemical Sciences, 28*(12), 646–654.

Schürmann, P., & Buchanan, B. B. (2008). The ferredoxin/thioredoxin system of oxygenic photosynthesis. *Antioxidants & Redox Signaling, 10*(7), 1235–1274.

Sehrawat, A., Abat, J. K., & Deswal, R. (2013). RuBisCO depletion improved proteome coverage of cold responsive S-nitrosylated targets in *Brassica juncea. Frontiers in Plant Science, 4*, 342.

Sehrawat, A., & Deswal, R. (2014a). S-nitrosylation analysis in *Brassica juncea* apoplast highlights the importance of nitric oxide in cold-stress signaling. *Journal of Proteome Research, 13*(5), 2599–2619.

Sehrawat, A., & Deswal, R. (2014b). Sub-proteome S-nitrosylation analysis in *Brassica juncea* hints at the regulation of Brassicaceae specific as well as other vital metabolic pathway(s) by nitric oxide and suggests post-translational modifications cross-talk. *Nitric Oxide: Biology and Chemistry/Official Journal of the Nitric Oxide Society, 43*, 97–111.

Seth, D., & Stamler, J. S. (2011). The SNO-proteome: causation and classifications. *Current Opinion in Chemical Biology, 15*(1), 129–136.

Shimoda, Y., Shimoda-Sasakura, F., Kucho, K., Kanamori, N., Nagata, M., Suzuki, A., et al. (2009). Overexpression of class 1 plant hemoglobin genes enhances symbiotic nitrogen fixation activity between *Mesorhizobium loti* and *Lotus japonicus. The Plant Journal: For Cell and Molecular Biology, 57*(2), 254–263.

Souza, J. M., Peluffo, G., & Radi, R. (2008). Protein tyrosine nitration-Functional alteration or just a biomarker? *Free Radical Biology and Medicine, 45*(4), 357–366.

Stamler, J. S., Lamas, S., & Fang, F. C. (2001). Nitrosylation: the prototypic redox-based signaling mechanism. *Cell, 106*(6), 675–683.

Stöhr, C., Strube, F., Marx, G., Ullrich, W. R., & Rockel, P. (2001). A plasma membrane-bound enzyme of tobacco roots catalyses the formation of nitric oxide from nitrite. *Planta, 212*(5–6), 835–841.

Stoimenova, M., Igamberdiev, A. U., Gupta, K. J., & Hill, R. D. (2007). Nitrite-driven anaerobic ATP synthesis in barley and rice root mitochondria. *Planta, 226*(2), 465–474.

Szabó, C., Ischiropoulos, H., & Radi, R. (2007). Peroxynitrite: biochemistry, pathophysiology and development of therapeutics. *Nature Reviews Drug Discovery, 6*(8), 662–680.

Tanou, G., Job, C., Rajjou, L., Arc, E., Belghazi, M., Diamantidis, G., et al. (2009). Proteomics reveals the overlapping roles of hydrogen peroxide and nitric oxide in the acclimation of citrus plants to salinity. *The Plant Journal: For Cell and Molecular Biology, 60*, 795–804.

Tun, N. N., Santa-Catarina, C., Begum, T., Silveira, V., Handro, W., Floh, E. I., et al. (2006). Polyamines induce rapid biosynthesis of nitric oxide (NO) in Arabidopsis thaliana seedlings. *Plant Cell Physiology, 47*(3), 346–354.

Vandelle, E., & Delledonne, M. (2011). Peroxynitrite formation and function in plants. *Plant Science, 181*(5), 534–539.

Wang, B. L., Tang, X. Y., Cheng, L. Y., Zhang, A. Z., Zhang, W. H., Zhang, F. S., et al. (2010). Nitric oxide is involved in phosphorus deficiency-induced cluster-root development and citrate exudation in white lupin. *The New Phytologist, 187*(4), 1112–1123.

Wawer, I., Bucholc, M., Astier, J., Anielska-Mazur, A., Dahan, J., Kulik, A., et al. (2010). Regulation of *Nicotiana tabacum* osmotic stress-activated protein kinase and its cellular partner GAPDH by nitric oxide in response to salinity. *The Biochemical Journal, 429*(1), 73–83.

Yamasaki, H., & Cohen, M. F. (2006). NO signal at the crossroads: polyamine-induced nitric oxide synthesis in plants? *Trends in Plant Science, 11*(11), 522–524.

Yu, M., Lamattina, L., Spoel, S. H., & Loake, G. J. (2014). Nitric oxide function in plant biology: a redox cue in deconvolution. *The New Phytologist, 202*(4), 1142–1156.

Yu, M., Yun, B. W., Spoel, S. H., & Loake, G. J. (2012). A sleigh ride through the SNO: regulation of plant immune function by protein S-nitrosylation. *Current Opinion in Plant Biology, 15*(4), 424–430.

Zaffagnini, M., Michelet, L., Sciabolini, C., Di Giacinto, N., Morisse, S., Marchand, C. H., et al. (2014). High-resolution crystal structure and redox properties of chloroplastic triosephosphate isomerase from *Chlamydomonas reinhardtii*. *Molecular Plant, 7*(1), 101–120.

Zaffagnini, M., Morisse, S., Bedhomme, M., Marchand, C. H., Festa, M., Rouhier, N., et al. (2013). Mechanisms of nitrosylation and denitrosylation of cytoplasmic glyceraldehyde-3-phosphate dehydrogenase from *Arabidopsis thaliana*. *The Journal of Biological Chemistry, 288*(31), 22777–22789.

CHAPTER EIGHT

The Functional Role of Nitric Oxide in Plant Mitochondrial Metabolism

Alok Kumar Gupta, Aprajita Kumari, Sonal Mishra, Aakanksha Wany, Kapuganti J. Gupta[1]
National Institute of Plant Genome Research, New Delhi, India
[1]Corresponding author: E-mail: jgk@nipgr.ac.in

Contents

Abstract

In recent years, mitochondrial nitric oxide (NO) production has attracted increasing attention. Mitochondria generate NO using nitrite as a substrate. Cytochrome c oxidase and other components of the electron transport chain also contribute to NO generation. Accumulating evidence indicates that mitochondria are scavengers of NO. Furthermore, several genes encoding mitochondrial proteins, as well as mitochondrial proteins, are regulated by NO. In this chapter, we provided an overview of the mechanisms of NO generation and scavenging in mitochondria and of the NO-dependent regulation of proteins and genes encoding mitochondrial proteins. In addition, the functional roles of NO in mitochondrial metabolism, such as inhibition of aconitase, production of ATP and induction of alternative oxidase are presented.

Advances in Botanical Research, Volume 77
ISSN 0065-2296
http://dx.doi.org/10.1016/bs.abr.2015.10.007

1. INTRODUCTION

Nitric oxide (NO) is a free radical molecule. After series of extensive studies, NO has been established as an important biomolecule that plays an important role in plant growth, development and stress physiology. Examples of such roles include plant microbe interactions, stomatal closure, regulation of respiration, resistance against various biotic and abiotic stresses, developmental programs such as cell wall synthesis and root development (Mur et al., 2013; Wendehenne & Hancock, 2011). Physiochemical nature of NO makes it as an excellent signal molecule. NO has diffusible capability in both liquid and lipid phase which makes it travel from site of production to site of reaction. Moreover NO is highly reactive in the presence of oxygen. The reactions of NO include (1) reaction with oxygen, forming NO_2; (2) NO_2 reaction with another molecule of NO leading to the production of N_2O_3; (3) reaction with superoxide anion (O_2^-) giving peroxynitrite ($ONOO^-$). NO half-life is very short, around 10 s, and determined by its concentration and immediate environment. For instance, at 10 μM concentration NO has a half-life of about 80 s whereas at 100 μM its half-life is about 8 s (Wink & Mitchell, 1998). That means the lower the concentration, the higher is its capacity of diffusion. Since NO reacts with oxygen and other metals, its half-life also depends on the concentration of these molecules.

Plants have several pathways for NO production. These are divided into oxidative and reductive pathways. Nearly half a dozen of pathways are known to operate in plants, located in different parts of cells, and activation of these pathways depends on the developmental stage and stress (Gupta, Fernie, Kaiser, & van Dongen, 2011). The well-studied pathway is the nitrate reductase (NR) pathway, this enzyme playing a major role in nitrogen metabolism. It is located in cytoplasm where it reduces nitrite to NO according to the following reaction:

$$NAD(P)H + 3H_3O^+ + 2NO_2^- \rightarrow NAD^+ + 2NO + 5H_2O.$$

NO production from NR requires high levels of nitrite ($K_{m\ nitrite} = 100\ \mu M$). Reduced cellular pH also leads to NO production by NR and post translational modifications play a role in production of NO from NR. Another enzyme is the plasma membrane-bound nitrite: NO reductase that is located in plasma membrane of roots and mitochondria and produces NO using nitrite as a substrate. Peroxisomal xanthine oxidoreductase also mediates conversion of nitrite to NO. NO synthase pathway is the most controversial oxidative pathway. This pathway is probably located in mitochondria or chloroplast or both. Polyamine and hydroxylamine pathways are poorly characterized pathways (Gupta et al., 2011). In this chapter, we particularly discuss the generation of NO by mitochondria and the impact of cellular NO on mitochondrial metabolism.

2. NITRIC OXIDE GENERATION IN MITOCHONDRIA

Previously, it was shown that electron transport chain of several ciliate protists and *Fusarium* can reduce nitrite to NO (Tielens, Rotte, van Hellemond, & Martin, 2002). It was also reported that skeletal muscle mitochondria are able to reduce nitrite to NO under decreased oxygen concentrations (Walters & Taylor, 1965). Later, Kozlov, Staniek, and Nohl (1999) found that nitrite in contact with actively respiring rat liver mitochondria accepts reducing equivalents from the ubiquinone cycle of the respiratory chain, leading to NO production. The green algae *Chlorella sorokiniana* is also capable of generating NO *in vivo* under anoxic conditions, the mitochondrial complex IV and alternative oxidase (AOX) being responsible for NO production in this single cell organism (Tischner, Planchet, & Kaiser, 2004). Similarly, tobacco cell suspensions are able to reduce nitrite to NO. This was the first evidence of NO production by mitochondria of higher plants (Planchet, Gupta, Sonoda, & Kaiser, 2005). Gupta, Stoimenova, and Kaiser (2005) conducted a detailed study and found that nitrite reduction to NO takes place in mitochondria from various species such as pea, barley, *Arabidopsis* and tobacco. They found that the K_m value for nitrite reduction to NO is 175 μM. This process occurs under hypoxia and anoxia. A K_i value of 0.05% was found, suggesting that this process occurs near anoxic conditions and in the absence of oxygen. Leaf slices of barley, pea and tobacco emitted 90–110 nmol NO g^{-1} FW h^{-1} during anoxic exposure, whereas root segments from these plants emitted only 2–30 nmol NO g^{-1} FW h^{-1} (Gupta et al. 2005; Planchet et al. 2005). These values correspond to 0.5–0.6 nmol NO mg^{-1} total root protein h^{-1}. Mitochondrial preparations from roots and cell suspensions produced 1–20 nmol NO mg^{-1} protein h^{-1}. Finally, Gupta and Kaiser (2010) found that this process occurs in membrane but not in the matrix of the mitochondria. Nitrite reduction to NO requires the interaction with a one electron donor of the respiratory chain. Inhibitory experiments show that complex III and complex IV of mitochondria are sites for NO production.

$$NO^{-} + 2H^{+} + e^{-} \rightarrow NO + H_2O$$

The mechanisms of nitrite reduction to NO by cytochrome c oxidase (COX) are still under investigation and several models are available for this mechanism (Figure 1).

Availability of oxygen, nitrite and NO determines the redox state of the COX centre which contains heme a_3 and copper B ($Fe_{a3}Cu_B$) that, in turn, depends on the redox state of cytochrome *c*. Fe^{2+} donates electron for nitrite reduction to NO. But this mechanism is still speculative (Figure 1).

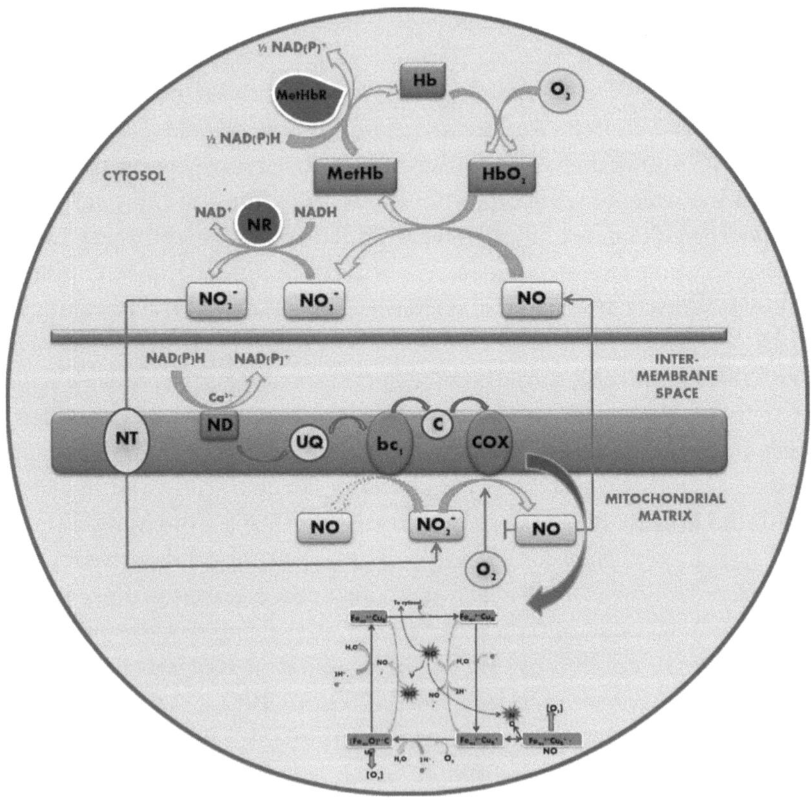

Figure 1 Role of mitochondria in the Hb/nitric oxide (NO) cycle under low oxygen conditions for production of ATP. Nitrite is transported to the mitochondria and its reduction takes place at the sites of cytochrome c oxidase (COX) and complex III (bc_1). The produced NO competes with O_2 for binding to COX. Since NO is a diffusible molecule, it goes to the cytosol where nonsymbiotic haemoglobin 1 converts it to nitrate (NO_3^{-}). Nitrate is then recycled and reduced by nitrate reductase (NR) to nitrite (NO_2^{-}). The overall process leads to the production of ATP. The bottom of the figure details reactions leading to NO production.

3. SCAVENGING OF NITRIC OXIDE BY MITOCHONDRIA

It has been reported that mitochondria scavenge significant amount of NO. For instance, Gupta et al. (2005) demonstrated that isolated barley root mitochondria scavenge 75–85% added NO. Surprisingly, mitochondria supplied with NADH scavenge more NO than without NADH. The mitochondria-induced scavenging of NO leads to nitrite. Moncada's group (Unitt, Hollis, Palacios-Callender, Frakich, & Moncada, 2010) reported that COX is responsible for scavenging and the optimal scavenging of NO by COX was achieved at 70 μM O_2, while at higher lower levels of oxygen, i.e. 15 μM O_2, NO scavenging was strongly suppressed. These authors also showed the involvement of the oxyferryl binuclear centre in this reaction. Long-term exposure of NO to mitochondria leads to slower respiration and nitrosylation of mitochondrial complex I (Brown & Borutaite, 2002) and probably of other proteins suggesting that a short-term scavenging of NO may help in prevention against nitrosative stress. Another scavenging mechanism is via $ONOO^-$ formation. Indeed, it has been shown that mitochondria also play a role in $ONOO^-$ formation (Radi, Cassina, Hodara, Quijano, & Castro, 2002). Oliveria de, Wulff, Saviani, and Salgado (2008) have shown that actively respiring mitochondria scavenge NO faster in superoxide dependent manner. For Instance, inhibition of complex III by antimycin A leads to increased superoxide and concomitant scavenging of NO.

4. PARTICIPATION OF MITOCHONDRIAL GENERATED NITRIC OXIDE IN CELL DEATH

In 1998, Delledonne et al. showed that NO plays a role in the development of cell death. NO reacts with superoxide, leading to the induction of cell death (Delledonne, Xia, Dixon, & Lamb, 1998; Zeidler et al., 2004). Both NOS and mitochondrial NO have been shown to play important role in this process (Modolo, Augusto, Almeida, Magalhaes, & Salgado, 2005; Zeidler et al., 2004). By using *Arabidopsis thaliana nia* 1 and *nia 2* mutants impaired in NRs expression, Modolo et al. (2005) found that nitrite is essential for NO production and subsequent cell death during the plant response to pathogen infection. Palmieri, Lindermayr, Bauwe, Steinhauser, and Durner (2010) further found that the nitrosylation of glycine decarboxylase (GDC) leads to an alteration in the redox status of mitochondria, causing cell death in tobacco cell suspensions treated with the elicitor harpin. The redox status was altered by a 60% reduction in GDC activity and a concomitant increase of the glycine/serine ratio.

5. AOX IN MITOCHONDRIA AND RELATION TO NO

AOX is a component of the mitochondrial electron transport chain. It plays an important role in prevention of over reduction of the ubiquinol pool but does not contribute for ATP production (Cvetkovska & Vanlerberghe, 2013). AOX also prevents excess of reactive oxygen species (ROS) production during various stress conditions (Maxwell, Wang, & McIntosh, 1999; Purvis & Shewfelt, 1993). Unlike COX, AOX is insensitive to NO and recent studies show that the AOX respiration acts to reduce the generation of ROS and reactive nitrogen species (RNS) in plant mitochondria by reducing the leak of single electrons from the electron transport chain (Cvetkovska & Vanlerberghe, 2012). AOX is induced under various conditions. For instance, phosphate deprivation leads to the production of NO and the induction of AOX (Royo, Moran, Ratcliffe, & Gupta, 2015). Also, treatment of *Arabidopsis* roots with the pathogenic *Fusarium oxysporum* led to increased production of NO and AOX (Gupta, Mur, & Brotman, 2014). These data suggest that NO production from nitrite takes place under normoxia to prevent excess of electron leakage. There is also evidence that AOX is involved in NO production from nitrite under hypoxia (Planchet et al., 2005; Tischner et al., 2004) but this needs further investigation.

6. NITROSYLATION AND NITRATION OF MITOCHONDRIAL PROTEINS

As discussed above, NO has a very short half-life (5–15 s) and displays high reactive potential (Hess, Matsumoto, Kim, Marshall, & Stamler, 2005). Due to these properties, NO plays a role in signaling via nitrosylation and tyrosine nitration of proteins. Indeed, NO can react with various intracellular and extracellular targets, including thiols (S-nitrosylation) and catalytic metal centres of proteins (metal-nitrosylation). Amongst these, there is an increasing interest in mitochondrial S-nitrosylated proteins (Foster & Stamler, 2004; Gupta, 2011; Piantadosi, 2012).

Several plant mitochondrial proteins are known to be targets of S-nitrosylation (Table 1). For instance, Palmieri et al. (2010) reported that the treatment of mitochondria extracts with S-nitrosoglutathione (GSNO)

Table 1 Nitrosylated Proteins in Mitochondria

S. No.	Biological Model	Protein Name	Effect of Nitric Oxide	References
1.	*Arabidopsis thaliana*	Gly decarboxylase subunits P1 and P2	Inhibition	Palmieri et al. (2010)
2.		Gly decarboxylase subunit H1	Inhibition	Palmieri et al. (2010)
3.		Gly decarboxylase subunit T (aminomethyltransferase)	Unknown	Palmieri et al. (2010)
4.		Ser hydroxymethyltransferase	Unknown	Palmieri et al. (2010)
5.		Lipoamide dehydrogenases 1 and 2	Unknown	Palmieri et al. (2010)
6.		NAD-malate dehydrogenase (mitochondrial isoform)	Unknown	Romero-Puertas et al. (2008)
7.		Subunits of complex I	Unknown	Burwell et al. (2006)
8.	*Pisum sativum*	PrxII F	Unknown	Camejo et al. (2013)
9.		Glycine dehydrogenase (GDC), P subunit	No significant effect	Camejo et al. (2013)
10.		ATP synthase F1 α subunit	No significant effect	Camejo et al. (2013)
11.		NADH-ubiquinone oxidoreductase subunit	No significant effect	Camejo et al. (2013)
12.		Heat shock 70-kDa protein mitochondrial	No significant effect	Camejo et al. (2013)
13.		Succinate dehydrogenase flavoprotein subunit	No significant effect	Camejo et al. (2013)
14.		ATP synthase β subunit	Activation	Camejo et al. (2013)
15.		Serine hydroxymethyltransferase	Activation	Camejo et al. (2013)
16.		Aminomethyltransferase (GDC, T subunit)	No significant effect	Camejo et al. (2013)
17.		NAD-dependent malate dehydrogenase (NAD MDH)	No significant effect	Camejo et al. (2013)
18.		AAC1 (ADP/ATP carrier 1)	No significant effect	Camejo et al. (2013)

(Continued)

Table 1 Nitrosylated Proteins in Mitochondria—cont'd

S. No.	Biological Model	Protein Name	Effect of Nitric Oxide	References
19.		Malate dehydrogenase, glyoxysomal	No significant effect	Camejo et al. (2013)
20.		F1 ATPase beta subunit	Activation	Camejo et al. (2013)
21.	Mammalian mitochondria	Aconitase	Inhibition	T.A. Prime et al. (2009), Chacko B.K. et al. (2010), Sun, Morgan, Shen, Steenbergen, and Murphy (2007), and Taylor and Moncada, 2010
22.		Citrate synthase	Unknown	Sun et al. (2007)
23.		2-Oxoglutarate dehydrogenase	Activation	T.A. Prime et al. (2009) and Chacko B.K. et al. (2010)
24.		Isocitrate dehydrogenase	Unknown	Chacko B.K. et al. (2010) and Sun et al. (2007)
25.		Malate dehydrogenase	Unknown	Konorev et al. (1996) and Sun et al. (2007)
26.		Succinyl CoA ligase	Unknown	Sun et al. (2007)
27.		Dihydrolipoyl dehydrogenase	Unknown	Konorev et al. (1996)
28.		NADH dehydrogenase subunit 3	Inhibition	Di Virgilio and Azzone (1982)
29.		Succinate dehydrogenases A	Unknown	Chacko B.K. et al. (2010) and Sun et al. (2007)

30.	Cytochrome b-c1 Complex subunit 1	Unknown	Chacko B.K. et al. (2010), Sun et al. (2007)
31.	Very long chain acyl CoA dehydrogenases	Unknown	Chacko B.K. et al. (2010) and Sun et al. (2007)
32.	Short chain acyl CoA dehydrogenase	Unknown	Chacko B.K. et al. (2010)
33.	Carnitine palmitoyl transferase 2	Unknown	Chacko B.K. et al. (2010)
34.	Enoyl CoA hydratase	Unknown	Chacko B.K. et al. (2010) and Sun et al. (2007)
35.	Electron transferring flavoprotein dehydrogenase	Unknown	Chacko B.K. et al. (2010) and Sun et al. (2007)
36.	Aldehyde dehydrogenase-2 (ALDH2)	Inhibition	Vinogradov A.D. (1998)
37.	Branched chain amino acid aminotransferase (BCAT2)	Inhibition	Reddy, Jones, Cross, Wong, & Van Der Vliet (2000)
38.	Heat shock protein 60 (Hsp60)	Activation	Zanella, Giordano, Muscari, Zini, & Guarnieri (2004)
39.	Heat shock protein 70 (mortalin)	Activation	Zanella et al. (2004)
40.	Dynamin-related protein-1 (Drp1)	Activation	Deocaris et al. (2006)
41.	Sarcosine dehydrogenase	Unknown	Konorev et al. (1996)
42.	Hydroxy methyl glutaryl CoA synthetase	Unknown	Konorev et al. (1996)
43.	Aspartate aminotransferase	Unknown	Sun et al. (2007)
44.	VDAC2	Unknown	Sun et al. (2007)

or GSH induced the S-nitrosylation of proteins identified by combining the biotin-switch method and nano-LC/MS/MS. These proteins include photorespiratory enzymes, notably three subunits of GDC, namely Gly decarboxylase subunits P1 and P2 and Gly decarboxylase subunit H but also Gly decarboxylase subunit T (aminomethyltransferase), serine hydroxymethyltransferase and lipoamide dehydrogenases 1 and 2. Romero-Puertas et al. (2008) identified NAD-malate dehydrogenase as a target for S-nitrosylation following GSNO treatment of leaf extracts. By combining combined separation techniques for complex I (blue-native gel electrophoresis, Superose 6 column chromatography) with detection methods for S-nitrosylated proteins (chemiluminescence, biotin-switch assay), Burwell, Nadtochiy, Tompkins, Young, and Brookes (2006) found that the 75-kDa subunit of complex I is S-nitrosylated in isolated rat heart mitochondria treated with S-nitrosoglutathione (10 μM–1 mM). The S-nitrosylation of complex I impaired several processes such as photorespiration, the tricarboxylic acid (TCA) cycle and ATP synthesis. Camejo et al. (2013) identified mitochondrial S-nitrosylated proteins in mitochondria isolated from control and salt-stress plants. These proteins included peroxiredoxin II F, Glycine dehydrogenase P subunit, ATP synthase F1 α and β subunits, NADH-ubiquinone oxidoreductase subunit, heat shock 70-kDa protein, succinate dehydrogenase flavoprotein subunit, serine hydroxymethyltransferase, aminomethyltransferase, NAD-dependent malate dehydrogenase, AAC1 (ADP/ATP Carrier 1) and malate dehydrogenase.

Tyrosine nitration is a covalent post-translational protein modification derived from the reaction of proteins with nitrating agents (Radi, 2013). Excess levels of reactive oxygen species in the presence of nitrogen oxides lead to the formation of nitrating species such as $ONOO^-$ that, in turn, leads to nitration of mitochondrial proteins. Protein tyrosine nitration is a biomarker for nitrosative stress (Radi, 2013). Several proteins become active and inactive upon nitration. For instance, nitration of NADP-isocitrate dehydrogenase and malate dehydrogenase induces a decrease in activities (Begara-Morales et al., 2013) while a similar treatment promotes the activation of superoxide dismutase (Sehrawat, Abat, & Deswal, 2013). The mitochondrial photorespiratory enzyme serine hydroxymethyltransferase and the subunit T of GDC (aminomethyltransferase) are also subjected to nitration (Lozano-Juste, Colom-Moreno, & León, 2011).

7. GENES ENCODING MITOCHONDRIAL PROTEINS ARE REGULATED BY NO

NO also influences the expression of nuclear and mitochondrial genes encoding mitochondria proteins. Huang, Von Rad, and Durner (2002) reported that NO acts as a transcriptional activator for the expression of the nuclear-encoded mitochondrial protein AOX1 in *Arabidopsis*. As already discussed, AOX plays a role in stress resistance. A cDNA-amplification fragment length polymorphism transcript profiling was performed by Polverari Molesini, Pezzotti, Buonaurio, Marte, & Delledonne (2003) in *A. thaliana* following treatment with the NO donor sodium nitroprusside (SNP). They reported the NO-dependent expression of 120 cDNAs including those encoding mitochondrial pyruvate dehydrogenase, glutathione peroxidase 6 and some other proteins involved in mitochondrial metabolism (Table 2). Besson-Bard et al. (2009) performed a microarray analysis from *A. thaliana* plantlets exposed to the toxic metal cadmium in presence or not of the mammalian NOS inhibitor Nw-nitro-L-Arg-methyl ester (L-NAME). This study identified around 2656 genes showing differential expression, out of which approximately three-fourth were NO-responsive transcripts. Some genes encoded proteins targeted to mitochondria such as the late embryogenesis abundant like protein 3 and 5, proteins from the cytochrome P450, the 2OG-Fe(II) oxygenase and the S-adenosyl-L-methionine-dependent methyltransferases families, 1-pyrroline-5-carboxylate dehydrogenase, the abscisic acid-responsive HVA22D protein, the retarded root growth protein, glyoxalase 2—5, the acyl-activating enzyme 3 and several other genes are listed in Table 2. Begara-Morales et al. (2014) performed a transcriptome analysis in *A. thaliana* plants for a comparative study of expressed transcripts in leaf and root tissues exposed to GSNO, an endogenous reservoir of NO in cells also acting as an S-nitrosylating agent. For this purpose, the authors used the paired-end RNA-seq technology developed by Illumina, thus providing the first original report in plants using such approach to investigate the GSNO signaling mechanism between different organs in plants. Several NO responsive transcripts related to mitochondrial metabolism were expressed such as those encoding 2-oxoglutarate, S-adenosyl-L-methionine-dependent methyltransferases and β-1,6-N-acetylglucosaminyltransferase.

Table 2 Nitric Oxide Responsive Mitochondrial Genes

S. No.	AGI ID	Annotation	Mitochondrial Localization	Encoded by Nuclear/Mitochondrial	References
1	AT4G02380	Late embryogenesis abundant like protein 5 (atlea5)	Yes	Nuclear	Besson-Bard et al. (2009)
2	AT4G37400	Cytochrome P450 family protein	Predicted/ extracellular region	Nuclear	Besson-Bard et al. (2009)
3	AT5G62530	1-Pyrroline-5-carboxylate dehydrogenase	Yes	Nuclear	Besson-Bard et al. (2009)
4	AT1G02820	Atlea3, late embryogenesis abundant 3, lea3	Predicted/ chloroplast	Nuclear	Besson-Bard et al. (2009)
5	AT4G24960	ABA – responsive-HVA22D protein	Yes	Nuclear	Besson-Bard et al. (2009)
6	AT2G41380	S-adenosyl-L-methionine-dependent methyltransferases superfamily protein	Yes	Nuclear	Besson-Bard et al. (2009)
7	AT1G69380	Retarded root growth, rrg protein	Yes	Nuclear	Besson-Bard et al. (2009)
8	AT2G31350	Glyoxalase 2–5	Yes	Nuclear	Besson-Bard et al. (2009)
9	AT2G26140	Ftsh protease 4, ftsh4	Yes	Nuclear	Besson-Bard et al. (2009)
10	AT4G02940	2OG-Fe(II) oxygenase family protein	Predicted	Nuclear	Besson-Bard et al. (2009)
11	AT3G48990	Acyl-activating enzyme 3	Yes	Nuclear	Besson-Bard et al. (2009)
12	AT5G26030	Ferrochelatase 1	Yes	Nuclear	Besson-Bard et al. (2009)
13	AT2G22500	Plant uncoupling mitochondrial protein 5	Yes	Nuclear	Besson-Bard et al. (2009)

14	AT2G24180	Cytochrome p450 71b6	Yes	Nuclear	Besson-Bard et al. (2009)
15	AT1G53580	Glyoxalase 2–3	Yes	Nuclear	Besson-Bard et al. (2009)
16	AT3G01290	Hypersensitive-induced reaction 2	Yes	Nuclear	Besson-Bard et al. (2009)
17	AT5G07440	Glutamate dehydrogenase 2	Yes	Nuclear	Besson-Bard et al. (2009)
18	AT1G64720	Membrane-related protein CP5	Yes	Nuclear	Besson-Bard et al. (2009)
19	AT1G59870	ATP-binding cassette g36	Yes	Nuclear	Besson-Bard et al. (2009)
20	AT4G33010	Glycine decarboxylase *p*-protein 1	Yes	Nuclear	Polverari et al. (2003)
21	AT1G54220	Mitochondrial pyruvate dehydrogenase subunit 2–3	Yes	Nuclear	Polverari et al. (2003)
22	AT2G30780	Tetratricopeptide repeat (TPR)-like superfamily protein	Yes	Nuclear	Polverari et al. (2003)
23	AT2G45210	Senescence-associated gene 201	Yes	Nuclear	Polverari et al. (2003)
24	AT4G13180	NAD(P)-binding Rossmann-fold superfamily protein	Predicted	Nuclear	Polverari et al. (2003)
25	AT4G11600	Glutathione peroxidase 6	Yes	Nuclear	Polverari et al. (2003)
26	AT4G34800	Small auxin upregulated RNA 4	Yes	Nuclear	Vidal et al. (2013)
27	AT3G25717	Rotundifolia like 16	Yes	Nuclear	Vidal et al. (2013)
28	AT1G13245	Rotundifolia like 17	Yes	Nuclear	Vidal et al. (2013)
29	AT1G68825	Rotundifolia like 15	Yes	Nuclear	Vidal et al. (2013)
30	AT3G29034	Unknown protein	Yes	Nuclear	Vidal et al. (2013)
31	AT2G31141	Unknown protein	Yes	Nuclear	Vidal et al. (2013)
32	AT2G23790	Protein of unknown function	Yes	Nuclear	Vidal et al. (2013)
	AT1G11655	Unknown protein	Yes	Nuclear	Vidal et al. (2013)
33	AT1G20070	Unknown protein	Yes	Nuclear	Begara-Morales et al. (2014)
34	AT1G74458	Unknown protein	Yes	Nuclear	Begara-Morales et al. (2014)

(Continued)

Table 2 Nitric Oxide Responsive Mitochondrial Genes—cont'd

S. No.	AGI ID	Annotation	Mitochondrial Localization	Encoded by Nuclear/ Mitochondrial	References
35	AT1G20270	2-Oxoglutarate (2OG) and Fe(II)-dependent oxygenase superfamily protein	Yes	Nuclear	Begara-Morales et al. (2014)
36	AT2G29995	Unknown protein	Yes	Nuclear	Begara-Morales et al. (2014)
37	AT5G16170	Beta-1,6-N-acetylglucosaminyltransferase family protein	Yes	Nuclear	Begara-Morales et al. (2014)
38	AT1G69520	S-adenosyl-L-methionine-dependent methyltransferases superfamily protein	Yes	Nuclear	Begara-Morales et al. (2014)
39	AT3G15352	ATCOX17; copper chaperone	Yes	Nuclear	Begara-Morales et al. (2014)
40	AT5G06750	Probable protein phosphatase 2C 68	Yes	Nuclear	Begara-Morales et al. (2014)
41	AT4G23230	Cysteine-rich receptor-like protein kinase 15	Yes	Nuclear	Begara-Morales et al. (2014)
43	AT3G22370	Alternative oxidase	Yes	Nuclear	Huang et al. (2002)

8. EFFECT OF NO ON TCA CYCLE VIA ACONITASE

Aconitase is an enzyme of TCA cycle. Due to its iron—sulphur (4Fe-4S) content, it is sensitive to NO. Aconitase mediates reversible isomerization of citrate to isocitrate. Two aconitase isoforms are located in plants, in the mitochondria and in the cytosol, depending on the isoform. It is known in animal systems that NO reversibly inhibits the cytosolic isoform of aconitase and, subsequently, enhances its mRNA-binding activity as an iron regulatory protein (Drapier, 1997).

Hypoxia leads to the inhibition of aconitase (Gupta et al., 2012). The NO-dependent inhibition of aconitase triggers an increase in citrate concentration which, in turn, results in the accumulation of amino acids biosynthesis. This process could contribute to protein synthesis required for survival under hypoxia. Indeed, although total protein synthesis is known to decline under hypoxia, some protective proteins need to be expressed for survival. Therefore, NO could contribute to cell survival under hypoxia. Increased levels of AOX in response to pathogen attack also helps in protecting against excess of ROS levels.

9. INCREASING ENERGY YIELD IN MITOCHONDRIA MEDIATED BY NITRITE REDUCTION TO NITRIC OXIDE

Addition of nitrite to roots mitochondria and nodules leads to an increased ratio of ADP/ATP, suggesting that NO production from nitrite participates to ATP synthesis (Stoimenova, Igamberdiev, Gupta & Hill, 2007; Horchani et al., 2011; Gupta & Igamberdiev, 2011). Stoimenova et al. (2007) demonstrated that isolated mitochondria from barley and rice exposed to hypoxia were able to oxidize reduced nicotinamide adenine dinucleotide (NADH) and reduced nicotinamide adenine dinucleotide phosphate (NADPH) and to promote ATP production. Either NADH/NADPH oxidation or synthesis of ATP was insensitive to rotenone, suggesting that complex I did not participate in ATP synthesis. In contrast, these reactions were sensitive to diphenylene iodonium (DPI), highlighting the involvement of alternative dehydrogenases. A same sensitivity was observed in presence of several mitochondrial electron transport chain inhibitors such as myxothiazol (complex III inhibitor), KCN (COX inhibitor) and oligomycin (ATP synthase inhibitor). It was also observed that the anaerobic ATP synthesis rate was 7—9 nmol min^{-1} mg^{-1} protein for barley and 15—17 nmol min^{-1} mg^{-1} protein for rice. This rate is approximately 3—5% of the aerobic mitochondrial ATP synthesis rate. Isolated rice mitochondria showed a more prolonged ATP synthesis than barley mitochondria. This observation could be attributed to the anoxic tolerance of rice but other mechanisms could also participate to this process.

10. CONCLUSION

Mitochondria play an important role in NO signaling via the production of NO. By scavenging NO, mitochondria also prevent cells against nitrosative stress. NO production in mitochondria under hypoxia or anoxia leads to increased production of energy and, by inhibiting aconitase, NO can increase citrate levels, resulting in the biosynthesis of amino acids. Mitochondrial NO plays a role in the induction of AOX and also mediates S-nitrosylation process in mitochondria which could lead to cell death. Several genes involved in mitochondria metabolism are regulated by NO. All together, these findings highlight the importance of mitochondria in NO signaling.

REFERENCES

Begara-Morales, J. C., Chakri, M., Sánchez-Calvo, B., Mata-Pérez, C., Leterrier, M., Palma, J. M., et al. (2013). Protein tyrosine nitration in pea roots during development and senescence. *Journal of Experimental Botany, 64*, 1121–1134.

Begara-Morales, J. C., Sánchez-Calvo, B., Luque, F., Leyva-Pérez, M., Leterrier, M., Corpas, F. J., et al. (2014). Differential transcriptomic analysis by RNA-Seq of GSNO-responsive genes between *Arabidopsis* roots and leaves. *Plant Cell Physiology, 55*, 1080–1095.

Besson-Bard, A., Gravot, A., Richaud, P., Auroy, P., Gaymard, F., Taconnat, L., et al. (2009). Nitric oxide contributes to cadmium toxicity in *Arabidopsis* by promoting cadmium accumulation in roots and by up regulating genes related to iron uptake. *Plant Physiology, 149*, 1302–1315.

Brown, G. C., & Borutaite, V. (2002). Nitric oxide inhibition of mitochondrial respiration and its role in cell death. *Free Radical Biology and Medicine, 33*, 1440–1450.

Burwell, L. S., Nadtochiy, S. M., Tompkins, A. J., Young, S., & Brookes, P. S. (2006). Direct evidence for S-nitrosation of mitochondrial complex I. *The Biochemical Journal, 394*, 627–634.

Camejo, D., Romero-Puertas Mdel, C., Rodríguez-Serrano, M., Sandalio, L. M., Lázaro, J. J., Jiménez, A., et al. (2013). Salinity-induced changes in S-nitrosylation of pea mitochondrial proteins. *Journal of proteomics, 79*, 87–99.

Chacko, B. K., Reily, C., Srivastava, A., Johnson, M. S., Ye, Y., Ulasova, E., Agarwal, A., Zinn, K. R., Murphy, M. P., Kalyanaraman, B., & Darley-Usmar, V. (2010). Prevention of diabetic nephropathy in Ins2(+/)(AkitaJ) mice by the mitochondria-targeted therapy MitoQ. *The Biochemical Journal, 432*, 9–19.

Cvetkovska, M., & Vanlerberghe, G. C. (2012). Alternative oxidase modulates leaf mitochondrial concentrations of superoxide and nitric oxide. *New Phytologist, 195*, 32–39.

Cvetkovska, M., & Vanlerberghe, G. C. (2013). Alternative oxidase impacts the plant response to biotic stress by influencing the mitochondrial generation of reactive oxygen species. *Plant Cell and Environment, 36*, 721–732.

Delledonne, M., Xia, Y., Dixon, R. A., & Lamb, C. (1998). Nitric oxide functions as a signal in plant disease resistance. *Nature, 394*, 585–588.

Deocaris, C. C., Kaul, S. C., & Wadhwa, R. (2006). On the brotherhood of the mitochondrial chaperones mortalin and heat shock protein 60. *Cell Stress and Chaperones, 11*, 116–128.

Di Virgilio, F., & Azzone, G. F. (1982). Activation of site I redox-driven H^+ pump by exogenous quinones in intact mitochondria. *The Journal of Biological Chemistry, 257*, 4106–4113.

Drapier, J. C. (1997). Interplay between NO and [Fe-S] clusters: relevance to biological systems. *Methods, 11*, 319–329.

Foster, M. W., & Stamler, J. S. (2004). New insights into protein S-nitrosylation: mitochondria as a model system. *The Journal of Biological Chemistry, 279*, 25891–25897.

Gupta, K. J., Fernie, A. R., Kaiser, W. M., & van Dongen, J. T. (2011). On the origins of nitric oxide. *Trends Plant Science, 16*, 160–168.

Gupta, K. J. (2011). Protein *S*-nitrosylation in plants: photorespiratory metabolism and NO signaling. *Science Signaling, 154*(4), jc1.

Gupta, K. J., & Igamberdiev, A. U. (2011). The anoxic plant mitochondrion as a nitrite: NO reductase. *Mitochondrion, 11*, 537–543.

Gupta, K. J., & Kaiser, W. M. (2010). Production and scavenging of nitric oxide by barley root mitochondria. *Plant and Cell Physiology, 51*(4), 576–584.

Gupta, K. J., Mur, L. A. J., & Brotman, Y. (2014). *Trichoderma asperelloides* suppresses nitric oxide generation elicited by *Fusarium oxysporum* in *Arabidopsis* roots. *Molecular Plant-Microbe Interactions, 27*(4), 307–314.

Gupta, K. J., Shah, J. K., Brotman, Y., Jahnke, K., Willmitzer, L., Kaiser, W. M., et al. (2012). Inhibition of aconitase by nitric oxide leads to induction of the alternative oxidase and to a shift of metabolism towards biosynthesis of amino acids. *Journal of Experimental Botany, 63*, 1773–1784.

Gupta, K. J., Stoimenova, M., & Kaiser, W. M. (2005). In higher plants, only root mitochondria, but not leaf mitochondria reduce nitrite to NO, in vitro and in situ. *Journal of Experimental Botany*, 2601–2609.

Hess, D. T., Matsumoto, A., Kim, S. O., Marshall, H. E., & Stamler, J. S. (2005). Protein *S*-nitrosylation: preview and parameters. *Nature Review. Molecular Cell Biology, 6*, 150–166.

Horchani, F., Prevot, M., Boscari, A., Evangelisti, E., Meilhoc, E., Bruand, C., et al. (2011). Both plant and bacterial nitrate reductases contribute to nitric oxide production in *Medicago truncatula* nitrogen-fixing nodules. *Plant Physiology, 155*, 1023–1036.

Huang, X., Von Rad, U., & Durner, J. (2002). Nitric oxide induces transcriptional activation of the nitric oxide-tolerant alternative oxidase in *Arabidopsis* suspension cells. *Planta, 215*, 914–923.

Konorev, E. A., Joseph, J., Tarpey, M. M., & Kalyanaraman, B. (1996). The mechanism of cardioprotection by S-nitrosoglutathione monoethyl ester in rat isolated heart during cardioplegic ischaemic arrest. *British Journal of Pharmacology, 119*, 511–518.

Kozlov, A. V., Staniek, K., & Nohl, H. (1999). Nitrite reductase activity is a novel function of mammalian mitochondria. *FEBS Letters, 454*, 127–130.

Lozano-Juste, J., Colom-Moreno, R., & León, J. (2011). *In vivo* protein tyrosine nitration in *Arabidopsis thaliana*. *Journal of Experimental Botany, 62*, 3501–3517.

Maxwell, D. P., Wang, Y., & McIntosh, L. (1999). The alternative oxidase lowers mitochondrial reactive oxygen production in plant cells. *Proceedings of the National Academy of Sciences of the United States of America, 96*, 8271–8276.

Modolo, L. V., Augusto, O., Almeida, I. M., Magalhaes, J. R., & Salgado, I. (2005). Nitrite as the major source of nitric oxide production by *Arabidopsis thaliana* in response to *Pseudomonas syringae*. *FEBS Letters, 579*, 3814–3820.

Mur, L. A. J., Mandon, J., Persijn, S., Cristescu, S. M., Moshkov, I. E., & Novikova, G. V. (2013). Nitric oxide in plants: an assessment of the current state of knowledge. *AoB plants, 5*, pls052.

Oliveira de, H. C., Wulff, A., Saviani, E. E., & Salgado, I. (2008). Nitric oxide degradation by potato tuber mitochondria: evidence for the involvement of external NAD(P)H dehydrogenases. *Biochemistry Biophysics Acta, 1777*, 470–476.

Palmieri, M. C., Lindermayr, C., Bauwe, H., Steinhauser, C., & Durner, J. (2010). Regulation of plant glycine decarboxylase by S-nitrosylation and glutathionylation. *Plant Physiology, 152*, 1514–1528.

Piantadosi, C. A. (2012). Regulation of mitochondrial processes by protein S-nitrosylation. *Biochimica et Biophysica Acta, 1820*, 712–721.

Planchet, E., Gupta, K. J., Sonoda, M., & Kaiser, W. M. (2005). Nitric oxide emission from tobacco leaves and cell suspensions: rate limiting factors and evidence for the involvement of mitochondrial electron transport. *Plant Journal, 41*, 732–743.

Polverari, A., Molesini, B., Pezzotti, M., Buonaurio, R., Marte, M., & Delledonne, M. (2003). Nitric oxide-mediated transcriptional changes in *Arabidopsis thaliana*. *Molecular Plant–Microbe Interactions, 16*, 1094–1105.

Prime, T. A., Blaikie, F. H., Evans, C., Nadtochiy, S. M., James, A. M., Dahm, C. C., et al. (2009). A mitochondria- targeted S-nitrosothiol modulates respiration, nitrosates thiols, and protects against ischemia-reperfusion injury. *Proc Natl Acad Sci USA, 106*, 10764–10769.

Purvis, A. C., & Shewfelt, R. L. (1993). Does the alternative pathway ameliorate chilling injury in sensitive plant tissues? *Plant Physiology, 88*, 712–718.

Radi, R., Cassina, A., Hodara, R., Quijano, C., & Castro, L. (2002). Peroxynitrite reactions and formation in mitochondria. *Free Radical Biology and Medicine, 33*, 1451–1464.

Radi, R. (February 19, 2013). Protein tyrosine nitration: biochemical mechanisms and structural basis of functional effects. *Accounts of Chemical Research, 46*(2), 550–559.

Reddy, S., Jones, A. D., Cross, C. E., Wong, P. S., & Van Der Vliet, A. (2000). Inactivation of creatine kinase by S-glutathionylation of the active-site cysteine residue. *The Biochemical Journal, 347*, 821–827.

Romero-Puertas, M. C., Campostrini, N., Mattè, A., Righetti, P. G., Perazzolli, M., Zolla, L., et al. (2008). Proteomic analysis of S-.proteins. *Arabidopsis thaliana* undergoing hypersensitive response. *Proteomics, 8*, 1459–1469.

Royo, B., Moran, J. F., Ratcliffe, R. G., & Gupta, K. J. (2015). Nitric oxide induces the alternative oxidase pathway in *Arabidopsis* seedlings deprived of inorganic phosphate. *Journal of Experimental Botany, 66*, 6273–6280.

Sehrawat, A., Abat, J. K., & Deswal, R. (2013). RuBisCO depletion improved proteome coverage of cold responsive S-nitrosylated targets in *Brassica juncea*. *Frontiers of Plant Science, 4*, 342.

Stoimenova, M., Igamberdiev, A. U., Gupta, K. J., & Hill, R. D. (2007). Nitrite-driven anaerobic ATP synthesis in barley and rice root mitochondria. *Planta, 226*, 465–474.

Sun, J., Morgan, M., Shen, R. F., Steenbergen, C., & Murphy, E. (2007). Preconditioning results in S-nitrosylation of proteins involved in regulation of mitochondrial energetics and calcium transport. *Circulation Research, 101*, 1155–1163.

Taylor, C. T., & Moncada, S. (2010). Nitric oxide, cytochrome C oxidase, and the cellular response to hypoxia. *Arteriosclerosis, Thrombosis and Vascular Biology, 30*, 643–647.

Tielens, A. G. M., Rotte, C., van Hellemond, J. J., & Martin, W. (2002). Mitochondria as we don't know them. *Trends Biochemical Science, 27*, 564–572.

Tischner, R., Planchet, E., & Kaiser, W. M. (2004). Mitochondrial electron transport as a source for nitric oxide in the unicellular green alga *Chlorella sorokiniana*. *FEBS Letters, 576*, 151–155.

Unitt, D. C., Hollis, V. S., Palacios-Callender, M., Frakich, N., & Moncada, S. (2010). Inactivation of nitric oxide by cytochrome *c* oxidase under steady-state oxygen conditions. *Biochemical Biophysics Acta Bioenergetics, 1797*, 371–377.

Vidal, E. A., Moyano, T. C., Krouk, G., Katari, M. S., Tanurdzic, M., McCombie, W. R., et al. (2013). Integrated RNA-seq and sRNA-seq analysis identifies novel nitrate-responsive genes in *Arabidopsis thaliana* roots. *BMC Genomics*. http://dx.doi.org/10.1186/1471-2164-14-701.

Vinogradov, A. D. (1998). Catalytic properties of the mitochondrial NADH-ubiquinone oxidoreductase (complex I) and the pseudo-reversible active/inactive enzyme transition. *Biochimica et Biophysica Acta, 1364*, 169–185.

Walters, C. L., & Taylor, A. M. (1965). The reduction of nitrite by skeletal muscle mitochondria. *Biochemical Biophysics Acta, 96*, ,522–524.

Wendehenne, D., & Hancock, J. T. (2011). New frontiers in nitric oxide biology in plant. *Plant Science, 181*, 507–508.

Wink, D. A., & Mitchell, J. B. (1998). Chemical biology of nitric oxide: insights into regulatory, cytotoxic, and cytoprotective mechanisms of nitric oxide. *Free Radical Biology and Medicine, 25*, 434–456.

Zanella, B., Giordano, E., Muscari, C., Zini, M., & Guarnieri, C. (2004). Nitric oxide synthase activity in rat cardiac mitochondria. *Basic Research in Cardiology, 99*, 159–164.

Zeidler, D., Zähringer, U., Dubery, I., Hartung, I., Bors, W., Hutzler, P., et al. (2004). Innate immunity in *Arabidopsis thaliana* lipopolysaccharides activate nitric oxide synthase (NOS) and induce defense genes. *Proceedings of the National Academy of Sciences of the United States of America, 102*, 15811–15816.

CHAPTER NINE

Nitric Oxide and Reactive Oxygen Species in PCD Signaling

Vittoria Locato*, Annalisa Paradiso[§], Wilma Sabetta[¶], Laura De Gara*, Maria Concetta de Pinto[§,1]

*Centro Integrato di Ricerca, Università Campus Bio-Medico di Roma, Rome, Italy
[§]Dipartimento di Biologia, Università degli Studi di Bari "Aldo Moro", Bari, Italy
[¶]Spin-off SINAGRI srl, Università degli Studi di Bari "Aldo Moro", Bari, Italy
[1]Corresponding author: E-mail: mariaconcetta.depinto@uniba.it

Contents

Abstract

Programmed cell death (PCD) is a process of genetically encoded and actively controlled cellular suicide. PCD is associated with different phases of plant life and also represents a defence mechanism that plants activate against different kinds of biotic and abiotic stresses. Reactive oxygen species (ROS) and nitric oxide (NO) have been proposed as key factors in the control of both developmentally and environmentally induced PCD. In this chapter, we give an overview of ROS and NO interplay in the signaling leading to PCD induced in senescence, self-incompatibility, hypersensitive response and under cadmium and heat stress. The data presented indicate that ROS and NO interact through different pathways, depending on the level and the timing of production of these reactive species, the cellular redox state and the plant species. Despite the complicate network of signals, some common points of NO/ROS crosstalk in the different kinds of PCD can be identified. First, during PCD, ROS and NO biosynthesis is mutually regulated with feedback mechanisms. The NO-dependent *S*-nitrosylation of proteins controlling ROS levels represents a key point of interaction

Advances in Botanical Research, Volume 77
ISSN 0065-2296
http://dx.doi.org/10.1016/bs.abr.2015.10.008

of the two species. Furthermore, NO and ROS interplay could amplify the induction of PCD executors, such as caspase-3-like. Finally, NO and/or ROS might control the ubiquitin—proteasome pathway involved in PCD induction.

1. INTRODUCTION

Programmed cell death (PCD) is a process of genetically encoded, actively controlled cellular suicide, through which cells are selectively eliminated with the involvement of specific proteases and nucleases. The importance of PCD in different phases of plant life is widely recognized. Indeed, PCD is associated with the development of female and male germ lines, incompatible pollen—pistil interaction, seed development and germination. PCD is also a key event during vegetative development, occurring during differentiation of xylem tracheary elements, lateral and adventitious root emergence, aerenchyma formation, abscission, dehiscence and senescence processes (Van Hautegem, Waters, Goodrich, & Nowack, 2015, and references therein). Moreover, plant PCD represents defence mechanisms against different kinds of biotic and abiotic stresses (Coll, Epple, & Dangl, 2011; Petrov, Hille, Mueller-Roeber, & Gechev, 2015).

Reactive oxygen species (ROS) and nitric oxide (NO) have been proposed as key factors in the control of both developmentally and environmentally induced PCD (Gross, Durner, & Gaupels, 2013; Petrov et al., 2015).

ROS, including hydrogen peroxide (H_2O_2), superoxide anion (O_2^-), hydroxyl radical and singlet oxygen, are produced in plants as a result of oxygen reduction during a number of metabolic processes. For a long time, these highly reactive intermediates of oxygen reduction, being able to damage biological molecules, have been considered only as unwelcome by-products of metabolism (Moller, Jensen, & Hansson, 2007). However, ROS can also play a positive role in metabolism by acting as signal molecules able to control different developmental processes and to activate defence responses (Mittler, Vanderauwera, Gollery, & Van Breusegem, 2004; Mittler et al., 2011).

Plants possess several mechanisms aimed at protecting or repairing their metabolism from ROS-induced damages. Superoxide dismutase (SOD), catalase (CAT) and ascorbate peroxidase (APX) are key enzymatic ROS scavengers that act together with antioxidant metabolites, such as ascorbate

(ASC) and glutathione (GSH), to regulate redox homeostasis, modulating timing and extent of ROS accumulation. ROS levels can increase by upregulation of ROS-producing enzymes, such as NADPH oxidase and peroxidases, alteration of electron flow, or downregulation of the antioxidant systems. ROS signals might be transduced by redox modifications of key proteins, like transcription factors and protein kinases (Foyer & Noctor, 2009; Mittler et al., 2011). Moreover, antioxidant molecules, whose redox state is affected by ROS, also participate to the ROS signaling pathway (Mittler et al., 2004). Depending on the type and the amount of ROS produced and the efficiency of the scavenging systems, plant cells can activate pathways aimed at restoring cellular redox balance or leading to cell death (de Pinto, Locato, & De Gara, 2012; de Pinto, Paradiso, Leonetti, & De Gara, 2006).

NO is a freely diffusible, gaseous free radical, acting as a signaling molecule. NO can influence different aspects of plant growth and development, such as seed germination, lateral root initiation, flowering and stomatal closure and can participate to the responses to abiotic and biotic stress (Yu, Lamattina, Spoel, & Loake, 2014). A major route by which NO can function as signal is *S*-nitrosylation, a redox-based posttranslation modification, consisting in the addition of an NO moiety to a reactive cysteine residue (Cys) to form an S-nitrosothiol (Astier et al., 2011). NO is also able to react with GSH, thus producing S-nitrosoglutathione (GSNO), which also acts as an endogenous trans-nitrosylating agent. GSNO is also considered as a physiological NO donor and, being more stable than NO, it acts as a long-distance NO transporter (Malik, Hussain, Yun, Spoel, & Loake, 2011). In some cases, the effects of NO are the result of its chemical interaction with ROS. Indeed, NO reacts with O_2^- to form the reactive nitrogen species peroxynitrite ($ONOO^-$), which can act as a signaling molecule mainly through the posttranslational modification of proteins by tyrosine nitration (Vandelle & Delledonne, 2011).

ROS have been proposed as key PCD inducers whereas the role of NO is less clear, since this molecule seems to be able to act both as a promoter or suppressor of PCD (Petrov et al., 2015; Wang, Chen, Loake, & Chu, 2010). However, a complex networking reactions between ROS, NO and antioxidants might be responsible for the fate of plant cells (Figure 1).

First, it is known that a feedback regulation might control ROS and NO biosynthesis. H_2O_2 is a recognized inducer of NO synthesis in various plants, acting through the enhancement of nitric oxide synthase (NOS)-like or nitrate reductase (NR) activities. H_2O_2 induces NO generation in mung

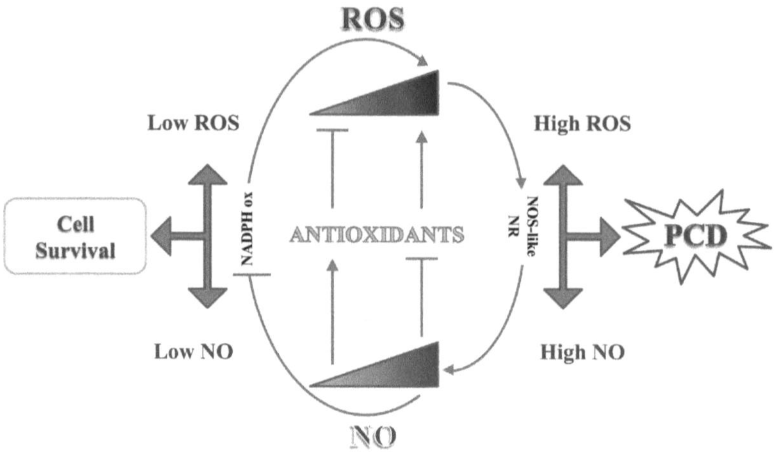

Figure 1 ROS and NO interactions in determining cell fate. ROS stimulate NO production, activating NOS-like or NR activities. On the other hand, NO can control ROS levels by inhibiting NADPH oxidase or by controlling antioxidant enzymes. In particular, low NO levels promote antioxidants, decreasing ROS; high NO levels inhibit antioxidants, increasing ROS. As a consequence low levels of NO/ROS might promote protection form oxidative damages and guarantee cell survival, whereas high NO/ROS levels contribute to induce PCD. More details are given in the text. ROS, Reactive oxygen species; PCD, programmed cell death; NO, nitric oxide; NR, nitrate reductase.

bean leaves, *Arabidopsis* roots, *Vicia faba* guard cells and tobacco BY-2 cells (He, Xu, She, Song, & Zhao, 2005; Lin et al., 2012; Lum, Butt, & Lo, 2002; de Pinto et al., 2006; Wang, Ries, Wu, Yang, & Crawford, 2010). Calcium (Ca^{2+}) might mediate the H_2O_2-induced NO generation, which is inhibited by Ca^{2+} channel blocker (Gonzalez et al., 2012). On the other hand, NO influences ROS levels, controlling their production and degradation. In *Arabidopsis* infected with avirulent pathogens, NO can nitrosylate NADPH oxidase, limiting ROS production (Yun et al., 2011). In *Arabidopsis* cells, NO production, occurring after cadmium treatment, inhibits APX activity with a consequent increase in H_2O_2 (De Michele et al., 2009). Furthermore, an NO-dependent decrease in APX activity, due to *S*-nitrosylation of the enzyme, seems to be responsible of high H_2O_2 levels in tobacco BY-2 cells after heat shock (de Pinto et al., 2013). Transgenic *Nicotiana tabacum* plants expressing the rat neuronal NOS accumulate high NO levels and show CAT inhibition, which causes an increase in H_2O_2 (Chun et al., 2012). However, many authors reported that NO can increase the activity of antioxidant enzymes, reducing ROS levels and protecting plants from different kinds of abiotic stress

(Li et al., 2008; Shing et al., 2009; Zhang et al., 2007). This contradictory results might be explained by the dose-dependent effect of NO on the cellular redox state. It has been proposed that low/high NO levels can stimulate/inhibit antioxidant systems. The relative amount of NO and ROS within cells or tissues is also a determinant for activating different responses (Gross et al., 2013 and references therein).

Overall, the above reported data indicate that NO and ROS levels could be finely regulated and that low NO-ROS levels can promote cell survival, whereas high NO-ROS levels are responsible of PCD induction (Figure 1).

2. PCD INDUCTION BY NO AND/OR H_2O_2

The first indication on the role of H_2O_2 as a key PCD inducer was obtained in the hypersensitive response (HR) occurring in soybean cells infected with the avirulent pathogen *Pseudomona syringae* pv. *glycinea*. The treatment of soybean cells with diphenyleneiodonium (DPI), a NADPH oxidase inhibitor, suppresses the pathogen-induced H_2O_2 burst and prevents cell death (Levine, Tenhaken, Dixon, & Lamb, 1994). Moreover, addition of exogenous H_2O_2 is able to trigger HR-PCD in a Ca^{2+}-dependent manner (Levine, Pennell, Alvarez, Palmer, & Lamb, 1996). Some years later, it was shown that in soybean cells NO is a further essential messenger in cell death execution occurring during HR (Delledonne, Xia, Dixon, & Lamb, 1998). The decrease of NO level, obtained by the addition of the NO scavenger cPTIO and by inhibitors of NOS activity, reduces HR-PCD. Furthermore, the treatment of soybean cells with the NO donor sodium nitroprusside (SNP) in combination with ROS is able to trigger PCD (Delledonne et al., 1998). Similarly, in tobacco BY-2 cells only the simultaneous application of SNP and glucose/glucose oxidase, an H_2O_2-generating donor system, but not each individual donor alone causes PCD. The PCD induced by NO and H_2O_2 in BY-2 cells requires the suppression of the antioxidant systems responsible for cellular redox balance. It has been proposed that the decrease in antioxidant defence is part of the strategy contributing to generate the oxidative burst that characterizes PCD (de Pinto, Tommasi, & De Gara, 2002). Consistently, *Arabidopsis* mutants, having a 10–25% of wild-type ASC content, spontaneously activate processes that typically occur during HR-PCD, such as localized cell death and the expression of pathogenesis-related proteins (Pavet et al., 2005).

In contrast with both ROS and NO requirement for PCD induction, it was shown that in tobacco BY-2 cells the sole exogenous application of H_2O_2 at given concentrations is able to induce PCD (Houot et al., 2001). This apparent contradiction has been successively explained. The metabolic responses activated in cells greatly differ depending both on the intensity of the oxidative stress generated and the different timing of ROS production. The production of H_2O_2 over a threshold value induces cell death in tobacco BY2 cells. Nevertheless, the kind of death (necrosis versus PCD) strongly depends on the timing of H_2O_2 production. The prolonged production of ROS for several hours induces a necrotic process. On the other hand, an amount of H_2O_2 comparable with that inducing cell necrosis, but given as a single pulse, triggers PCD (de Pinto et al., 2006). Direct addition of H_2O_2 probably simulates the oxidative burst induced by various forms of biotic or abiotic stress against which PCD is frequently activated (de Pinto et al., 2012). Interestingly, only the direct addition of H_2O_2 activates NO synthesis through the enhancement of an NOS-like activity; the treatment with cPTIO, before the exposure to H_2O_2, significantly blocked cell death, confirming that at least in tobacco BY-2 cells, H_2O_2 and NO are team players in the PCD-signaling pathway (de Pinto et al., 2006, 2013). In this H_2O_2-induced PCD, NO plays a key role in the control of cytosolic APX (cAPX), a key enzyme involved in the removal of H_2O_2. In particular, the precocious *S*-nitrosylation of cAPX seems to be responsible for the immediate decrease in the enzyme activity and for the ubiquitination of cAPX, suggesting that a tight relationship exists between NO and H_2O_2 in PCD induction (de Pinto et al., 2013). A role for thylakoidal APX (tAPX) has been also proposed in the signaling of NO-induced PCD in *Arabidopsis*. Indeed, after treatment with SNP, tAPX overexpressing plants show a reduction of cell death compared to wild-type plants subjected to the same treatment. On the other hand, the tAPX antisense plants show enhanced symptoms of damage when NO-induced cell death is triggered. The use of these transgenic lines confirms that H_2O_2 acts in partnership with NO in PCD induction (Murgia et al., 2004; Tarantino et al., 2005).

Another well-studied NO–ROS interaction is that occurring in H_2O_2-induced cell death in rice leaves (Lin et al., 2012). The rice *noe1* (*nitricoxideexcess1*) mutant accumulates high NO level. The NOE1 gene encodes the rice catalase OsCATC. Leaves of *noe1* mutants accumulate high H_2O_2 levels which, by activating NR, promote NO production. In this system, the removal of NO reduces cell death. Leaf cell death in *noe1* plants is also alleviated by overexpression of OsGSNO reductase (GSNOR) that reduces the intracellular S-nitrosothiol (SNO) levels, suggesting that both NO and SNOs are important mediators in the H_2O_2-induced cell death (Lin et al., 2012; Wang, Lin, Loake, & Chu, 2013). Interestingly, different proteins are nitrosylated in the *noe1* plants, among which glyceraldehyde 3-phosphate dehydrogenase (GAPDH) and thioredoxin, which are reported to be involved in PCD in animals (Hara et al., 2005; Sumbayev, 2003). In the light-driven leaf cell death in rice, NO might function downstream of H_2O_2 since no changes in H_2O_2 content occur in either *GSNOR*-overexpressing or *GSNOR*-RNAi transgenic lines in the *noe1* background (Lin et al., 2012). However, in maize leaves NO treatment leads to a rapid cell death and induces H_2O_2 accumulation through the enhancement of NADPH oxidase. Moreover, pharmacological studies show that in this case H_2O_2 is required, but not sufficient, to induce cell death, suggesting that some other key signaling components can be induced by NO (Kong, Zhang, Pan, Zhou, & Li, 2013).

All these studies imply that NO and H_2O_2 are involved in PCD induction and that they could interact through different pathways, depending on plant species and redox state. In the following paragraphs the involvement of ROS and NO in the signaling pathways leading to different kinds of developmentally and environmentally induced PCD will be discussed.

3. NO AND ROS SIGNALING DURING SENESCENCE

Senescence, the final stage of plant development, can be considered as a form of plant PCD (Lim, Woo, & Nam, 2003; Quirino, Noh, Himelblau, & Amasino, 2000). This developmental process is characterized by remarkable changes in cellular metabolism and nutrient remobilization, resulting from alterations in the dynamic equilibrium between anabolic and catabolic processes (Balazadeh, Parlitz, Mueller-Roeber, & Meyer, 2008; Lim et al., 2003).

ROS accumulation is a common event in the cell death pathway in senescent organs (Smykowski, Zimmermann, & Zentgraf, 2010; Zentgraf & Hemleben, 2008). Many literature data report an increase of different ROS during natural and artificially induced senescence. For instance, in cucumber leaves, during the premature senescence induced by high-temperature treatments, accumulation of large amounts of O_2^- and H_2O_2 is responsible for the increased lipid peroxidation and the decreased protein and chlorophyll contents (Zhao, Nishimura, Fukumoto, & Li, 2011). During the senescence process, among the involved ROS a key role has been attributed to H_2O_2. This molecule can function as a signal to promote senescence in different plant species and it seems to be part of a complex regulatory network (Bieker, Riester, Stahl, Franzaring, & Zentgraf, 2012). The accelerated senescence, observed in *Arabidopsis* plants overexpressing the Bax-like apoptotic protein CRS, is due to H_2O_2 accumulation that acts as a positive mediator for cell death by enhancing metacaspase expression (Cui et al., 2013). H_2O_2, in addition to its role in oxidizing macromolecules, has been proposed to function as a signal able to induce the expression of senescence-associated genes (SAGs) (Navabpour et al., 2003). Moreover, NAC transcription factors ORS1, JUB1 and ATAF1, key regulators of leaf senescence, have been also found to be rapidly and strongly induced by H_2O_2 treatment (Balazadeh et al., 2011; Garapati, Xue, Munne-Bosch, & Balazadeh, 2015; Wu et al., 2012). Thus, during senescence ROS might promote cell death by directly oxidizing target macromolecules and by driving the expression of senescence-related genes.

NO is also involved in the control of senescence-induced PCD; however, literature data show that this molecule can both accelerate and slow down this process.

In *Arabidopsis*, a senescence-like phenotype has been shown for plants expressing an NO degrading dioxygenase and for *nos1/noa1* mutants, deficient in NO (Mishina, Lamb, & Zeier, 2007; Niu & Guo, 2012), suggesting that NO may alleviate ROS toxicity, acting as a leaf senescence delaying factor. *Arabidopsis* mutants *dnd1*, lacking the cation-permeable channel CNGC2, show senescence-associated phenotypes, such as loss of chlorophyll, H_2O_2 production, lipid peroxidation, expression of SAGs and tissue necrosis. Interestingly, basal levels of NO in *dnd1* plants are lower than that observed in wild-type plants and the application of an NO donor rescues the *dnd1* senescence-related phenotypes. This provides evidences that during senescence, CNGC2 is involved in Ca^{2+} influx and impacts downstream basal NO production, confirming that NO acts as a negative regulator in leaf senescence signaling (Ma, Smigel, et al., 2010). The protective effect of NO during senescence can be also linked to its role as negative regulator of the chlorophyll catabolic pathway and as a positive factor in maintaining the stability of thylakoid membranes (Liu & Guo, 2013).

On the other hand, NO can also promote senescence. During the progression of senescence, due to the exogenous addition of high concentration of the cytokinin 6-benzylaminopurine (BA), an increase of NO levels occurs. Either an NOS inhibitor or an NO scavenger prevents cell death, indicating that NO synthesis is a key player in BA-induced PCD/senescence (Carimi et al., 2005). In *Arabidopsis*, during age-dependent senescence the accumulation of the ferritin isoform AtFer1, required for its iron-detoxification function when ROS accumulate, occurs via an NO-mediated pathway (Murgia et al., 2007). In sunflower seedlings, during salt-triggered senescence, the level of endogenous NO sharply increases (David, Yadav, & Bhatla, 2010). Recently, it has been reported that in *Arabidopsis* NO can induce premature cotyledon senescence (Du et al., 2014).

A possible explanation of this contrasting data is that the role of NO may differ in senescence, probably in function of its concentration. This is supported by a study of leaf senescence in sunflower, which shows that depending on the doses NO can either delay or accelerate senescence (Prochazkova & Wilhelmova, 2011).

4. NO AND ROS INTERPLAY IN SELF-INCOMPATIBILITY

Self-incompatibility (SI) is an important genetically controlled mechanism in higher plants, which is able to prevent self-fertilization, thus promoting out-cross pollination and genetic diversity of populations (Franklin-Tong, 2008). Two main systems of SI, gametophitic (GSI) and sporophytic have been identified in plants. The triggering of PCD via ROS and/or NO during SI has been shown in several species (Bosch & Franklin-Tong, 2008; Jiang et al., 2014; Rogers, 2006; Roldan, Rojas, & Goldraij, 2012; Serrano, Pelliccione, & Olmedilla, 2010; Wang, Wu, et al., 2010; Wilkins et al., 2011). However, NO and ROS interaction has been better clarified in some cases of GSI.

The most common type of GSI is under the control of the *S-RNase* gene, responsible for pistil specificity and the multiple *S-locus F-box* gene, responsible for pollen specificity (Kear & McClure, 2012; Kubo et al., 2010). The growth of self-pollen or genetically related pollen is blocked by the action of extracellular S-RNase able to penetrate into pollen tube and thus to degrade its RNAs (Luu, Qin, Morse, & Cappadocia, 2000). In *Pyrus pyrifolia*, a species characterized by this type of GSI, S-RNases specifically disrupt tip-localized ROS production, causing mitochondrial alterations. The decrease in ROS leads to reduction in Ca^{2+} current, depolymerization of actin cytoskeleton and nuclear DNA degradation (Wang, Wu, et al., 2010). Recently, it has been shown that anti-calmodulin treatment permits a partially rescue of the inhibition of self-incompatible pollen tube growth in pear, indicating that calmodulin acts upstream ROS regulation and actin filament depolymerization (Jiang et al., 2014).

Interactions among ROS and NO have been proposed in the SI of *Olea europaea* L. Although the olive SI determinants have not yet been characterized, it has been suggested that this SI belongs to the GSI under the control of *S-RNase* genes, since it shows its typical features such as bicellular pollen, wet stigma and RNAse activity (Serrano & Olmedilla, 2012). A possible bidirectional exchange of signals between pollen and stigma regulates ROS and NO production. The H_2O_2 content decreases in stigma papillae after pollen arrival, whereas the levels of O_2^- and NO increase after pollination. Peroxidase and NADPH oxidase seem to be involved in O_2^- production whereas an NOS-like activity is responsible of the increase of NO. Biochemical evidences strongly suggest that both O_2^- and NO are essential for triggering PCD in this SI. During olive pollination the concomitant increase in $ONOO^-$ concentration, generated by the rapid reaction between O_2^- and NO, is responsible of the increase in protein nitration, strongly associated with PCD. Indeed, the treatment with $ONOO^-$ scavengers drastically reduces the percentage of papillar cell death and the number of pollen grains undergoing PCD (Serrano, Romero-Puertas, Rodriguez-Serrano, Sandalio, & Olmedilla, 2012).

A crosstalk of NO and ROS in PCD occurring during SI has been also shown in *Papaver rhoeas*. In this plant, GSI is due to the molecular interaction between the pistil surface protein PrsS, and the pollen trans-membrane protein PrpS (Wheeler et al., 2009). This interaction causes an intracellular Ca^{2+} increase, which is, in turn, responsible for a signaling cascade that culminates in PCD and rejection of self-pollen (Wilkins, Poulter, & Franklin-Tong, 2014). The rapid increase in cytosolic Ca^{2+} and K^+ levels, due to the PrsS–PrpS interaction, causes a network of signaling events, leading to mitogen-activated protein kinases (MAPKs) activation, ROS and NO production, cytoskeleton alteration and culminating in PCD (Wilkins et al., 2011). In particular, the increase in cytosolic ROS occurs in few minutes after the SI induction, upstream the transient increases in NO. A combined pretreatment with ROS and NO scavengers, such as DPI and cPTIO, drastically reduces the formation of actin punctate foci and the activation of a caspase-3-like, key markers of the SI response in *Papaver* (Wilkins et al., 2011). The effect of NO and ROS on actin polymerization might be due to protein posttranslation modifications mediated by these species. Indeed, the carbonylation and *S*-nitrosylation of actin causes severe disorders in actin cytoskeleton structure and function (Rodriguez-Serrano et al., 2014).

5. NO AND ROS CROSSTALK DURING HYPERSENSITIVE RESPONSE

HR is probably the form of plant PCD in which the interplay between ROS and NO in signaling has been most widely studied. HR is a defence mechanism that plants activate against avirulent pathogens at the site of attack, to inhibit further pathogen entry and spreading (Greenberg & Yao, 2004). As previously mentioned, ROS are important signals mediating HR and defence gene activation (Levine et al., 1994). A biphasic ROS overproduction has been indicated as typical of the HR. The early ROS increase seems to be independent of the specific pathogen recognition, whereas the late ROS accumulation is typical of the incompatible interactions (Grant et al., 2000; Lamb & Dixon, 1997). ROS production during HR principally involves NADPH oxidases, encoded by the respiratory burst oxidase homologue (RBOH) gene family, localized into the plant cell plasma membrane (Torres, Dangl, & Jones, 2002). The pathogen recognition causes a cytosolic Ca^{2+} increase that, on one hand, can directly bind to the EF motifs on the cytosolic side of NADPH oxidase and, on the other hand, activates a class of Ca^{2+}-dependent protein kinases that modulate NADPH oxidase activity by phosphorylation (Blume, Nurnberger, Nass, & Scheel, 2000; Keller et al., 1998; Kobayashi et al., 2007). In most cases a lack of RBOH expression leads to very low levels of ROS production, resulting in the alteration of different plant responses in terms of cell death and pathogen resistance (Marino, Dunand, Puppo, & Pauly, 2012).

During the incompatible plant—pathogen interaction NO also acts as a trigger for HR. The pharmacological modulation of intracellular levels of NO, by exogenous treatments with NO donors or with NO scavengers and inhibitors of the mammalian NOS activates and delays/suppresses HR, respectively (Delledonne et al., 1998). Many literature data report that a key factor for the HR is the balance between intracellular NO and ROS levels and that the two species work together to induce cell death (Delledonne, Zeier, Marocco, & Lamb, 2001; Kulik et al., 2015; Mur, Kenton, & Draper, 2005).

Interestingly, during plant–pathogen interaction, ROS can control NO production and vice versa. The production of NO induced by the elicitor cryptogein in tobacco cells is regulated through a ROS-dependent pathway involving NADPH oxidase, whereas NO downregulates the level of H_2O_2 (Kulik et al., 2015). On the other hand, NO is required for the proper production of H_2O_2 in grape cells elicited by *Botrytis cinerea* endopolygalacturonase 1 and in *Arabidopsis* leaves elicited by oligogalacturonides (Rasul et al., 2012; Vandelle, Poinssot, Wendehenne, Bentejac, & Pugin, 2006).

Once again, the data highlight discrepancies that might be due to the specificity of the utilized plant models or can reflect a complex signaling network that has not been fully determined. Nevertheless, different mechanisms explaining the crosstalk of NO and ROS as cell death mediators have been proposed (Trapet et al., 2015; Figure 2).

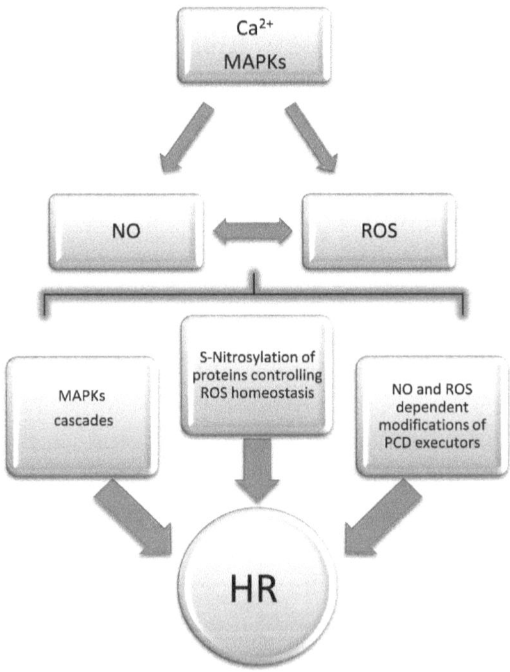

Figure 2 Possible NO/ROS-dependent downstream signaling during the hypersensitive response. NO and ROS seems to be regulated by a Ca^{2+}-dependent MAPKs cascade. Furthermore, NO and ROS, whose biosyntheses are mutually controlled, could interact in inducing HR-PCD through three different downstream signaling mechanisms: activation of MAPKs cascade, *S*-nitrosylation of proteins involved in ROS homeostasis, oxidation and *S*-nitrosylation of PCD executors. MAPKs, mitogen-activated protein kinases; ROS, reactive oxygen species; PCD, programmed cell death; NO, nitric oxide; HR, hypersensitive response.

First, calcium flux could represent a molecular link between NO and ROS; indeed, the production of both species are calcium-dependent and both NO and ROS account for calcium mobilization in plant signaling (Lecourieux, Ranjeva, & Pugin, 2006; Trapet et al., 2015). A possible link between calcium and downstream NO and ROS production in response to pathogen perception could be represented by MAPKs signaling cascade. It has been shown that in *Nicotiana benthamiana* MAPKs regulate the RBOH-dependent ROS and NOA1-mediated NO production (Asai, Ohta, & Yoshioka, 2008). MAPK cascades could also act downstream NO and ROS signals. For instance, xylanase perception in tomato cell suspensions activates a protein kinase pathway required for NO formation and *S*-nitrosylation-dependent mechanisms involved in downstream signaling, leading to ROS production (Lanteri, Lamattina, & Laxalt, 2011).

Another mechanism linking NO with ROS is the *S*-nitrosylation of proteins controlling ROS homeostasis during the HR. An example is the *S*-nitrosylation of peroxiredoxin IIE (PrxII), occurring in *Arabidopsis* upon infection by the avirulent bacterial pathogen *P. syringae* pv. tomato (Romero-Puertas et al., 2007). Peroxiredoxins are thiol-based enzymes involved in ROS removal as well as in $ONOO^-$ detoxification (Rhee, Chae, & Kim, 2005). The formation of $ONOO^-$ is highly competitive with O_2^- dismutation performed by SOD; thus hyperaccumulation of NO can also restrain H_2O_2 release into the cell (Ferrer-Sueta & Radi, 2009). During the HR, *S*-nitrosylation of Cys-121 of the active site of PrxII causes the loss of the detoxification activity of the protein, thus regulating the antioxidant function of this enzyme and contributing to HR (Romero-Puertas et al., 2007). Another well-described example of NO-dependent regulation of ROS homeostasis is the *S*-nitrosylation of AtRBOHD in *Arabidopsis* plants challenged by the avirulent bacteria *P. syringae*. In this plant—pathogen interaction, the early burst of ROS and NO initiates HR, but at later stages of this defence response, when the SNO levels exceed a certain threshold, the AtRBOHD is *S*-nitrosylated. The *S*-nitrosylation, occurring at Cys-890, causes a decrease in Flavin adenine dinucleotide binding and, thus, inhibits AtRBOHD activity. Accordingly, mutation of this Cys to an alanine residue compromises *S*-nitrosylation of AtRBOHD, leading to an increase in ROS production and cell death in response to the pathogen attack. This NO-dependent regulation of AtRBOHD represents a feedback regulative mechanism helping plants to control excessive ROS production and HR expansion (Yun et al., 2011).

A further link between ROS and NO in HR might be represented by proteins involved in PCD execution that can be subjected to posttranslation modification by both reactive species. Two examples of these kind of proteins are GAPDH and CDC48 (Cell Division Cycle 48), both reported, at least in animal systems, as PCD executioners through the ubiquitin—proteasome pathway (Baek et al., 2013; Braun & Zischka, 2008; Hara et al., 2005). GAPDH is a target of *S*-nitrosylation and H_2O_2-mediated oxidation in plants (Hancock et al., 2005; Lindermayr, Saalbach, & Durner, 2005; Zaffagnini et al., 2013). In *Arabidopsis* undergoing HR, *S*-nitrosylation of GAPDH occurs (Romero-Puertas et al., 2008). In animal system it has been reported that *S*-nitrosylated GAPDH interacts with Siah1, an E3 ubiquitin ligase and the resulting complex is translocated into the nucleus, where it promotes the ubiquitination and degradation of proteins, thus facilitating cell death (Hara et al., 2005). CDC48 is a member of the AAA+ ATPases family, containing two domains involved in the binding and hydrolysis of ATP. This chaperone-like enzyme uses energy from ATP hydrolysis for remodelling ubiquitinated proteins, in order to make possible their degradation via the proteasome (Baek et al., 2013). In tobacco cells elicited by cryptogein, CDC48 is *S*-nitrosylated on Cys-526; the *S*-nitrosylation causes the inhibition of the ATPase activity of CDC48 (Astier et al., 2012). It has been reported that in *Drosophila* CDC48 the oxidation of the same Cys inhibits protein activity (Noguchi et al., 2005). Although not demonstrated in plants, it is possible to speculate that *S*-nitrosylation and H_2O_2-mediated oxidation of GAPDH and CDC48 could regulate HR through the control of ubiquitin-mediated proteolysis (Trapet et al., 2015). This hypothesis is supported by papers indicating the involvement of components of the ubiquitin pathways in the regulation of plant defence (Kulik et al., 2015; Trujillo & Shirasu, 2010).

6. NO AND ROS INVOLVEMENT IN PCD INDUCED BY ABIOTIC STRESS

A common factor of environmental stress is an increase in reactive species that are integral part of the signaling pathway activating defence mechanisms. Many abiotic stress such as drought, flooding, heavy metals and extreme temperatures are known to induce PCD, which is directly or indirectly mediated by ROS and NO (Petrov et al., 2015; Qiao, Li, & Fan, 2014). Here, we give a description of ROS—NO interplay and downstream signaling events leading to cadmium- and heat stress-dependent PCD.

6.1 Cadmium-Induced PCD

In plants, cadmium (Cd) negatively affects crucial physiological processes, such as uptake and transport of water and nutrients, nitrogen metabolism, photosynthesis and respiration (see Costs and Benefits of Nitric Oxide Generation in Plants Exposed to Cadmium (Arasimowicz-Jelonek, Floryszak-Wieczorek, & Izbianska, 2016)). As a consequence, Cd inhibits seed germination and decreases plant growth (Sanità di Toppi & Gabrielli, 1999). It has been shown that Cd can induce cell death which, depending on the metal concentration, can be PCD or necrosis (Behboodi & Samadi, 2004; De Michele et al., 2009; Fojtova & Kovarik, 2000).

Cd-induced PCD (Cd-PCD) shows typical apoptotic-like markers (Table 1; Arasimowicz-Jelonek et al., 2012; De Michele et al., 2009; Ma, Xu, et al., 2010; Ye, Li, & Xing, 2013). Moreover, Cd-PCD resembles an accelerated senescence in *Arabidopsis* cells, as evidenced by the expression of SAG12 (De Michele et al., 2009) and an increase in the expression of the *Hsr203J* gene, an early marker for HR, in tobacco BY-2 cells (Ma, Xu, et al., 2010).

In tobacco BY-2 cells, Cd-PCD requires a biphasic production of ROS, with the two waves differing in their nature and subcellular localization. A first transient NADPH oxidase-dependent accumulation of H_2O_2 is followed by the accumulation of O_2^- in mitochondria and both waves are

Table 1 PCD Markers and NO-Dependent Mechanisms of Cd Cytotoxicity during Cd-PCD in Different Experimental Systems

Experimental System	PCD Markers	NO-Dependent Mechanisms in Cd-Cytotoxicity	References
Arabidopsis cell suspension	Chromatin condensation TUNEL positive nuclei	*S*-nitrosylation-dependent inhibition of phytochelatins	De Michele et al. (2009)
Tobacco BY-2 cell suspension	Chromatin condensation TUNEL positive nuclei	Increase of Cd uptake and accumulation	Ma, Xu, et al. (2010)
Lupine roots	TUNEL positive nuclei Positive comet assay	Increase of Cd uptake and accumulation	Arasimowicz-Jelonek et al. (2012)
Arabidopsis plants	Chromatin condensation	MPK6-mediated activation of caspase-3-like	Ye et al. (2013)

PCD, programmed cell death; NO, nitric oxide; Cd-PCD, Cd-induced PCD TUNEL, Terminal deoxynucleotidyl transferase dUTP nick end labeling.

needed to induce PCD (Garnier et al., 2006). On the other hand, in *Arabidopsis* cells H_2O_2 production seems to increase as a late event, in an NO-dependent way. Indeed, inhibition of NO synthesis results in partial prevention of H_2O_2 production and cell death. In this case, NO might regulate H_2O_2 levels by APX inhibition (De Michele et al., 2009). An early boosted NO production is required to initiate Cd-PCD in lupine roots. NO generation is accompanied by the activation of NADPH-oxidase and subsequent O_2^- accumulation. In this case, O_2^-, rather than H_2O_2, functions as the molecule that synergizes with NO to activate PCD (Arasimowicz-Jelonek et al., 2012).

The role of NO in plant response to Cd seems to be different if the source of the molecule is exogenous or endogenous. Indeed, NO released by NO-donors functions as an antioxidant and significantly reduces Cd-induced oxidative damage (Saxena & Shekhawat, 2013 and references therein), whereas endogenous NO production contributes to Cd toxicity (Besson-Bard et al., 2009; Groppa, Zawoznik, Tomaro, & Benavides, 2008) and is required for Cd-PCD (Arasimowicz-Jelonek et al., 2012; De Michele et al., 2009; Ma, Xu, et al., 2010). This further underlines the relevance of the relative amount of ROS and NO, and their subcellular compartmentalization, in the activation of a certain metabolic response.

Different mechanisms by which NO can exert its effect in Cd-PCD have been proposed (Table 1). NO might play a role in the regulation of Cd cytotoxicity by direct *S*-nitrosylation of phytochelatins (PCs). Indeed, as both Cd and NO are able to bind Cys residues of PCs, *S*-nitrosylation of PCs might reduce the efficiency of Cd detoxication (De Michele et al., 2009; Elviri et al., 2010). A second mechanism by which NO can regulate Cd-PCD is related to the ability of this molecule to stimulate Cd uptake, thus promoting Cd accumulation (Arasimowicz-Jelonek et al., 2012; Ma, Xu, et al., 2010). In this case NO might regulate the expression of genes mediating Cd uptake or blocking Cd detoxication. In *Arabidopsis*, Cd enters root cells through the IRT1 transporter. Thus, the competition for Cd and Fe absorption through IRT1 leads to a decline in the iron cellular content, which is perceived as a signal for NO synthesis (Besson-Bard et al., 2009; Graziano & Lamattina, 2007). NO promotes the upregulation of iron acquisition-related genes expression, among which *IRT1* and as a consequence amplifies Cd uptake (Besson-Bard et al., 2009). Finally, NO can contribute to Cd-PCD by MPK6-mediated caspase-3-like activation (Ye et al., 2013). It has been shown that in Cd-PCD, the activity of caspase-3-like, an executor of PCD, increases in both *Arabidopsis* and tomato cells (Iakimova, Woltering, Kapchina-Toteva, Harren, & Cristescu, 2008; Ye et al., 2013). Interestingly, the reduction of NO levels with cPTIO significantly reduces the activity of Cd-induced caspase-3-like (Ye et al., 2013). A MAPK cascade is involved in Cd-induced cell death and NO seems to be required for activation of MAPKs (Pagnussat, Lanteri, Lombardo, & Lamattina, 2004; Zhang et al., 2007). Consistently, the pretreatment of *Arabidopsis* plants with cPTIO suppresses the Cd-induced activation of MAPKs. Moreover, the impairment of MPK6, obtained using the MAPK inhibitor PD98059 or by utilizing the *mpk-6* mutant, suppresses Cd-induced caspase-3-like activity and rescues Cd-PCD (Ye et al., 2013).

6.2 Heat Stress-Induced PCD

Heat stress (HS) can cause several apoptosis-like characters and triggers PCD in plant cells (Vacca et al., 2004; Zhang, Li, Xing, & Gao, 2009; Zuppini, Bugno, & Baldan, 2006). However, as shown in tobacco BY-2 cells, HS-induced PCD is induced only when temperature rises above a specific threshold (Locato, Gadaleta, De Gara, & de Pinto, 2008; Vacca et al., 2004). The production of ROS and NO, as well as the metabolic changes observed in cells subjected to HS at temperature inducing PCD are considerably different to those occurring in response to less severe stimuli (Locato et al., 2008; Locato, de Pinto, & De Gara, 2009; Marsoni et al., 2010).

In tobacco BY-2 cells, the exposure to temperature inducing PCD causes a rapid and drastic production of ROS. Both H_2O_2 and O_2^- are required for a successful PCD since the addition of ASC or SOD to the cultures prevents cell death. NADPH oxidase could have a key role in the oxidative burst occurring during HS (Konigshofer, Tromballa, & Loppert, 2008; Miller et al., 2009). However, in tobacco BY-2 cells undergoing HS-induced PCD (HS-PCD), mitochondria might be responsible for the early ROS production, due to a prompt impairment of the mitochondrial oxidative metabolism (Vacca et al., 2004; Valenti et al., 2007). An important role for mitochondria in ROS production during HS-PCD has been also shown in *Arabidopsis* (Zhang et al., 2009).

Interestingly, in tobacco BY-2 cells undergoing HS-PCD, an impairment of L-galactone-γ-lactone dehydrogenase (GLDH), the last enzyme of ASC biosynthesis, occurs (Valenti et al., 2007). GLDH is an integral part of plant mitochondrial complex I, the redox state of which affects GLDH catalysis (Millar et al., 2003). GLDH inhibition might be due to complex I alteration. As a consequence of GLDH inhibition, a decrease in the amount of available ASC occurs, contributing to the oxidative burst needed for PCD (Locato et al., 2008; Vacca et al., 2004).

In tobacco BY-2 cells undergoing HS-PCD, mitochondria have also a key role for PCD execution; indeed they promptly release functionally active cytochrome c (cyt c) in a ROS-dependent manner. The cyt c release is needed for caspase-3-like activation. Interestingly, after 2 h from the HS, a decrease in mitochondrial cyt c fraction and a parallel increase in the cytosolic one occur, whereas after longer periods from the HS the amount of cyt c remains constant in the mitochondria and strongly decreases in the cytosol. This behaviour might be explained by the late activation of caspase-3-like activity, which might be responsible of cyt c proteolysis and execution of PCD program (Vacca et al., 2006).

The caspase-3-like activation has also been observed in HS-induced *Arabidopsis* death (Li, Yue, & Xing, 2012), confirming a key role for this protein in HS-PCD. Caspase-3-like activation is due to a vacuole-localized cysteine protease called γVPE, exhibiting caspase-1-like activity. The model proposes that HS treatment causes ROS production, cytosolic Ca^{2+} increase and activation of a Ca^{2+}-calmodulin cascade that, in turn, activates MPK6 protein. The MPK6 protein upregulates the transcript level of the inactive γVPE that, in the vacuole, can be self-processed into activated VPEs, which disrupts vacuole and promotes caspase-3-like activation (Li et al., 2012).

Exposure to temperature inducing PCD also causes a rapid increase in NO production in tobacco BY-2 cells (Locato et al., 2008). The addition of cPTIO in the culture medium, before exposure to HS, significantly blocks cell death, thus confirming a key role of NO for PCD triggering (de Pinto et al., 2013). Interestingly, as above reported, NO promotes caspase-3-like activation via MPK6 in Cd-PCD (Ye et al., 2013), suggesting that a similar mechanism can also be active in HS-PCD. The finding that H_2O_2-mediated activation of MPK6 modulates NO biosynthesis (Wang, Du, Li, Ren, & Song, 2010) makes the induction of caspase-3-like an event that could be amplified by NO and ROS interaction.

Another link between NO and ROS in HS-PCD is the occurrence of *S*-nitrosylation of cAPX. It is worth noting that *S*-nitrosylation of cAPX is responsible for the immediate decrease in the enzyme activity and acts as a signal for ubiquitin-dependent protein degradation. When NO production is reduced by treatment with cPTIO, cAPX ubiquitination and degradation are remarkably prevented (de Pinto et al., 2013). These data are coherent with the ROS-dependent proteasome activation during HS-induced PCD. A pretreatment of tobacco BY-2 cells with the proteasome inhibitor MG132, not only prevents HS-PCD, but also permits a recovery of cAPX activity (Vacca et al., 2007). Collectively, the data suggest that NO and ROS are involved in ubiquitin–proteasome-dependent cAPX degradation. Furthermore, in tobacco BY-2 cells undergoing HS-PCD, a later decrease in cAPX transcript also occurs (Locato et al., 2008). The decline in cAPX has a feedback-like effect on ROS production. Consistently, in tobacco BY-2 cells *en route* to HS-PCD a biphasic production of H_2O_2 occurs (Locato et al., 2008).

The above reported interaction among NO and ROS and the downstream events leading to HS-PCD are schematized in Figure 3.

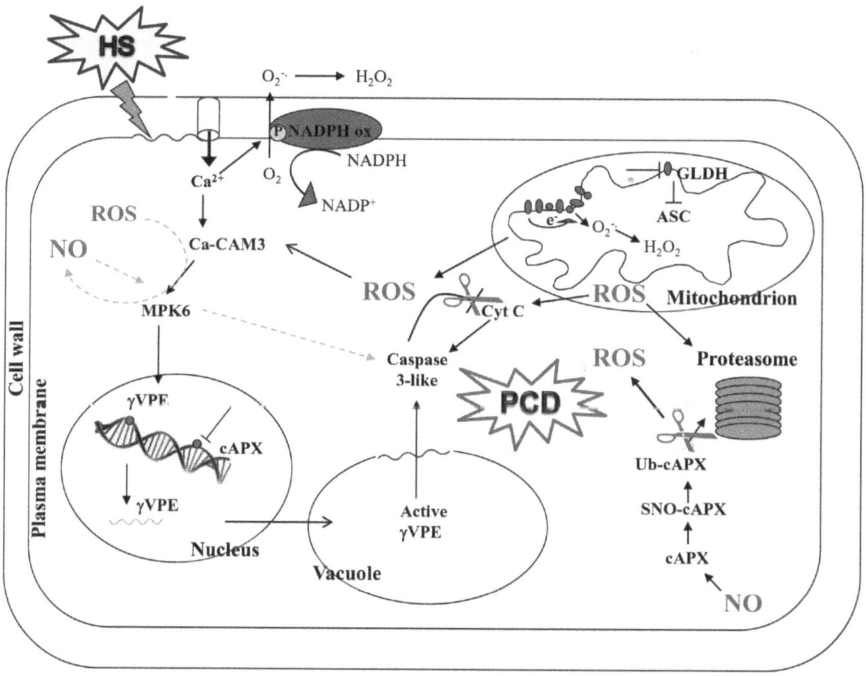

Figure 3 Schematic representation of NO- and ROS-dependent signaling involved in HS-PCD. After HS, ROS can be produced by NADPH oxidase or by an impairment of the mitochondrial oxidative metabolism. In mitochondria, the inhibition of GLDH activity causes a decrease in ASC content, contributing to the ROS accumulation. Moreover, ROS-induced cyt c release from mitochondria is responsible of caspase-3-like activation. Caspase-3-like can be also activated by NO trough a MPK6-dependent release of active γVPE. Furthermore, NO can *S*-nitrosylate APX, contributing to its ubiquitination. The ubiquinated APX could be degraded by the ROS-dependent activation of the proteasome. APX can also decrease due to inhibition of its transcription. More details are given in the text. ROS, reactive oxygen species; PCD, programmed cell death; NO, nitric oxide; HS, heat stress; GLDH, L-galactone-γ-lactone dehydrogenase; ASC, ascorbate; APX, ascorbate peroxidase.

7. CONCLUSIONS

In the last decades, evidences that NO and ROS are relevant partners in controlling metabolic pathways are strongly increased. The data reported in this chapter show that NO and ROS are involved in PCD induction, acting through different pathways, depending on the level and the timing of production of these reactive species, the cellular redox state and the kind of PCD induced. Thus, we are very far to have a complete and clear picture on NO–ROS interaction in PCD signaling, which is further complicated by the differences in the plant model utilized and the possible lack of knowledge of some key points of the network. However, some common factor in the different kinds of analyzed PCD can be identified. First, ROS and NO biosyntheses are mutually regulated with feedback mechanisms and the production of both species might rely on a Ca^{2+}-dependent activation of MAPKs. The NO-dependent *S*-nitrosylation of proteins controlling ROS homeostasis represents another key point of interaction of the two species during PCD. This is the case of NADPH-oxidase and PrxII during HR and cAPX in H_2O_2- and HS-induced PCD. Apart from the relations at the biosynthetic level, NO and ROS can also crosstalk in downstream signaling, leading to PCD. NO and ROS interplay could amplify the induction of PCD executors, probably via MAPKs, as is the case of NO- and/or ROS-dependent activation of caspase-3-like, observed in SI, Cd- and HS-induced PCD.

A further interesting point is the emerging evidence of an involvement of an NO- and/or ROS-dependent ubiquitin–proteasome pathway in PCD induction. At the moment only few data are reported in the literature, like the NO-dependent ubiquitination of cAPX and the ROS-dependent activation of proteasome in HS-induced PCD. The possibility that *S*-nitrosylation and oxidation of GAPDH and CDC48 could regulate HR trough the control of ubiquitin-mediated proteolysis highlights that this topic might deserve more attention in the future.

REFERENCES

Arasimowicz-Jelonek, M., Floryszak-Wieczorek, J., Deckert, J., Rucinska-Sobkowiak, R., Gzyl, J., Pawlak-Sprada, S., et al. (2012). Nitric oxide implication in cadmium-induced programmed cell death in roots and signaling response of yellow lupine plants. *Plant Physiology and Biochemistry, 58*, 124–134.

Arasimowicz-Jelonek, M., Floryszak-Wieczorek, J., & Izbianska, K. (2016). Costs and benefits of nitric oxide generation in plants exposed to cadmium. *Nitric Oxide and Signaling in Plants, 77*, 97–121.

Asai, S., Ohta, K., & Yoshioka, H. (2008). MAPK signaling regulates nitric oxide and NADPH oxidase-dependent oxidative bursts in *Nicotiana benthamiana*. *The Plant Cell, 20*, 1390–1406.

Astier, J., Besson-Bard, A., Lamotte, O., Bertoldo, J., Bourque, S., Terenzi, H., et al. (2012). Nitric oxide inhibits the ATPase activity of the chaperone-like AAA+ ATPase CDC48, a target for S-nitrosylation in cryptogein signalling in tobacco cells. *Biochemical Journal, 447*, 249–260.

Astier, J., Rasul, S., Koen, E., Manzoor, H., Besson-Bard, A., Lamotte, O., et al. (2011). S-nitrosylation: an emerging post-translational protein modification in plants. *Plant Science, 181*, 527–533.

Baek, G. H., Cheng, H., Choe, V., Bao, X., Shao, J., Luo, S., et al. (2013). Cdc48: a swiss army knife of cell biology. *Journal of Amino Acids, 2013*, 183421.

Balazadeh, S., Kwasniewski, M., Caldana, C., Mehrnia, M., Zanor, M. I., Xue, G. P., et al. (2011). ORS1, an H_2O_2-responsive NAC transcription factor, controls senescence in *Arabidopsis thaliana*. *Molecular Plant, 4*, 346–360.

Balazadeh, S., Parlitz, S., Mueller-Roeber, B., & Meyer, R. C. (2008). Natural developmental variations in leaf and plant senescence in *Arabidopsis thaliana*. *Plant Biology (Stuttg), 10*(Suppl. 1), 136–147.

Behboodi, B. S., & Samadi, L. (2004). Detection of apoptotic bodies and oligonucleosomal DNA fragments in cadmium-treated root apical cells of *Allium cepa* Linnaeus. *Plant Science, 167*, 411–416.

Besson-Bard, A., Gravot, A., Richaud, P., Auroy, P., Duc, C., Gaymard, F., et al. (2009). Nitric oxide contributes to cadmium toxicity in *Arabidopsis* by promoting cadmium accumulation in roots and by up-regulating genes related to iron uptake. *Plant Physiology, 149*, 1302–1315.

Bieker, S., Riester, L., Stahl, M., Franzaring, J., & Zentgraf, U. (2012). Senescence-specific alteration of hydrogen peroxide levels in *Arabidopsis thaliana* and oilseed rape spring variety *Brassica napus* L. cv. Mozart. *Journal of Integrative Plant Biology, 54*, 540–554.

Blume, B., Nurnberger, T., Nass, N., & Scheel, D. (2000). Receptor-mediated increase in cytoplasmic free calcium required for activation of pathogen defense in parsley. *The Plant Cell, 12*, 1425–1440.

Bosch, M., & Franklin-Tong, V. E. (2008). Self-incompatibility in *Papaver*: signalling to trigger PCD in incompatible pollen. *Journal of Experimental Botany, 59*, 481–490.

Braun, R. J., & Zischka, H. (2008). Mechanisms of Cdc48/VCP-mediated cell death: from yeast apoptosis to human disease. *Biochimica et Biophysica Acta-Molecular Cell Research, 1783*, 1418–1435.

Carimi, F., Zottini, M., Costa, A., Cattelan, I., De Michele, R., Terzi, M., et al. (2005). NO signalling in cytokinin-induced programmed cell death. *Plant Cell and Environment, 28*, 1171–1178.

Chun, H. J., Park, H. C., Koo, S. C., Lee, J. H., Park, C. Y., Choi, M. S., et al. (2012). Constitutive expression of mammalian nitric oxide synthase in tobacco plants triggers disease resistance to pathogens. *Molecules and Cells, 34*, 463–471.

Coll, N. S., Epple, P., & Dangl, J. L. (2011). Programmed cell death in the plant immune system. *Cell Death and Differentiation, 18*, 1247–1256.

Cui, M. H., Ok, S. H., Yoo, K. S., Jung, K. W., Yoo, S. D., & Shin, J. S. (2013). An *Arabidopsis* cell growth defect factor-related protein, CRS, promotes plant senescence by increasing the production of hydrogen peroxide. *Plant and Cell Physiology, 54*, 155–167.

David, A., Yadav, S., & Bhatla, S. C. (2010). Sodium chloride stress induces nitric oxide accumulation in root tips and oil body surface accompanying slower oleosin degradation in sunflower seedlings. *Physiologia Plantarum, 140*, 342–354.

De Michele, R., Vurro, E., Rigo, C., Costa, A., Elviri, L., Di Valentin, M., et al. (2009). Nitric oxide is involved in cadmium-induced programmed cell death in *Arabidopsis* suspension cultures. *Plant Physiology, 150*, 217–228.

de Pinto, M. C., Locato, V., & De Gara, L. (2012). Redox regulation in plant programmed cell death. *Plant, Cell and Environment, 35*, 234–244.

de Pinto, M. C., Locato, V., Sgobba, A., Romero-Puertas, M. D., Gadaleta, C., Delledonne, M., et al. (2013). S-Nitrosylation of ascorbate peroxidase is part of programmed cell death signaling in tobacco Bright Yellow-2 cells. *Plant Physiology, 163*, 1766–1775.

de Pinto, M. C., Paradiso, A., Leonetti, P., & De Gara, L. (2006). Hydrogen peroxide, nitric oxide and cytosolic ascorbate peroxidase at the crossroad between defence and cell death. *The Plant Journal, 48*, 784–795.

de Pinto, M. C., Tommasi, F., & De Gara, L. (2002). Changes in the antioxidant systems as part of the signaling pathway responsible for the programmed cell death activated by nitric oxide and reactive oxygen species in tobacco Bright-Yellow 2 cells. *Plant Physiology, 130*, 698–708.

Delledonne, M., Xia, Y., Dixon, R. A., & Lamb, C. (1998). Nitric oxide functions as a signal in plant disease resistance. *Nature, 394*, 585–588.

Delledonne, M., Zeier, J., Marocco, A., & Lamb, C. (2001). Signal interactions between nitric oxide and reactive oxygen intermediates in the plant hypersensitive disease resistance response. *Proceedings of the National Academy of Sciences of the United States of America, 98*, 13454–13459.

Du, J., Li, M., Kong, D., Wang, L., Lv, Q., Wang, J., et al. (2014). Nitric oxide induces cotyledon senescence involving co-operation of the NES1/MAD1 and EIN2-associated ORE1 signalling pathways in *Arabidopsis*. *Journal of Experimental Botany, 65*, 4051–4063.

Elviri, L., Speroni, F., Careri, M., Mangia, A., di Toppi, L. S., & Zottini, M. (2010). Identification of in vivo nitrosylated phytochelatins in *Arabidopsis thaliana* cells by liquid chromatography-direct electrospray-linear ion trap-mass spectrometry. *Journal of Chromatography A, 1217*, 4120–4126.

Ferrer-Sueta, G., & Radi, R. (2009). Chemical biology of peroxynitrite: kinetics, diffusion, and radicals. *ACS Chemical Biology, 4*, 161–177.

Fojtova, M., & Kovarik, A. (2000). Genotoxic effect of cadmium is associated with apoptotic changes in tobacco cells. *Plant, Cell and Environment, 23*, 531–537.

Foyer, C. H., & Noctor, G. (2009). Redox regulation in photosynthetic organisms: signaling, acclimation, and practical implications. *Antioxidants and Redox Signaling, 11*, 861–905.

Franklin-Tong, V. (2008). In V. Franklin-Tong (Ed.), *Self-incompatibility in flowering plants – Evolution, diversity, and mechanisms*. Berlin Heidelberg: Springer-Verlag.

Garapati, P., Xue, G. P., Munne-Bosch, S., & Balazadeh, S. (2015). Transcription factor ATAF1 in *Arabidopsis* promotes senescence by direct regulation of key chloroplast maintenance and senescence transcriptional cascades. *Plant Physiology, 168*, 1122–1139.

Garnier, L., Simon-Plas, F., Thuleau, P., Agnel, J. P., Blein, J. P., Ranjeva, R., et al. (2006). Cadmium affects tobacco cells by a series of three waves of reactive oxygen species that contribute to cytotoxicity. *Plant, Cell and Environment, 29*, 1956–1969.

Gonzalez, A., Cabrera, M. A., Henríquez, M. J., Contreras, R. A., Morales, B., & Moenne, A. (2012). Cross talk among calcium, hydrogen peroxide, and nitric oxide and activation of gene expression involving calmodulins and calcium-dependent protein kinases in *Ulva compressa* exposed to copper excess. *Plant Physiology, 158*, 1451–1462.

Grant, M., Brown, I., Adams, S., Knight, M., Ainslie, A., & Mansfield, J. (2000). The RPM1 plant disease resistance gene facilitates a rapid and sustained increase in cytosolic calcium that is necessary for the oxidative burst and hypersensitive cell death. *The Plant Journal, 23*, 441–450.

Graziano, M., & Lamattina, L. (2007). Nitric oxide accumulation is required for molecular and physiological responses to iron deficiency in tomato roots. *The Plant Journal, 52*, 949–960.

Greenberg, J. T., & Yao, N. (2004). The role and regulation of programmed cell death in plant-pathogen interactions. *Cellular Microbiology, 6*, 201–211.

Groppa, M. D., Zawoznik, M. S., Tomaro, M. L., & Benavides, M. P. (2008). Inhibition of root growth and polyamine metabolism in sunflower (*Helianthus annuus*) seedlings under cadmium and copper stress. *Biological Trace Element Research, 126*, 246–256.

Gross, F., Durner, J., & Gaupels, F. (2013). Nitric oxide, antioxidants and prooxidants in plant defence responses. *Frontiers in Plant Science, 4*, 419.

Hancock, J. T., Henson, D., Nyirenda, M., Desikan, R., Harrison, J., Lewis, M., et al. (2005). Proteomic identification of glyceraldehyde 3-phosphate dehydrogenase as an inhibitory target of hydrogen peroxide in *Arabidopsis*. *Plant Physiology and Biochemistry, 43*, 828–835.

Hara, M. R., Agrawal, N., Kim, S. F., Cascio, M. B., Fujimuro, M., Ozeki, Y., et al. (2005). S-nitrosylated GAPDH initiates apoptotic cell death by nuclear translocation following Siah1 binding. *Nature Cell Biology, 7*, 665–674.

He, J. M., Xu, H., She, X. P., Song, X. G., & Zhao, W. M. (2005). The role and the interrelationship of hydrogen peroxide and nitric oxide in the UV-B-induced stomatal closure in broad bean. *Functional Plant Biology, 32*, 237–247.

Houot, V., Etienne, P., Petitot, A. S., Barbier, S., Blein, J. P., & Suty, L. (2001). Hydrogen peroxide induces programmed cell death features in cultured tobacco BY-2 cells, in a dose-dependent manner. *Journal of Experimental Botany, 52*, 1721–1730.

Iakimova, E. T., Woltering, E. J., Kapchina-Toteva, V. M., Harren, F. J., & Cristescu, S. M. (2008). Cadmium toxicity in cultured tomato cells—role of ethylene, proteases and oxidative stress in cell death signaling. *Cell Biology International, 32*, 1521–1529.

Jiang, X., Gao, Y., Zhou, H., Chen, J., Wu, J., & Zhang, S. (2014). Apoplastic calmodulin promotes self-incompatibility pollen tube growth by enhancing calcium influx and reactive oxygen species concentration in *Pyrus pyrifolia*. *Plant Cell Reports, 33*, 255–263.

Kear, P. J., & McClure, B. (2012). How did flowering plants learn to avoid blind date mistakes? Self-incompatibility in plants and comparisons with nonself rejection in the immune response. *Advances in Experimental Medicine and Biology, 738*, 108–123.

Keller, T., Damude, H. G., Werner, D., Doerner, P., Dixon, R. A., & Lamb, C. (1998). A plant homolog of the neutrophil NADPH oxidase gp91phox subunit gene encodes a plasma membrane protein with Ca^{2+} binding motifs. *The Plant Cell, 10*, 255–266.

Kobayashi, M., Ohura, I., Kawakita, K., Yokota, N., Fujiwara, M., Shimamoto, K., et al. (2007). Calcium-dependent protein kinases regulate the production of reactive oxygen species by potato NADPH oxidase. *The Plant Cell, 19*, 1065–1080.

Kong, X., Zhang, D., Pan, J., Zhou, Y., & Li, D. (2013). Hydrogen peroxide is involved in nitric oxide-induced cell death in maize leaves. *Plant Biology (Stuttg), 15*, 53–59.

Konigshofer, H., Tromballa, H. W., & Loppert, H. G. (2008). Early events in signalling high-temperature stress in tobacco BY2 cells involve alterations in membrane fluidity and enhanced hydrogen peroxide production. *Plant, Cell and Environment, 31*, 1771–1780.

Kubo, K., Entani, T., Takara, A., Wang, N., Fields, A. M., Hua, Z., et al. (2010). Collaborative non-self recognition system in S-RNase-based self-incompatibility. *Science, 330*, 796–799.

Kulik, A., Noirot, E., Grandperret, V., Bourque, S., Fromentin, J., Salloignon, P., et al. (2015). Interplays between nitric oxide and reactive oxygen species in cryptogein signalling. *Plant, Cell and Environment, 38*, 331–348.

Lamb, C., & Dixon, R. A. (1997). The oxidative burst in plant disease resistance. *Annual Review in Plant Physiology and Plant Molecular Biology, 48*, 251–275.

Lanteri, M. L., Lamattina, L., & Laxalt, A. M. (2011). Mechanisms of xylanase-induced nitric oxide and phosphatidic acid production in tomato cells. *Planta, 234*, 845–855.

Lecourieux, D., Ranjeva, R., & Pugin, A. (2006). Calcium in plant defence-signalling pathways. *New Phytologist, 171*, 249–629.

Levine, A., Pennell, R. I., Alvarez, M. E., Palmer, R., & Lamb, C. (1996). Calcium-mediated apoptosis in a plant hypersensitive disease resistance response. *Current Biology, 6*, 427–437.

Levine, A., Tenhaken, R., Dixon, R., & Lamb, C. (1994). H_2O_2 from the oxidative burst orchestrates the plant hypersensitive disease resistance response. *Cell, 79*, 583–593.

Li, Q. Y., Niu, H. B., Yin, J., Wang, M. B., Shao, H. B., Deng, D. Z., et al. (2008). Protective role of exogenous nitric oxide against oxidative-stress induced by salt stress in barley (*Hordeum vulgare*). *Colloids and Surfaces B-Biointerfaces, 65*, 220–225.

Li, Z., Yue, H., & Xing, D. (2012). MAP Kinase 6-mediated activation of vacuolar processing enzyme modulates heat shock-induced programmed cell death in *Arabidopsis*. *New Phytologist, 195*, 85–96.

Lim, P. O., Woo, H. R., & Nam, H. G. (2003). Molecular genetics of leaf senescence in *Arabidopsis*. *Trends in Plant Science, 8*, 272–278.

Lin, A., Wang, Y., Tang, J., Xue, P., Li, C., Liu, L., et al. (2012). Nitric oxide and protein S-nitrosylation are integral to hydrogen peroxide-induced leaf cell death in rice. *Plant Physiology, 158*, 451–464.

Lindermayr, C., Saalbach, G., & Durner, J. (2005). Proteomic identification of *S*-nitrosylated proteins in *Arabidopsis*. *Plant Physiology, 137*, 921–930.

Liu, F., & Guo, F. Q. (2013). Nitric oxide deficiency accelerates chlorophyll breakdown and stability loss of thylakoid membranes during dark-induced leaf senescence in *Arabidopsis*. *PLoS One, 8*, e56345.

Locato, V., Gadaleta, C., De Gara, L., & de Pinto, M. C. (2008). Production of reactive species and modulation of antioxidant network in response to heat shock: a critical balance for cell fate. *Plant, Cell and Environment, 31*, 1606–1619.

Locato, V., de Pinto, M. C., & De Gara, L. (2009). Different involvement of the mitochondrial, plastidial and cytosolic ascorbate-glutathione redox enzymes in heat shock responses. *Physiologia Plantarum, 135*, 296–306.

Lum, H. K., Butt, Y. K., & Lo, S. C. (2002). Hydrogen peroxide induces a rapid production of nitric oxide in mung bean (*Phaseolus aureus*). *Nitric Oxide, 6*, 205–213.

Luu, D. T., Qin, X., Morse, D., & Cappadocia, M. (2000). S-RNase uptake by compatible pollen tubes in gametophytic self-incompatibility. *Nature, 407*, 649–651.

Ma, W., Smigel, A., Walker, R. K., Moeder, W., Yoshioka, K., & Berkowitz, G. A. (2010). Leaf senescence signaling: the Ca^{2+}-conducting *Arabidopsis* cyclic nucleotide gated channel2 acts through nitric oxide to repress senescence programming. *Plant Physiology, 154*, 733–743.

Ma, W., Xu, W., Xu, H., Chen, Y., He, Z., & Ma, M. (2010). Nitric oxide modulates cadmium influx during cadmium-induced programmed cell death in tobacco BY-2 cells. *Planta, 232*, 325–335.

Malik, S. I., Hussain, A., Yun, B. W., Spoel, S. H., & Loake, G. J. (2011). GSNOR-mediated de-nitrosylation in the plant defence response. *Plant Science, 181*, 540–544.

Marino, D., Dunand, C., Puppo, A., & Pauly, N. (2012). A burst of plant NADPH oxidases. *Trends in Plant Science, 17*, 9–15.

Marsoni, M., Cantara, C., de Pinto, M. C., Gadaleta, C., De Gara, L., Bracale, M., et al. (2010). Exploring the soluble proteome of Tobacco Bright Yellow-2 cells at the switch towards different cell fates in response to heat shocks. *Plant, Cell and Environment, 33*, 1161–1175.

Millar, A. H., Mittova, V., Kiddle, G., Heazlewood, J. L., Bartoli, C. G., Theodoulou, F. L., et al. (2003). Control of ascorbate synthesis by respiration and its implications for stress responses. *Plant Physiology, 133*, 443–447.

Miller, G., Schlauch, K., Tam, R., Cortes, D., Torres, M. A., Shulaev, V., et al. (2009). The plant NADPH oxidase RBOHD mediates rapid systemic signaling in response to diverse stimuli. *Science Signaling, 2*, ra45.

Mishina, T. E., Lamb, C., & Zeier, J. (2007). Expression of a nitric oxide degrading enzyme induces a senescence programme in *Arabidopsis*. *Plant, Cell and Environment, 30*, 39–52.

Mittler, R., Vanderauwera, S., Gollery, M., & Van Breusegem, F. (2004). Reactive oxygen gene network of plants. *Trends in Plant Science, 9*, 490–498.

Mittler, R., Vanderauwera, S., Suzuki, N., Miller, G., Tognetti, V. B., Vandepoele, K., et al. (2011). ROS signaling: the new wave? *Trends in Plant Science, 16*, 300–309.

Moller, I. M., Jensen, P. E., & Hansson, A. (2007). Oxidative modifications to cellular components in plants. *Annual Review of Plant Biology, 58*, 459–481.

Mur, L. A. J., Kenton, P., & Draper, J. (2005). In planta measurements of oxidative bursts elicited by avirulent and virulent bacterial pathogens suggests that H_2O_2 is insufficient to elicit cell death in tobacco. *Plant, Cell and Environment, 28*, 548–561.

Murgia, I., Tarantino, D., Vannini, C., Bracale, M., Carravieri, S., & Soave, C. (2004). *Arabidopsis thaliana* plants overexpressing thylakoidal ascorbate peroxidase show increased resistance to Paraquat-induced photooxidative stress and to nitric oxide-induced cell death. *The Plant Journal, 38*, 940–953.

Murgia, I., Vazzola, V., Tarantino, D., Cellier, F., Ravet, K., Briat, J. F., et al. (2007). Knock-out of ferritin AtFer1 causes earlier onset of age-dependent leaf senescence in *Arabidopsis*. *Plant Physiology and Biochemistry, 45*, 898–907.

Navabpour, S., Morris, K., Allen, R., Harrison, E., A-H-Mackerness, S., & Buchanan-Wollaston, V. (2003). Expression of senescence-enhanced genes in response to oxidative stress. *Journal of Experimental Botany, 54*, 2285–2292.

Niu, Y. H., & Guo, F. Q. (2012). Nitric oxide regulates dark-induced leaf senescence through EIN2 in *Arabidopsis*. *Journal of Integrative Plant Biology, 54*, 516–525.

Noguchi, M., Takata, T., Kimura, Y., Manno, A., Murakami, K., Koike, M., et al. (2005). ATPase activity of p97/valosin-containing protein is regulated by oxidative modification of the evolutionally conserved cysteine 522 residue in Walker A motif. *The Journal of Biological Chemistry, 280*, 41332–41341.

Pagnussat, G. C., Lanteri, M. L., Lombardo, M. C., & Lamattina, L. (2004). Nitric oxide mediates the indole acetic acid induction activation of a mitogen-activated protein kinase cascade involved in adventitious root development. *Plant Physiology, 135*, 279–286.

Pavet, V., Olmos, E., Kiddle, G., Mowla, S., Kumar, S., Antoniw, J., et al. (2005). Ascorbic acid deficiency activates cell death and disease resistance responses in *Arabidopsis*. *Plant Physiology, 139*, 1291–1303.

Petrov, V., Hille, J., Mueller-Roeber, B., & Gechev, T. S. (2015). ROS-mediated abiotic stress-induced programmed cell death in plants. *Frontiers in Plant Science, 6*, 69.

Prochazkova, D., & Wilhelmova, N. (2011). Nitric oxide, reactive nitrogen species and associated enzymes during plant senescence. *Nitric Oxide, 24*, 61–65.

Qiao, W. H., Li, C. N., & Fan, L. M. (2014). Cross-talk between nitric oxide and hydrogen peroxide in plant responses to abiotic stresses. *Environmental and Experimental Botany, 100*, 84–93.

Quirino, B. F., Noh, Y. S., Himelblau, E., & Amasino, R. M. (2000). Molecular aspects of leaf senescence. *Trends in Plant Science, 5*, 278–282.

Rasul, S., Dubreuil-Maurizi, C., Lamotte, O., Koen, E., Poinssot, B., Alcaraz, G., et al. (2012). Nitric oxide production mediates oligogalacturonide-triggered immunity and resistance to *Botrytis cinerea* in *Arabidopsis thaliana*. *Plant, Cell and Environment, 35*, 1483–1499.

Rhee, S. G., Chae, H. Z., & Kim, K. (2005). Peroxiredoxins: a historical overview and speculative preview of novel mechanisms and emerging concepts in cell signaling. *Free Radical Biology and Medicine, 38*, 1543–1552.

Rodriguez-Serrano, M., Pazmino, D. M., Sparkes, I., Rochetti, A., Hawes, C., Romero-Puertas, M. C., et al. (2014). 2,4-Dichlorophenoxyacetic acid promotes *S*-nitrosylation and oxidation of actin affecting cytoskeleton and peroxisomal dynamics. *Journal of Experimental Botany, 65*, 4783–4793.

Rogers, H. J. (2006). Programmed cell death in floral organs: how and why do flowers die? *Annals of Botany, 97*, 309–315.

Roldan, J. A., Rojas, H. J., & Goldraij, A. (2012). Disorganization of F-actin cytoskeleton precedes vacuolar disruption in pollen tubes during the in vivo self-incompatibility response in *Nicotiana alata*. *Annals of Botany, 110*, 787–795.

Romero-Puertas, M. C., Campostrini, N., Matte, A., Righetti, P. G., Perazzolli, M., Zolla, L., et al. (2008). Proteomic analysis of S-nitrosylated proteins in *Arabidopsis thaliana* undergoing hypersensitive response. *Proteomics, 8*, 1459–1469.

Romero-Puertas, M. C., Laxa, M., Matte, A., Zaninotto, F., Finkemeier, I., Jones, A. M., et al. (2007). S-nitrosylation of peroxiredoxin II E promotes peroxynitrite-mediated tyrosine nitration. *The Plant Cell, 19*, 4120–4130.

Sanità di Toppi, L., & Gabrielli, R. (1999). Response to cadmium in higher plants. *Environmental and Experimental Botany, 41*, 105–130.

Saxena, I., & Shekhawat, G. S. (2013). Nitric oxide (NO) in alleviation of heavy metal induced phytotoxicity and its role in protein nitration. *Nitric Oxide-Biology and Chemistry, 32*, 13–20.

Serrano, I., & Olmedilla, A. (2012). Histochemical location of key enzyme activities involved in receptivity and self-incompatibility in the olive tree (*Olea europaea* L.). *Plant Science, 197*, 40–49.

Serrano, I., Pelliccione, S., & Olmedilla, A. (2010). Programmed-cell-death hallmarks in incompatible pollen and papillar stigma cells of *Olea europaea* L. under free pollination. *Plant Cell Reports, 29*, 561–572.

Serrano, I., Romero-Puertas, M. C., Rodriguez-Serrano, M., Sandalio, L. M., & Olmedilla, A. (2012). Peroxynitrite mediates programmed cell death both in papillar cells and in self-incompatible pollen in the olive (*Olea europaea* L.). *Journal of Experimental Botany, 63*, 1479–1493.

Singh, H. P., Kaur, S., Batish, D. R., Sharma, V. P., Sharma, N., & Kohli, R. K. (2009). Nitric oxide alleviates arsenic toxicity by reducing oxidative damage in the roots of *Oryza sativa* (rice). *Nitric Oxide, 20*, 289–297.

Smykowski, A., Zimmermann, P., & Zentgraf, U. (2010). G-Box binding factor1 reduces CATALASE2 expression and regulates the onset of leaf senescence in *Arabidopsis*. *Plant Physiology, 153*, 1321–1331.

Sumbayev, V. V. (2003). S-nitrosylation of thioredoxin mediates activation of apoptosis signal-regulating kinase 1. *Archives of Biochemistry and Biophysics, 415*, 133–136.

Tarantino, D., Vannini, C., Bracale, M., Campa, M., Soave, C., & Murgia, I. (2005). Antisense reduction of thylakoidal ascorbate peroxidase in *Arabidopsis* enhances paraquat-induced photooxidative stress and nitric oxide-induced cell death. *Planta, 221*, 757–765.

Torres, M. A., Dangl, J. L., & Jones, J. D. (2002). *Arabidopsis* gp91phox homologues AtrbohD and AtrbohF are required for accumulation of reactive oxygen intermediates in the plant defense response. *Proceedings of the National Academy of Sciences of the United States of America, 99*, 517–522.

Trapet, P., Kulik, A., Lamotte, O., Jeandroz, S., Bourque, S., Nicolas-Frances, V., et al. (2015). NO signaling in plant immunity: a tale of messengers. *Phytochemistry, 112*, 72–79.

Trujillo, M., & Shirasu, K. (2010). Ubiquitination in plant immunity. *Current Opinion in Plant Biology, 13*, 402–408.
Vacca, R. A., de Pinto, M. C., Valenti, D., Passarella, S., Marra, E., & De Gara, L. (2004). Production of reactive oxygen species, alteration of cytosolic ascorbate peroxidase, and impairment of mitochondrial metabolism are early events in heat shock-induced programmed cell death in tobacco Bright-Yellow 2 cells. *Plant Physiology, 134*, 1100–1112.
Vacca, R. A., Valenti, D., Bobba, A., Merafina, R. S., Passarella, S., & Marra, E. (2006). Cytochrome c is released in a reactive oxygen species-dependent manner and is degraded via caspase-like proteases in tobacco Bright-Yellow 2 cells en route to heat shock-induced cell death. *Plant Physiology, 141*, 208–219.
Vacca, R. A., Valenti, D., Bobba, A., de Pinto, M. C., Merafina, R. S., De Gara, L., et al. (2007). Proteasome function is required for activation of programmed cell death in heat shocked tobacco Bright-Yellow 2 cells. *FEBS Letters, 581*, 917–922.
Valenti, D., Vacca, R. A., de Pinto, M. C., De Gara, L., Marra, E., & Passarella, S. (2007). In the early phase of programmed cell death in Tobacco Bright Yellow 2 cells the mitochondrial adenine nucleotide translocator, adenylate kinase and nucleoside diphosphate kinase are impaired in a reactive oxygen species-dependent manner. *Biochimica et Biophysica Acta-Bioenergetics, 1767*, 66–78.
Van Hautegem, T., Waters, A. J., Goodrich, J., & Nowack, M. K. (2015). Only in dying, life: programmed cell death during plant development. *Trends in Plant Science, 20*, 102–113.
Vandelle, E., & Delledonne, M. (2011). Peroxynitrite formation and function in plants. *Plant Science, 181*, 534–539.
Vandelle, E., Poinssot, B., Wendehenne, D., Bentejac, M., & Pugin, A. (2006). Integrated signaling network involving calcium, nitric oxide, and active oxygen species but not mitogen-activated protein kinases in BcPG1-elicited grapevine defenses. *Molecular Plant-Microbe Interactions, 19*, 429–440.
Wang, C. L., Wu, J., Xu, G. H., Gao, Y. B., Chen, G., Wu, J. Y., et al. (2010). S-RNase disrupts tip-localized reactive oxygen species and induces nuclear DNA degradation in incompatible pollen tubes of *Pyrus pyrifolia*. *Journal of Cell Science, 123*, 4301–4309.
Wang, P., Du, Y., Li, Y., Ren, D., & Song, C. P. (2010). Hydrogen peroxide-mediated activation of MAP kinase 6 modulates nitric oxide biosynthesis and signal transduction in *Arabidopsis*. *The Plant Cell, 22*, 2981–2998.
Wang, Y., Chen, C., Loake, G. J., & Chu, C. (2010). Nitric oxide: promoter or suppressor of programmed cell death? *Protein and Cell, 1*, 133–142.
Wang, Y., Lin, A., Loake, G. J., & Chu, C. (2013). H_2O_2-induced leaf cell death and the crosstalk of reactive nitric/oxygen species. *Journal of Integrative Plant Biology, 55*, 202–208.
Wang, Y., Ries, A., Wu, K., Yang, A., & Crawford, N. M. (2010). The *Arabidopsis* prohibitin gene PHB3 functions in nitric oxide-mediated responses and in hydrogen peroxide-induced nitric oxide accumulation. *The Plant Cell, 22*, 249–259.
Wheeler, M. J., de Graaf, B. H., Hadjiosif, N., Perry, R. M., Poulter, N. S., Osman, K., et al. (2009). Identification of the pollen self-incompatibility determinant in *Papaver rhoeas*. *Nature, 459*, 992–995.
Wilkins, K. A., Bancroft, J., Bosch, M., Ings, J., Smirnoff, N., & Franklin-Tong, V. E. (2011). Reactive oxygen species and nitric oxide mediate actin reorganization and programmed cell death in the self-incompatibility response of *papaver*. *Plant Physiology, 156*, 404–416.
Wilkins, K. A., Poulter, N. S., & Franklin-Tong, V. E. (2014). Taking one for the team: self-recognition and cell suicide in pollen. *Journal of Experimental Botany, 65*, 1331–1342.
Wu, A., Allu, A. D., Garapati, P., Siddiqui, H., Dortay, H., Zanor, M. I., et al. (2012). JUNGBRUNNEN1, a reactive oxygen species-responsive NAC transcription factor, regulates longevity in *Arabidopsis*. *The Plant Cell, 24*, 482–506.

Ye, Y., Li, Z., & Xing, D. (2013). Nitric oxide promotes MPK6-mediated caspase-3-like activation in cadmium-induced *Arabidopsis thaliana* programmed cell death. *Plant, Cell and Environment, 36*, 1–15.

Yu, M., Lamattina, L., Spoel, S. H., & Loake, G. J. (2014). Nitric oxide function in plant biology: a redox cue in deconvolution. *New Phytologist, 202*, 1142–1156.

Yun, B. W., Feechan, A., Yin, M., Saidi, N. B., Le Bihan, T., Yu, M., et al. (2011). *S*-nitrosylation of NADPH oxidase regulates cell death in plant immunity. *Nature, 478*, 264–268.

Zaffagnini, M., Morisse, S., Bedhomme, M., Marchand, C. H., Festa, M., Rouhier, N., et al. (2013). Mechanisms of nitrosylation and denitrosylation of cytoplasmic glyceraldehyde-3-phosphate dehydrogenase from *Arabidopsis thaliana. The Journal of Biological Chemistry, 288*, 22777–22789.

Zentgraf, U., & Hemleben, V. (2008). Molecular cell biology: are reactive oxygen species regulators of leaf senescence? *Progress in Botany, 69*, 117–138.

Zhang, A., Jiang, M., Zhang, J., Ding, H., Xu, S., Hu, X., et al. (2007). Nitric oxide induced by hydrogen peroxide mediates abscisic acid-induced activation of the mitogen-activated protein kinase cascade involved in antioxidant defense in maize leaves. *New Phytologist, 175*, 36–50.

Zhang, L., Li, Y., Xing, D., & Gao, C. (2009). Characterization of mitochondrial dynamics and subcellular localization of ROS reveal that HsfA2 alleviates oxidative damage caused by heat stress in *Arabidopsis. Journal of Experimental Botany, 60*, 2073–2091.

Zhao, X., Nishimura, Y., Fukumoto, Y., & Li, J. (2011). Effect of high temperature on active oxygen species, senescence and photosynthetic properties. *Environmental and Experimental Botany, 70*, 212–216.

Zuppini, A., Bugno, V., & Baldan, B. (2006). Monitoring programmed cell death triggered by mild heat shock in soybean-cultured cells. *Functional Plant Biology, 33*, 617–627.

CHAPTER TEN

Nitric Oxide: Jack-of-All-Trades of the Nitrogen-Fixing Symbiosis?

Imène Hichri[*,§,¶], Eliane Meilhoc[||,#], Alexandre Boscari[*,§,¶], Claude Bruand[||,#], Pierre Frendo[*,§,¶], Renaud Brouquisse[*,§,¶,1]

[*]Institut Sophia Agrobiotech (ISA), INRA, Sophia Antipolis, France
[§]Institut Sophia Agrobiotech (ISA), CNRS, Sophia Antipolis, France
[¶]Institut Sophia Agrobiotech (ISA), Université Nice Sophia Antipolis, Sophia Antipolis, France
[||]Laboratoire des Interactions Plantes-Microorganismes (LIPM), INRA, Castanet-Tolosan, France
[#]Laboratoire des Interactions Plantes-Microorganismes (LIPM), CNRS, Castanet-Tolosan, France
[1]Corresponding author: E-mail: renaud.brouquisse@sophia.inra.fr

Contents

Abstract

The symbiotic interaction between legumes and bacteria of Rhizobium type leads to the formation of new organs, called nodules, which provides a niche for bacterial nitrogen (N_2) fixation. In the nodules, bacteria differentiate into bacteroids able to fix atmospheric N_2 through nitrogenase activity. As nitrogenase is strongly inhibited by oxygen, N_2-fixation is made possible thanks to microaerophilic conditions prevailing in the nodules, characterized by a balance between low oxygen contents and adjusted nitric oxide (NO) contents. NO was shown to be produced by both the plant and bacterial partners during symbiosis, from early interaction steps between the plant and the

Advances in Botanical Research, Volume 77
ISSN 0065-2296
http://dx.doi.org/10.1016/bs.abr.2015.10.014

bacteria to N_2-fixing and senescence steps in mature nodules. NO is required for an optimal establishment of the symbiotic interaction. Transcriptomic analysis at early stage of the symbiosis showed that NO regulates about 2030 genes including genes involved in the induction of cell dedifferentiation and organogenesis, and in the repression of plant defense reactions, favouring the establishment of the plant–microbe interaction. In mature nodules, NO was found to act as both an inhibitor of N_2-fixation, a regulator of nitrogen metabolism, a beneficial metabolic intermediate for the maintenance of the energy metabolism under hypoxic conditions and a signal triggering nodule senescence. The present review provides an overview of NO sources and degradation pathways and of its multifaceted functions throughout the whole symbiotic process. It additionally analyses NO crosstalks with reactive oxygen species and hormones and presents particularly promising issues to decipher the roles of NO in N_2-fixing symbioses.

1. INTRODUCTION

Nitrogen (N) constitutes one of the primary macronutrients essential for plant growth and development, together with phosphate and potassium. It plays a key role in the production of amino acids, chlorophyll and nucleotides used in DNA/RNA synthesis and energy transfer compounds such as ATP. Plants absorb nitrogen from the soil mainly as ammonium (NH_4^+) and nitrate (NO_3^-) ions. However, the amount of mineral N present in the soil is generally a limiting factor for plant growth.

The *Fabaceae* plant group, or legumes, has the remarkable ability to establish a symbiotic interaction with soil bacteria, named rhizobia, leading to the formation of a new organ called the nodule. Although nodules are generally observed on plant roots, some legume species bear nodules on both roots and stems (Masson-Boivin, Giraud, Perret, & Batut, 2009). The primary function of nodules is to give favourable conditions for the reduction of atmospheric nitrogen (N_2) in ammonia (NH_3) by the bacteria through nitrogenase activity, and its inclusion in plant metabolism. Rhizobia cannot reduce N_2 without a plant host. Although the majority of the N_2-fixing plants belongs to the legume family, there are few exceptions with Parasponian interaction with rhizobia (Behm, Geurts, & Kiers, 2014) and actinorhizal plants in symbiotic association with *Frankia* bacteria (Svistoonoff, Hocher, & Gherbi, 2014).

The establishment of the symbiosis involves a molecular dialog between the two partners. In regards to the legume–rhizobium symbiosis, the plant secreted (iso)flavonoids are recognized by the bacteria, resulting in the induction of the synthesis pathway of bacterial nodulation factors (NF), which are lipochito-oligosaccharides. The NF recognition by the plant partner induces the root hair curling around bacteria and the cell divisions in the root inner cortex, resulting in the nodule primordium formation. Concomitantly, bacteria penetrate the root hair after degradation of the cell wall, migrate through the epidermis and the outer cortex inside a dedicated structure, the infection thread, and are finally released into the host cells by endocytosis. The resulting organelle, named symbiosome, is constituted by the bacteria (called bacteroid from this stage) surrounded by the peribacteroid membrane. The differentiation of the two symbiotic partners gives rise to a functional N_2-fixing nodule.

Nodules can be of determinate or indeterminate type. Indeterminate nodules (Medicago, clover, lupine) have a persistent meristem which generates the cylindrical shape of the nodules. These nodules can be divided into four main different zones: a meristematic zone (zone I) which allows the growth of the nodule via cell division, the infection and differentiation zone (zone II) in which the bacteria enter the plant cells and where the differentiation of both symbiotic partners begins, the N_2-fixing zone (zone III) where the symbiosomes reduce the N_2 in NH_3, and the senescence zone (zone IV) in which the cells of the two symbiotic partners die (Figure 1). In contrast, determinate nodules (soybean, pea, broad bean) lose their meristematic activity soon after initiation. The nodules are spherical and the cell expansion is mainly responsible for the nodule growth. In determinate nodule, the various stages of development are present successively during the lifetime of the nodule (Sprent, 2007).

Nitric oxide (NO) is a gaseous reactive and signaling molecule involved in multiple physiological processes such as plant growth, tissue and organ development, stress response, and plant–microorganism interactions (see other chapters of the present issue), including legume–rhizobium symbiosis (Hichri et al., 2015; Meilhoc, Boscari, Bruand, Puppo, & Brouquisse, 2011; Puppo, Pauly, Boscari, Mandon, & Brouquisse, 2013). Evidence for NO presence in N_2-fixing nodules came from the detection of NO complexed to leghaemoglobin (Lb), the major haemoprotein of legume nodules, in soybean and cowpea (Maskall, Gibson, & Dart, 1977). Such a complex was subsequently reported in several legumes forming determinate and indeterminate nodules (Kanayama & Yamamoto, 1991; Mathieu, Moreau, Frendo, Puppo, & Davies, 1998; Sanchez et al., 2010). Using either NO-specific fluorescent dyes, or NO biosensor rhizobium strains, NO

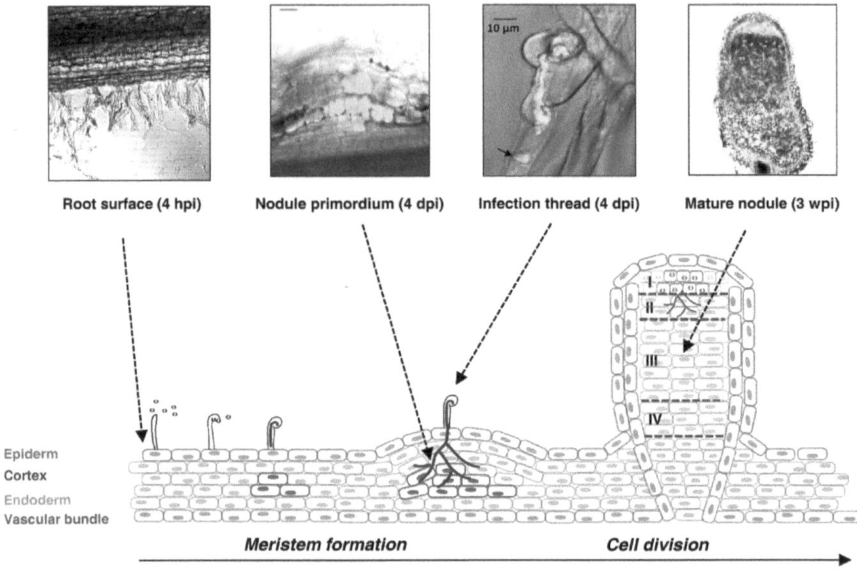

Figure 1 NO production during the legume–*Rhizobium* symbiotic interaction. NO production sites are detailed for each step of the infection process: on root surface (4 hpi), nodule primordium (4 dpi), infection thread (4 dpi) and mature nodules where NO is present in the infection zone, the N_2-fixing zone, and the inter-zone III–IV (3 wpi). hpi, hours post-inoculation; dpi, days post-inoculation; wpi, weeks post-inoculation. *The picture of infection thread (4 dpi) is reprinted from Figure 1 in del Giudice et al. (2011) with kind permission from Wiley-Blackwell.* (See colour plate)

was detected in *Lotus japonicus* and *Medicago truncatula* roots as early as 4 h after inoculation with their bacterial partners (Nagata et al., 2008), in inoculation thread and nodule primordia of 4-days inoculated *M. truncatula* roots (del Giudice et al., 2011), and in the N_2-fixing zone of mature nodules (Baudouin, Pieuchot, Engler, Pauly, & Puppo, 2006). The presence of NO was nevertheless not systematically observed in nodules of all legume species as free NO was not detected in *Arachis hypogaea* (peanut or groundnut) nodules. Instead, the detection of nitrosothiols and S-nitrosylated proteins clearly indicates that NO is present in these nodules (Maiti, Sarkar, & Ghosh, 2012). Consequently, NO is produced from early interaction steps between the two partners to the onset of senescence (Figure 1), raising the question of its physiological and signaling functions during the symbiotic process. The present review provides an overview of the multifaceted roles of NO during N_2-fixing symbiosis and proposes several lines of investigation for analyzing its role in the field of plant–microbe interactions.

2. NO IN PLANT AND BACTERIA

2.1 NO Sources

The origin of NO at the different steps of the symbiotic interaction is still a matter of debate. It can be suspected that only the plant is responsible for NO synthesis in roots right after inoculation with the symbiont or during the formation of nodule primordia. Several enzymatic pathways for NO production have been identified in plants (for a review see Mur et al. (2013)) but the sources of NO at early steps of the interaction remain obscure. The reductive pathway involving nitrate reductase (NR) is probably the best characterized. During early interaction, root treatment with an NR inhibitor indeed mimics the addition of NO scavenger on the transcriptional regulation of genes involved in the nodulation process (Boscari, del Giudice, et al., 2013). In addition, the mammalian NO synthase (NOS) inhibitor, L-NAME, was ineffective suggesting the probable involvement of NR rather than an NOS-like enzyme in the synthesis of NO at this stage (Boscari, del Giudice, et al., 2013). Three putative NR-encoding genes have been identified in the genome of *M. truncatula* and preliminary data suggest that essentially NR1 is involved in NO production (Boscari, unpublished results). At later stages of the symbiotic interaction (mature nodules), both plant and bacteria are involved in NO production (Figure 2). Indeed Horchani et al. (2011) showed that in nodules of *M. truncatula*, up to 35%

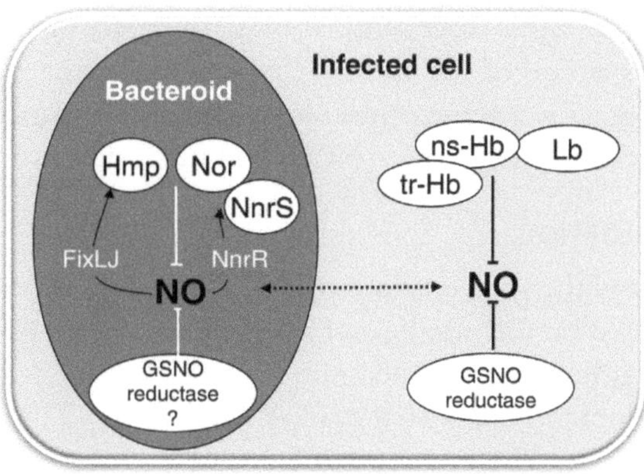

Figure 2 Bacterial and plant systems to modulate NO level in legume root nodules. Hmp expression is regulated by the two-component system FixLJ while the NO specific regulator NnrS controls Nor and NnrS expression. tr-Hb, truncated haemoglobin; Lb, leghaemoglobin; ns-Hb, non symbiotic haemoglobin.

of the detected NO was produced by the symbiont. Comparable results revealing NO production by the bacterial partner have been reported in *Bradyrhizobium japonicum*/*Glycine max* (soybean) nodules (Sanchez et al., 2010). In these later stages NR could also be a major source of NO allowing the reduction of NO_3^- to NO_2^- (nitrite) subsequently reduced to NO via the mitochondrial electron transfer chain (ETC) (Horchani et al., 2011). Although NR activity seems a predominant source of NO, it is not known whether the three NRs encoded by the genome of *M. truncatula* are effective at this stage. Other sources of NO in plants such as polyamine oxidases (PAOx) have been considered since the addition of spermine triggered an increase in NO content whereas that of gazatine (a polyamine oxydase inhibitor) strongly reduced it (Hichri et al., 2015). In rhizobia either under free-living conditions or as bacteroids inside nodules, the main source of NO production is the denitrification pathway which includes the *nap* (nitrate reductase), *nir* (nitrite reductase), *nor* (NO reductase) and *nos* (nitrous reductase) genes (Horchani et al., 2011; Sanchez et al., 2010). A *Sinorhizobium meliloti* strain mutated for *nirK* (nitrite reductase) lost more than 95% of its NO production capacity when grown in microaerobic conditions in the presence of NO_3^-, indicating that nitrite reductase is indeed the major bacterial source of NO in these conditions (Meilhoc, unpublished data). Nevertheless, a NOS-like activity was also detected in *S. meliloti* grown in aerobic conditions although no corresponding gene has been identified in its genome (Pii, Crimi, Cremonese, Spena, & Pandolfini, 2007).

Hence the sources of NO are various but should be carefully analyzed at each step of the symbiotic interaction as they might greatly vary depending upon the stage considered. While NO produced by the plant might be considered as a signal, a defense molecule or a metabolic compound, the interest for the bacteroid to produce NO is still puzzling.

2.2 NO Degradation

The roles of NO in the symbiotic interaction depend upon its concentration and localization within the nodule and, thus, NO level is a finely tuned balance between synthesis and degradation. NO degradation was first ascribed to plant haemoglobins (Hbs) which oxidize NO in NO_3^- in the presence of O_2 (Figures 2 and 3). Three major families of Hbs have been described in plants: non-symbiotic haemoglobins (nsHbs, essentially Class 1 Hb), leghaemoglobins (Lbs, Class 2 Hb) and truncated haemoglobins (trHbs, Class 3 Hb) (for review (Gupta, Hebelstrup, Mur, & Igamberdiev, 2011; Hill, 2012)). Lbs are abundant (up to mM range) in legumes and although having

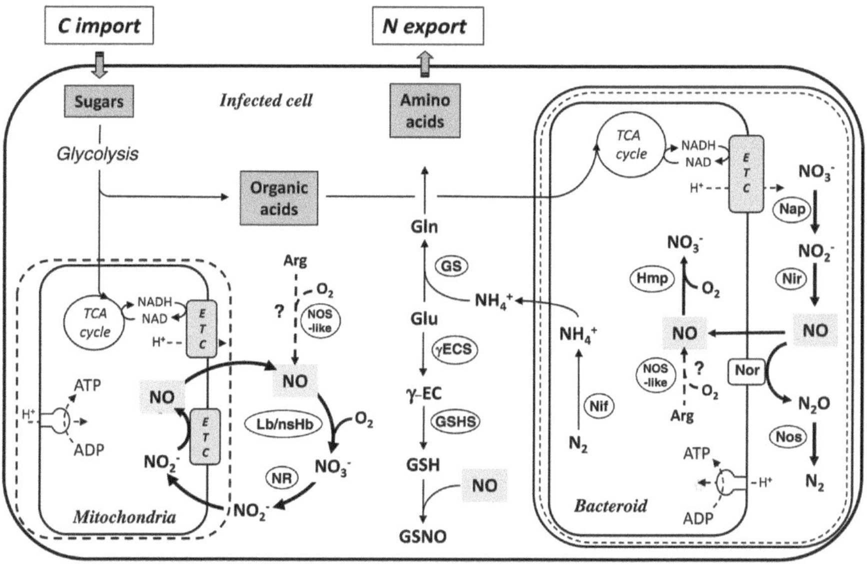

Figure 3 Schematic representation of NO involvement in multiple biological processes within N_2-fixing nodules. NO regulates N_2-fixation, mitochondrial respiration and both carbon and nitrogen metabolism. NO is synthesized both by the bacterial and plant partners. In bacteria, NO arises out of the denitrification pathway, where NO_3^- is respectively reduced into NO_2^-, NO, N_2O and N_2, by nitrate reductase (Nap), nitrite reductase (Nir), NO reductase (Nor) and nitrous oxide reductase (Nos). In the cytosol, the flavohaemoglobin Hmp oxidizes NO into NO_3^-. In the plant partner, NO mainly comes from and regulates the Hb-NO respiration. In the cytosol, NR (nitrate reductase) reduces NO_3^- to NO_2^-, which is transported to the mitochondria. NO_2^- is subsequently reduced by the mitochondrial ETC (electron transfert chain) into NO, which will passively diffuse into the cytosol where Lb (leghaemoglobin) or nsHb (non-symbiotic haemoglobin) will oxidize it back into NO_3^-. This cycle allows mitochondrial respiration to occur and energy (ATP) regeneration. A putative involvement of a plant and a bacterial NOS (nitric oxide synthase)-like enzyme in NO production remains speculative. Arg, Arginine; γ-EC, γ-glutamyl cysteine; γ-ECS, γ-glutamyl cysteine synthetase; GSHS, glutathione synthetase; Gln, glutamine; Glu, glutamic acid; GSNO, S-nitrosoglutathione; Nif, nitrogenase; TCA, tricarboxylic acid.

a key role in facilitating O_2 transfer to N_2-fixing bacteria in root nodules, they now appear to also have a function in NO scavenging (Gupta et al., 2011), together with nsHbs and trHbs. Indeed, purified recombinant class1 nsHb from alfalfa displayed an NO degradation activity (Seregelyes et al., 2004) and the overexpression of class1 nsHb led to a reduction of NO level in nodules of *L. japonicus* (Shimoda et al., 2009). The different plant Hbs might display specificity in their expression pattern and their role at each step of the interaction. As an illustration, it is remarkable that *L. japonicus* class 1 nsHb displayed an expression burst only 4 hpi with its symbiont, which correlated with a rise in NO production (Nagata et al., 2008). In the same way, it is worth noting that *M. truncatula* possesses as many as 17 putative haemoglobin encoding genes in its genome (Young et al., 2011). RNA sequencing of laser captured microdissection of the different zones of *M. truncatula* nodules indicate that these genes have slightly different expression patterns, raising the issue of their likely role (Roux et al., 2014).

Plant Hbs could be expected to be sufficient for controlling NO level in nodules. However, four proteins involved in the control of NO level have also been identified in *S. meliloti* (Figure 2). Indeed, the flavohaemoglobin (Hmp), the respiratory NO reductase (Nor) and two proteins of the NnrS family (NnrS1 and NnrS2) have been shown to be involved (directly or indirectly) in NO degradation and to be essential in maintaining efficient nitrogen-fixing symbiosis (Blanquet et al., 2015; Cam et al., 2012; Meilhoc, Blanquet, Cam, & Bruand, 2013). In many bacteria, Hmp oxidizes NO in NO_3^- while Nor catalyzes the reduction of NO into N_2O (nitrous oxide) in denitrifying bacteria. In *B. japonicum*, a globin (Bjgb) is also involved in NO detoxification under free-living conditions (Cabrera et al., 2011) whereas Nor is involved in NO degradation in *B. japonicum* and in *Rhizobium etli* (Gomez-Hernandez et al., 2011; Sanchez et al., 2011). Therefore, not only plant but also bacterial proteins control NO level inside nodules and we anticipate that other bacterial proteins such as thioredoxins or GSNO (S-nitroso-glutathione) reductase might indirectly participate to NO detoxification/degradation/signalization *in planta* (Lee et al., 2010). Hence, two questions remain: why do bacteria need as many different systems to reduce NO levels? Is there a specific role of bacterial NO degradation proteins which might explain why plant Hbs are not sufficient to control NO level inside nodules? These questions have been partly answered by a recent work on *M. truncatula* nodules showing that bacterial NO degrading proteins are not only dedicated to protect bacterial proteins against NO deleterious effects, but also protect plant proteins from being inactivated by NO-mediated post translational modifications (Blanquet et al., 2015).

2.3 The Mechanisms of NO Action

Chemically, NO is classically considered in its radical form (NO•) with a single electron in its $2p$-π orbital. It can either donate, or accept, one electron to form nitrosonium cation (NO^+) and nitroxyl anion (NO^-), respectively. These reactive nitrogen species (RNS) have different chemical properties which explain their reactivity with a large panel of molecules ranging from dioxygen to protein, lipids and nucleic acids (Stamler, Singel, & Loscalzo, 1992). The best characterized NO-dependent modification is the S-nitrosylation (or transnitrosylation) of Cys residues in proteins. This involves the addition of an NO group to specific Cys thiols which results in the modification of protein activity, stability or cellular compartmentation (Hess, Matsumoto, Kim, Marshall, & Stamler, 2005; Hess, Matsumoto, Nudelman, & Stamler, 2001). In plant and bacteria, several hundred proteins involved in a wide variety of cellular functions have been found to be regulated by S-nitrosylation such as glyceraldehyde dehydrogenase (GAPDH), NADPH oxidase, peroxiredoxin II E or the transcription factor (TF) TGA1 (Astier et al., 2012; Waszczak et al., 2015). NO is also known to nitrosylate transition metals, such as the haem moiety of Hbs (Hill, 2012). A number of enzymes, including guanylate cyclase, catalase, aconitase and cytochrome oxidase, are regulated by metal nitrosylation which generally results in the loss of their activity (Besson-Bard, Pugin, & Wendehenne, 2008). NO may also react with superoxide ($O_2^{\bullet -}$) to form peroxynitrite ($ONOO^-$). $ONOO^-$ has the ability to modify a variety of amino acids, such as oxidation of sulfur-containing amino acid (Cys, Met) and nitration of aromatic amino acids (Tyr, Trp, Phe and His), often resulting in modulation of protein function (Alvarez & Radi, 2003). Tyrosine nitration is considered as a selective mechanism and is emerging as an important feature of stress response, signal transduction, cytoskeletal organization, or developmental processes in plants (Vandelle & Delledonne, 2011). Interestingly, a number of proteins such as ascorbate peroxidase (Correa-Aragunde, Foresi, & Lamattina, 2015) or glutamine synthetase (Melo, Silva, Ribeiro, Seabra, & Carvalho, 2011) are regulated both by S-nitrosylation and tyrosine nitration, indicating that NO-dependent regulation occurs in interaction with other ROS/RNS through a complex redox-dependent regulatory network.

Biological effects of NO are also related to changes in gene expression through NO-dependent processes. Over the past 10 years, hundreds of bacterial genes (Hyduke, Jarboe, Tran, Chou, & Liao, 2007; Meilhoc, Cam, Skapski, & Bruand, 2010) and thousands of plant genes (Besson-Bard et al., 2009; Boscari, del Giudice, et al., 2013; Palmieri et al., 2008) displaying differential NO-dependent expression have been identified. In bacteria, most of the NO-regulated genes are regulated through either the two-component regulator FixLJ, or the regulator NnrR (Meilhoc et al., 2011). In plants, the promoter analysis of 28 NO-regulated genes led to the identification of eight families of transcription factor binding sites (TFBS) which turned to be binding sites for stress-related TFs (Palmieri et al., 2008). Recently, Gibbs and coworkers showed that the transcriptional response to hypoxic stress in plants occurs through NO-dependent processes controlling the stability of specific ethylene responsive factors-VII (ERF-VII) TFs (Gibbs et al., 2014). NO-mediated turnover of ERF-VII coordinates diverse NO responses throughout plant development, including seed germination, hypocotyl growth and stomatal movements (Gibbs et al., 2014). Whether ERF-VII bind to TFBS of NO-regulated genes is still unknown and is an open field of investigation for deciphering NO signaling in plants.

3. NO ROLES IN NITROGEN-FIXING SYMBIOSIS

3.1 Symbiosis Establishment

The establishment of N_2-fixing symbiosis involves numerous steps including the cross recognition between the two partners, the entry of bacteria into the plant and the processes required for the development of nodule primordia. Numbers of recent works described the production of NO at these different stages and shed some light on the requirement of this signaling molecule during the initiation of the interaction.

Inoculation of legumes with their symbionts triggers a production of NO within few hours post-inoculation (hpi). Upon *L. japonicus* inoculation with its symbiont *Mesorhizobium loti*, a transient production of NO is observed at the root surface 4 hpi (Figure 1), which then decreases to its basal level after 10 and 24 hpi (Nagata et al., 2008). Besides, inoculation of *L. japonicus* with non-symbiont rhizobium does not induce such an NO accumulation, indicating that NO production results from the specific recognition of the bacterial partner by the plant partner (Nagata et al., 2008). Such NO production was initially described as a defense-inducible signal in plant interaction with bacterial pathogens (Delledonne, Xia, Dixon, & Lamb, 1998; Durner, Wendehenne, & Klessig, 1998). However, when *L. japonicus* is infected with the plant pathogens *Ralstonia solanacearum* and *Pseudomonas syringae*, a continuous NO production was observed for at least 24 h (Nagata et al., 2008). The decrease in NO level observed in *L. japonicus*, after its initial accumulation during the 4 h following infection with *M. loti*, was assigned to the expression of the *LjHB1* gene encoding nsHb1 (Nagata et al., 2008, 2009; Murakami et al., 2011). Treatments of *L. japonicus* roots with NO-donor (S-nitroso-N-acetyl-D,L-penicillamine – SNAP) and NO-scavenger (c-PTIO) resulted in the induction and repression of *LjHB1*, respectively (Nagata et al., 2008; Shimoda et al., 2005). These observations suggest that at early step of the symbiotic interaction, the burst of NO specifically induces the expression of Hb1 which, in return, down-regulates the level of NO to lower plant defense response and allows the reception of the symbiont in the roots (Murakami et al., 2011). Uchiumi's team demonstrated that in *L. japonicus/M. loti* symbiosis, NO production was induced by the lipopolysaccharide covering the cell surface of *M. loti* (Murakami et al., 2011; Nagata et al., 2009).

Using both the cell-permeable NO-specific fluorescent probe diaminofluorescein 2-diacetate (DAF-2DA) and an NO biosensor bacterial strain, NO production was detected during the first steps of the interaction between *M. truncatula* and *S. meliloti* (del Giudice et al., 2011). The detection of NO was observed along the infection threads but only on the plant cell side which suggests that NO production at this step is mainly due to the plant partner, and/or that bacterial system(s) prevent(s) NO accumulation inside bacteria (Meilhoc et al., 2010). NO was also detected in dividing cortical cells of the nodule primordia (del Giudice et al., 2011).

NO can also act as a signal molecule in controlling, either directly and/or indirectly, nodule number in Medicago plants (Pii et al., 2007). Both the scavenging of NO by c-PTIO, and the overexpression of the bacterial flavohaemoglobin *hmp* by the plant partner, led to a reduced efficiency of nodulation in the *M. truncatula/S. meliloti* interaction (del Giudice et al., 2011). In the same way, treatments of determinate nodules such as those of soybean with the NOS inhibitor Nω-nitro-arginine (L-NNA) during early interaction with *B. japonicum* also resulted in a 70% reduction of nodule number, and the addition of the NO-donor DETA-NO reverted the reduction (Leach, Keyster, Du Plessis, & Ludidi, 2010). In contrast, Uchiumi's team reported an increase in the number of nodules formed on *L. japonicus* transgenic hairy roots overexpressing either *LjHb1* or *Alnus firma AfHb1*, compared to those formed on control roots (Shimoda et al., 2009). Discrepancies between these studies could be due to the choice of enzymes with NO-detoxifications activities, Hmp or Hb, and the level of NO resulting from these activities. Taken together these results underline the importance of NO in symbiotic interaction and the need for the tight setting-up of NO concentrations for successful establishment of the symbiotic relationship.

Increasing evidence implied that NO could modulate gene expression and transcriptomic studies have been performed to identify the NO-responsive genes during the symbiotic interaction. A first study conducted on *M. truncatula* roots treated with two NO donors, sodium nitroprusside (SNP) and GSNO, enabled the identification of 999 putative NO-responsive genes in which 290 were also regulated during nodule development (Ferrarini et al., 2008). More recently, a transcriptomic analysis using RNA-Seq technology was performed with *M. Truncatula* inoculated roots (4 days post-inoculation) treated or not with c-PTIO (Boscari, del Giudice, et al., 2013). Differential expression analysis revealed 2030 genes affected by the NO-depletion, among which many genes encoding TFs and proteins involved in nodule development, such as cyclin-like proteins, peptidases or ribosomal protein families. These results suggest that NO detected in nodule primordia not yet invaded by the rhizobial cells could be required for the dedifferentiation of cortical cells and the induction of cell division during nodule formation. The reduced expression of genes involved in nodule development (*MtCRE1*, *MtCCS52A*) in NO-depleted roots reinforces the hypothesis that NO could regulate the nodulation process (del Giudice et al., 2011). It has been proposed that cytokinin is a key differentiation signal for nodule organogenesis (Frugier, Kosuta, Murray, Crespi, & Szczyglowski, 2008). Thus, NO could control the nodulation process through cytokinin perception mechanisms. Another striking finding is the number of genes affected by c-PTIO, and probably regulated by NO, involved in terpene, flavonoid and phenylpropanoid pathways and in defense response (PR-proteins, cytochrome P450). These observations show that NO could be involved in the regulation of plant defense reactions to enable the establishment of the beneficial plant–microbe interactions as it was proposed to occur in mycorrhizal symbiosis (Espinosa, Garrido, Ortega, Casimiro, & Alvarez-Tinaut, 2014).

3.2 Mature Nodules, Nitrogen-Fixation

Toxic effects versus useful signaling/metabolic roles of NO in N_2-fixing nodules have been a matter of debate over the last years. NO was first reported to be a potent inhibitor of *B. japonicum* nitrogenase activity in vitro (Trinchant & Rigaud, 1982). Likewise, upon NO-donor treatments, NO inhibited in vivo nitrogenase activity in soybean, *A. firma*, *L. japonicus* or *M. truncatula* nodules (Cam et al., 2012; Kato, Kanahama, & Kanayama, 2010; Shimoda et al., 2009; Trinchant & Rigaud, 1982). In addition, *L. japonicus* nodules over-expressing nsHb1 from either *L. japonicus* or *A. firma*, exhibited lower NO levels and higher N_2 fixation efficiency than control ones (Shimoda et al., 2009), whereas *M. truncatula* nodules inoculated with an *S. meliloti hmp* deficient mutant exhibited a higher NO content and a reduced N_2 fixation as compared to the wild-type (WT) strain (Cam et al., 2012). Thus, it clearly appears that an excess of NO inhibits the symbiotic fixation of N_2. It is so far unclear whether in vivo N_2-fixation was reduced through either a direct nitrogenase inhibition by NO, or an indirect regulation of nitrogen metabolism by NO, or both. Nitrogenase displays at least three putative S-nitrosylation sites (Xue et al., 2010) and different nitrogenase subunits have been identified among the S-nitrosylated proteins found in *M. truncatula* mature nodules (Puppo et al., 2013), suggesting that NO may inhibit nitrogenase activity through S-nitrosylation. In addition, it was demonstrated in soybean that the NO formed in response to flooding conditions has a negative effect on the expression of *B. japonicum nifH* and *nifD* genes, which could be partially restored by a c-PTIO treatment (Sanchez et al., 2010). This means that nitrogenase may be targeted by NO both at transcriptional and post-translational levels.

During the N_2-fixing process, NH_4^+ generated by bacteroid nitrogenase is assimilated to glutamate (Glu) to form glutamine (Gln) through the activity of the plant glutamine synthetase (GS) (Figure 3). In *M. truncatula*, MtGS1a is responsible for 90% of the total nodule GS activity (Carvalho et al., 2000). MtGS1a from *M. truncatula* nodules was shown to be inactivated by NO through tyrosine nitration (Melo et al., 2011), resulting in a reduction of N_2-fixation. In addition, it has been reported that GS nitration is enhanced in *hmp* mutant nodules (Blanquet et al., 2015), indicating that NO originating from both the plant and the bacterial partner may contribute to GS1a inhibition. Glu is also a precursor for the biosynthesis of glutathione (GSH) which is both a major antioxidant compound and the synthesis precursor of GSNO (Figure 3). Consequently, NO-mediated inactivation of the GS activity could promote a redirection of the Glu pool for GSH and GSNO synthesis (Melo et al., 2011). The upregulation by NO of *M. truncatula* γ-glutamylcysteine synthetase (γ-ECS) and glutathione synthetase (GSHS) expression, both genes involved in GSH synthesis, supports this hypothesis (Innocenti et al., 2007). Aside nitrogenase and GS, Puppo et al. (2013) reported that about 80 S-nitrosylated proteins were identified in *M. truncatula* mature nodules. Most of these proteins are related to nitrogen, carbon and energy metabolism both in plant and bacterial partners. The activity of most of these enzymes, related to the tricarboxylic acid cycle, glycolysis and N_2-assimilation, is inhibited by NO donors (Boscari, del Giudice, et al., 2013; Igamberdiev, Ratcliffe, & Gupta, 2014), indicating that NO acts as a global down regulator of primary metabolism.

Functional nodules are characterized by a microoxic environment, raising the question of energy supply within this organ, and a metabolic role for NO has been proposed in relation to ATP regeneration in the nodule (Horchani et al., 2011). NO production is a hallmark of plant response to hypoxia, and this production is supposed to be linked to a cyclic respiration process, known as 'Hb-NO cycle', with improved capacity for the plant to maintain cell energy status under low oxygen (Gupta & Igamberdiev, 2011). In this Hb-NO respiration (Figure 3), NO_3^- is first reduced to NO_2^- in the cytosol by NR. Subsequently, NO_2^- is translocated into the mitochondrial matrix where it functions as a terminal acceptor for ETC, allowing the regeneration of ATP. The NO resulting from the reduction of NO_2^- diffuses freely from the matrix to the cytosol, where it is oxidized back to NO_3^- by nsHb (Figure 3). This latter can operate at nanomolar oxygen level, i.e. two orders of magnitude lower than the Km of cytochrome oxidase for oxygen (Igamberdiev et al., 2014). Thus, under hypoxic environment, plant mitochondria preserve their capacity to oxidize external NADH and retain limited power for ATP synthesis, complementing glycolytic ATP production. Similar function is provided by the denitrification pathway in the bacteroid, where NO_3^- is reduced to N_2 via NO formation. Different lines of evidence support the existence of such Hb-NO respiration in legume nodules. First, in mature *G. max* and *M. truncatula* nodules, the plant NR and ETC and the bacterial denitrification pathway are involved in NO production, particularly under hypoxic conditions (Horchani et al., 2011; Meakin et al., 2007; Sanchez et al., 2010). Second, the energy status of the nodules depends either partly, or almost entirely, on NR functioning under normoxic, or hypoxic conditions, respectively (Horchani et al., 2011). Third, Lbs have the capacity to efficiently react with NO to produce NO_3^- (Herold & Puppo, 2005) and the NO generated at the ETC level may therefore be oxidized into NO_3^- by Lbs and/or ns-Hbs. The capacity of plant Hbs and bacterial Hmp to oxidize NO into NO_3^- and their high affinity for NO favour the recycling of NO through both the Hb-NO respiration and the denitrification pathway, both keeping NO concentration below toxic level and maintaining a minimal energy status under hypoxia. Interestingly, in *L. japonicus* nodules, N_2 fixation was stimulated by low concentration (0.1 mM) of the NO-donor SNP, but inhibited in presence of 1 and 10 mM of SNP, illustrating that low but significant NO contents are beneficial to N_2 fixation (Kato et al., 2010). Overall, these observations suggest that NO could play a dual function in the nodule, as a useful intermediate necessary for maintaining basal energy metabolism on the one hand, and as a down-regulator of N and carbon metabolism to reduce energy demand under microoxic environment on the other hand.

3.3 Senescence, Symbiotic Breaking Off

Few weeks-aged nodules enter a senescence process characterized by the death of both plant tissues and bacteroids, and therefore a decline of N_2 fixation. A similar process of nodule senescence can also be induced prematurely by adverse environmental conditions such as a NO_3^- treatment or a dark stress (Matamoros et al., 1999). Interestingly, an increase of NO levels in *M. truncatula* nodules, resulting either from the use of *S. meliloti* strains deficient in NO degradation (*hmp, norB, nnrS1 or nnrS2*) or from the exogenous addition of an NO donor, led to a premature senescence of the nodules, the severity of which correlated with the NO levels inside the nodules (Blanquet et al., 2015). Conversely, nodules induced by *S. meliloti* strains overexpressing either *hmp* or *nnrS1* contained a low level of NO and displayed a delayed senescence phenotype. Dark-induced senescence was also inhibited in nodules induced by a *hmp*-overexpressing strain, which suggests that NO is also involved in dark-induced nodule senescence (Cam et al., 2012). Altogether, these results suggest that NO is a key player of nodule senescence. Whether nodule senescence induced by NO_3^- is also dependent on NO is currently unknown, but it is known that NO_3^- addition leads to an increase of NO production in nodules through NO_3^- reduction, which could be the cause of the premature senescence (Horchani et al., 2011).

Two evidences suggest that NO might be more abundant in senescing nodule tissues. First, NO detection in *M. truncatula* nodules using an NO biosensor *S. meliloti* strain suggested that NO is present at its highest level in cells located in the center of the basal part of the nodule (Cam et al., 2012), where senescence was described to start (Perez Guerra et al., 2010). Second, Lb derivatives particularly abundant in aging soybean nodules were reported to display nitrated haemes, suggesting that RNS could be more abundant during nodule senescence (Navascues et al., 2012). However, nothing is known about the origin of NO at the onset of senescence, in particular whether it is synthesized by plant cells, bacteroids, or both.

Although the mode of action of NO during nodule senescence is presently unknown, several hypotheses can be formulated. First, as described above (III-2), NO is known to be an inhibitor of global N_2 fixation (Boscari, Meilhoc, et al., 2013). NO is also a potent inhibitor of respiration as it has been shown to inhibit both mitochondrial and bacterial terminal respiratory oxidases, and generate an increase in reactive oxygen species (ROS) and RNS (Arjona, Wikstrom, & Adelroth, 2015; Igamberdiev et al., 2014; Sarti et al., 2012). Senescence could be therefore an indirect consequence of N_2 fixation or respiration inhibition. Alternatively, NO might act as a signal to trigger a specific senescence pathway.

At the molecular level, NO causes post-translational modifications on proteins, including the S-nitrosylation of cysteine and the nitration of tyrosine, which can affect the activity of proteins (for a review Astier and Lindermayr (2012)). As mentioned above (III-2), several symbiotic targets are either known or predicted, whose NO-mediated modification might directly or indirectly trigger nodule senescence: (1) nitrogenase, which multiple subunits possess several potential S-nitrosylation sites (Xue et al., 2010), (2) various enzymes of the TCA cycle, essential to carbon metabolism in symbiotic nodules, which are known to be the target of S-nitrosylation in *Mycobacterium tuberculosis* (Rhee, Erdjument-Bromage, Tempst, & Nathan, 2005) and were found to be S-nitrosylated in *S. meliloti in planta* (Puppo et al., 2013); (3) Lb, which plays an essential role as O_2 transporter, and has been shown in several plant systems to be the target of post-translational modifications mediated by RNS including metal nitrosylation (Mathieu et al., 1998), haeme nitration (Navascues et al., 2012) and tyrosine nitration (Sainz et al., 2015), (4) cysteine proteases, which are well-known markers of legume nodule senescence (Perez Guerra et al., 2010; Pierre et al., 2014) and may be putative targets for NO-mediated post-translational modifications. Exploration of S-nitrosylated and tyrosine nitrated plant and bacterial proteins from nodules at the onset of senescence will provide critical information about molecular targets of NO, helping us to decipher the role of NO in nodule senescence.

3.4 Redox State and Crosstalk with ROS—RNS—GSH

ROS and RNS are key players of the redox regulation process. H_2O_2 and NO are both able to react with protein targets and modulate their activity. Common S-sulfenylated and S-nitrosylated protein targets have been identified in root nodules (Puppo et al., 2013). Many of these proteins are involved in the carbon and nitrogen primary metabolism and may regulate the general energetic cell metabolism of both symbionts. The most featuring example in *M. truncatula* is, as described above, GS1a which is subjected to NO-mediated inactivation through tyrosine 167 (Melo et al., 2011). GS1a is also subjected to methionine sulfoxidation during nodule senescence

(Matamoros et al., 2013) showing that this key enzyme is regulated by both ROS and RNS. Moreover, as already mentioned, NO may also react with $O_2^{\cdot -}$ to form $ONOO^-$. $ONOO^-$ has the ability to modify proteins by oxidation of sulfur-containing amino acid and nitration of aromatic amino acids. In this context, nodule GS has also been shown to be regulated by tyrosine nitration (Melo et al., 2011), indicating that direct interaction between ROS and RNS may participate to a complex redox-dependent regulation as in plant pathogen interactions (Kulik et al., 2015; Vandelle & Delledonne, 2011). The potential regulatory role of NO on NADPH oxidase (Yun et al., 2011) could also be involved in the crosstalk between ROS and RNS in root nodule by modifying ROS production.

RNS are also involved the cellular redox regulation through interactions with constituents of the antioxidant defense. NO treatment regulates the expression of the glutathione peroxidases *Gpx1* and *Gpx3* in *L. japonicus* nodules (Matamoros et al., 2015). Moreover, Gpx3 nitrosylation results in enzyme activity inhibition in vitro indicating a post-translational control of this enzyme (Matamoros et al., 2015). In contrast, NO treatment of soybean root nodules resulted in an increase in ascorbate peroxidase activity (Keyster, Klein, Egbichi, Jacobs, & Ludidi, 2011). NO was also found to modify antioxidant level in the symbiotic partners. Indeed, GSNO and SNP treatments increase the GSH level in *M. truncatula* (Innocenti et al., 2007) and in *S. meliloti* (Maiti et al., 2012). Numerous redox associated systems are adjusted during the different steps of nodule formation and functioning (Frendo, Matamoros, Alloing, & Becana, 2013; Puppo et al., 2013; Ribeiro, Alloing, Mandon, & Frendo, 2014) showing that the symbiotic association requires a fine regulation of the various steps to lead to a functional symbiosis. Deciphering the cross-talk between the different players involved in the nodule redox regulation promises to be an exciting objective in the coming years.

3.5 NO and Hormone Crosstalk

Key phytohormones and/or rhizobia-synthesized hormones such as abscisic acid (ABA), auxin, cytokinins (CKs), ethylene, jasmonic acid (JA), brassinosteroids and many more regulate a broad range of symbiotic processes from the infection steps (nodule initiation) to nodule development and senescence, nodule number, size, morphology or positioning, N fixation or response to environmental stresses and ultimately legume plant growth (Breakspear et al. (2014) reviewed in Ferguson and Mathesius (2014)). Since NO is necessary for nodule formation (del Giudice et al., 2011), N fixation

(Horchani et al., 2011; Sanchez et al., 2010) and nodule senescence (Cam et al., 2012), a putative link between NO accumulation and hormonal regulation remains intriguing. Indeed, the crosstalk between NO and phytohormones during nodule establishment and functioning is still poorly understood. However, treatment of inoculated *M. truncatula* roots with the NO scavenger cPTIO differently affected the expression of auxin, brassinosteroid, JA, ethylene and salicylic acid-related genes, strongly supporting an interference of NO in hormonal regulation of the symbiosis (Boscari, del Giudice, et al., 2013).

In the *M. truncatula/S. meliloti* system, auxins have recently been reported to play a role in root-hair infection, by virtue of their involvement in cell division and expansion (Breakspear et al., 2014), but auxins role during primordium initiation/development seems rather more critical in indeterminate (*Medicago* species) than in determinate-type nodules such as those of common bean (Pii et al., 2007; Turner et al., 2013). Indeed, inoculation of *M. truncatula* plants with an auxin-overproducing *S. meliloti* strain increased nodule indole acetic acid (IAA) and NO contents, and promoted nodules formation and root growth (Pii et al., 2007). The NO scavenger cPTIO reduced nodules number in plants inoculated with either WT or IAA-overproducing strain (Pii et al., 2007) and inhibited the expression of several auxin-related genes (Boscari, del Giudice, et al., 2013). Since NO depletion in *M. truncatula* inoculated roots resulted in the inhibition of several auxin-related genes, a direct transcriptional regulatory effect of NO on these genes, and consequently on nodule establishment, may be suspected (Boscari, del Giudice, et al., 2013).

CKs have been suggested to be key regulators of the nodulation and symbiosis processes in relation to NO (Frugier et al., 2008; del Giudice et al., 2011). The reduction of NO content in *M. truncatula* inoculated roots significantly decreased the expression of genes involved in nodule formation such as *MtCRE1* (del Giudice et al., 2011) encoding a CK receptor involved in early symbiotic interaction with *S. meliloti* and nodule organogenesis (Gonzalez-Rizzo, Crespi, & Frugier, 2006). In addition, NO was found to affect CK signaling via the inhibition of the histidine phosphotransfer protein activity through S-nitrosylation, as reported for Arabidopsis AHP1 (Feng et al., 2013). Finally, *L. japonicus enhanced nitrogen fixation 1* (*enf1*) mutants, showing low sensitivity to exogenous ABA and reduced endogenous ABA concentration, exhibited a higher number and weight of root nodules and, consequently, improved N fixation and plant growth (Tominaga et al., 2009). In fact, NO production was reduced in *enf1* nodules, which in turn may have increased N_2 fixation (Tominaga et al., 2009).

A part from symbiotic interaction and nodule organogenesis/development, NO and phytohormones interaction may interfere with oxidative stress management and thus nodule senescence. In 7 week-old *L. japonicus* nodules, the glutathione peroxidase *LjGpx1* is significantly induced by GSNO, but not by phytohormones, while *LjGpx3* is only induced by the ethylene precursor 1-aminocyclopropane-1-carboxylic acid (ACC) and by CK, but not by GSNO treatment (Matamoros et al., 2015). However, both enzymes are nitrosylated in vitro and in vivo, resulting in the inhibition of their activity. Consequently, both NO and phytohormones regulate antioxidative enzymes activity/expression within the nodule (Matamoros et al., 2015). Similarly, the haemoglobin encoding gene *LjGLB1-1* is specifically up-regulated by NO in *L. japonicus* nodules (45 days old) (Shimoda et al., 2005), but also by ABA and ACC (Bustos-Sanmamed et al., 2011).

Finally, since ethylene and JA play a positive role in nodule senescence (Van De Velde et al., 2006), and since NO was suggested to be a key regulator of nodule senescence delay (Cam et al., 2012), it is also tempting to consider that NO could also be part of the signaling pathway regulating ethylene and JA biosynthesis during the nodule senescence event. However, such hypothesis remains to be investigated.

4. CONCLUSIONS AND FUTURE DIRECTIONS

Considered together, the data presented in this review show that NO is produced all along the N_2-fixing symbiosis, from the first hours of the interaction between the plant and the bacteria to the breaking off of the association between the two partners. Its spatiotemporal accumulation, i.e. in the infection thread, nodule cortex, N_2-fixation zone or at the fixation to senescence onset, underlines the diversity of the physiological environments wherein it is produced. During the early steps of symbiosis, NR and NOS-like activities have been evidenced as NO sources, whereas bacterial denitrification pathway and the plant NR/mitochondrial ETC system appear to be the main sources of NO in N_2-fixing mature nodules. If NOS-like or PAOx activities have been found in mature nodules extracts or slices, their contribution to NO production remains to be determined in vivo. The control of how much, where and when NO is produced is a challenging issue during the whole symbiotic process. Indeed, NO is toxic for free Rhizobia and is a potent inhibitor of nitrogenase activity and symbiotic N_2 fixation. On another hand, NO production is increased in hypoxic nodules and this production is supposed to be linked – via a Hb/NO respiration process – with improved capacity of the nodules to maintain their energy status under hypoxic conditions. It has been also suggested that NO might be a developmental signal involved in the induction of nodule senescence. Hence, the questions are raised of the toxic effects versus signaling/metabolic functions of NO, and of the regulation of NO levels compatible with N_2 fixation. Regarding this point, the number of plant haemoglobins (nsHbs, Lbs) and of bacterial detoxication systems (Hmp, Nor, NnrS) acting together to control NO level has been particularly evidenced. However, the involvement of both plant and bacterial GSNO reductases and thioredoxins systems remains to be investigated to decipher their respective contribution to NO balance in legume roots, free bacteria and nodules.

Hb/NO pair appears to be a key component of symbiotic as well as pathogen interactions. During the early stage of the interaction the reaction against symbiotic rhizobia resembles plant defense response to pathogens. In both cases, infection by the microorganism first induces an NO-dependent expression of many plant defense-related genes. However, in the pathogen interaction, the repression or the maintaining of Hb expression at low level is accompanied by a prolonged NO production and the set-up of an elevated resistance response under the control of SA-dependent or JA/ethylene-dependent pathways (Wally, Mira, Hill, & Stasolla, 2013). In the symbiotic interaction, Hb genes are rapidly induced, triggering within the first hours of the interaction a decrease in NO level, the repression of plant defense system and the establishment of the host process (Boscari, Meilhoc, et al., 2013; Nagata et al., 2009). If the involvement of ns-Hbs during the early stages of symbiosis has been evidenced, the respective functions of the three classes of Hbs remains to be clarified. For instance, class-1 and class-2 Hbs bind O_2 very tightly, making it unlikely that they function as O_2 carriers, stores or sensors (Hill, 2012). Deciphering their respective role in the balance of NO level in N_2-fixing nodules and at the onset of senescence is a challenging issue. Although the amount of available data concerning NO role(s) during biological N_2-fixation in leguminous plants is increasing, little is known about non-legume plants. Indeed, even though Hbs have been identified in actinorhizal plants and a truncated Hb induced upon plant infection by Frankia, the role of NO remains to be studied in actinorhizal as well as in Parasponia–rhizobia symbiosis (for a review Santi, Bogusz, and Franche (2013)).

Considered together, the data presented in this review show that NO acts as a multifaceted regulator — as a Jack-of-all-trades — involved in numerous mechanisms of the symbiotic process such as symbiont recognition, modulation of plant defense reactions, cell division and organogenesis, in energy, nitrogen and carbon metabolisms, and in the setting up of senescence. NO interplays with other major regulators of plant growth and development such as hormones and ROS. Deciphering these interplays and the mechanisms of NO perception and signaling appears to be another promising issue for the future.

ACKNOWLEDGEMENTS

This work was supported by the 'Institut National de la Recherche Agronomique', the 'Centre National de la Recherche Scientifique', the University of Nice—Sophia Antipolis, The Institut National des Sciences Appliquées of Toulouse and the French Government (National Research Agency, ANR) through the LABEX SIGNALIFE (ANR-11-LABX-0028-01) and LABEX TULIP (ANR-10-LABX-41; ANR-11-IDEX-0002-02) programs. IH was supported by a post-doctoral fellowship from the Institut National de la Recherche Agronomique.

REFERENCES

Alvarez, B., & Radi, R. (2003). Peroxynitrite reactivity with amino acids and proteins. *Amino Acids, 25*, 295—311.

Arjona, D., Wikstrom, M., & Adelroth, P. (2015). Nitric oxide is a potent inhibitor of the cbb(3)-type heme-copper oxidases. *FEBS Letters, 589*, 1214—1218.

Astier, J., Besson-Bard, A., Lamotte, O., Bertoldo, J., Bourque, S., Terenzi, H., et al. (2012). Nitric oxide inhibits the ATPase activity of the chaperone-like AAA+ ATPase CDC48, a target for S-nitrosylation in cryptogein signalling in tobacco cells. *Biochemical Journal, 447*, 249—260.

Astier, J., & Lindermayr, C. (2012). Nitric oxide-dependent posttranslational modification in plants: an update. *International Journal of Molecular Sciences, 13*, 15193—15208.

Baudouin, E., Pieuchot, L., Engler, G., Pauly, N., & Puppo, A. (2006). Nitric oxide is formed in *Medicago truncatula-Sinorhizobium meliloti* functional nodules. *Molecular Plant—Microbe Interaction, 19*, 970—975.

Behm, J. E., Geurts, R., & Kiers, E. T. (2014). Parasponia: a novel system for studying mutualism stability. *Trends in Plant Science, 19*, 757—763.

Besson-Bard, A., Astier, J., Rasul, S., Wawer, I., Dubreuil-Maurizi, C., Jeandroz, S., et al. (2009). Current view of nitric oxide-responsive genes in plants. *Plant Science, 177*, 302—309.

Besson-Bard, A., Pugin, A., & Wendehenne, D. (2008). New insights into nitric oxide signaling in plants. *Annual Review of Plant Biology, 59*, 21—39.

Blanquet, P., Silva, L., Catrice, O., Bruand, C., Carvalho, H., & Meilhoc, E. (2015). *Sinorhizobium meliloti* controls NO-mediated post-translational modification of a *Medicago truncatula* nodule protein. *Molecular Plant—Microbe Interaction*. http://dx.doi.org/10.1094/MPMI-05-15-0118-R.

Boscari, A., del Giudice, J., Ferrarini, A., Venturini, L., Zaffini, A. L., Delledonne, M., et al. (2013). Expression dynamics of the *Medicago truncatula* transcriptome during the symbiotic interaction with *Sinorhizobium meliloti*: which role for nitric oxide? *Plant Physiology, 161*, 425–439.

Boscari, A., Meilhoc, E., Castella, C., Bruand, C., Puppo, A., & Brouquisse, R. (2013). Which role for nitric oxide in symbiotic N2-fixing nodules: toxic by-product or useful signaling/metabolic intermediate? *Frontiers in Plant Science, 4*, 384.

Breakspear, A., Liu, C., Roy, S., Stacey, N., Rogers, C., Trick, M., et al. (2014). The root hair "infectome" of *Medicago truncatula* uncovers changes in cell cycle genes and reveals a requirement for Auxin signaling in rhizobial infection. *The Plant Cell, 26*, 4680–4701.

Bustos-Sanmamed, P., Tovar-Mendez, A., Crespi, M., Sato, S., Tabata, S., & Becana, M. (2011). Regulation of nonsymbiotic and truncated hemoglobin genes of *Lotus japonicus* in plant organs and in response to nitric oxide and hormones. *The New Phytologist, 189*, 765–776.

Cabrera, J. J., Sanchez, C., Gates, A. J., Bedmar, E. J., Mesa, S., Richardson, D. J., et al. (2011). The nitric oxide response in plant-associated endosymbiotic bacteria. *Biochemical Society Transactions, 39*, 1880–1885.

Cam, Y., Pierre, O., Boncompagni, E., Herouart, D., Meilhoc, E., & Bruand, C. (2012). Nitric oxide (NO): a key player in the senescence of *Medicago truncatula* root nodules. *New Phytologist, 196*, 548–560.

Carvalho, H., Lescure, N., De Billy, F., Chabaud, M., Lima, L., Salema, R., et al. (2000). Cellular expression and regulation of the *Medicago truncatula* cytosolic glutamine synthetase genes in root nodules. *Plant Molecular Biology, 42*, 741–756.

Correa-Aragunde, N., Foresi, N., & Lamattina, L. (2015). Nitric oxide is a ubiquitous signal for maintaining redox balance in plant cells: regulation of ascorbate peroxidase as a case study. *Journal of Experimental Botany, 66*, 2913–2921.

Delledonne, M., Xia, Y., Dixon, R. A., & Lamb, C. (1998). Nitric oxide functions as a signal in plant disease. *Nature, 394*, 585–588.

Durner, J., Wendehenne, D., & Klessig, D. F. (1998). Defense gene induction in tobacco by nitric oxide, cyclic GMP, and cyclic ADP-ribose. *Proceedings of the National Academy of Sciences of the United States of America, 95*, 10328–10333.

Espinosa, F., Garrido, I., Ortega, A., Casimiro, I., & Alvarez-Tinaut, M. C. (2014). Redox activities and ROS, NO and phenylpropanoids production by axenically cultured intact olive seedling roots after interaction with a mycorrhizal or a pathogenic fungus. *PLoS One, 9*, e100132.

Feng, J., Wang, C., Chen, Q., Chen, H., Ren, B., Li, X., et al. (2013). S-nitrosylation of phosphotransfer proteins represses cytokinin signaling. *Nature Communications, 4*, 1529.

Ferguson, B. J., & Mathesius, U. (2014). Phytohormone regulation of legume–rhizobia interactions. *Journal of Chemical Ecology, 40*, 770–790.

Ferrarini, A., De Stefano, M., Baudouin, E., Pucciariello, C., Polverari, A., Puppo, A., et al. (2008). Expression of *Medicago truncatula* genes responsive to nitric oxide in pathogenic and symbiotic conditions. *Molecular Plant–Microbe Interaction, 21*, 781–790.

Frendo, P., Matamoros, M. A., Alloing, G., & Becana, M. (2013). Thiol-based redox signaling in the nitrogen-fixing symbiosis. *Frontiers of Plant Science, 4*, 376.

Frugier, F., Kosuta, S., Murray, J. D., Crespi, M., & Szczyglowski, K. (2008). Cytokinin: secret agent of symbiosis. *Trends in Plant Science, 13*, 115–120.

Gibbs, D. J., Md Isa, N., Movahedi, M., Lozano-Juste, J., Mendiondo, G. M., Berckhan, S., et al. (2014). Nitric oxide sensing in plants is mediated by proteolytic control of group VII ERF transcription factors. *Molecular Cell, 53*, 369–379.

del Giudice, J., Cam, Y., Damiani, I., Fung-Chat, F., Meilhoc, E., Bruand, C., et al. (2011). Nitric oxide is required for an optimal establishment of the *Medicago truncatula-Sinorhizobium meliloti* symbiosis. *The New Phytologist, 191*, 405–417.

Gomez-Hernandez, N., Reyes-Gonzalez, A., Sanchez, C., Mora, Y., Delgado, M. J., & Girard, L. (2011). Regulation and symbiotic role of nirK and norC expression in *Rhizobium etli*. *Molecular Plant–Microbe Interaction, 24*, 233–245.

Gonzalez-Rizzo, S., Crespi, M., & Frugier, F. (2006). The *Medicago truncatula* CRE1 cytokinin receptor regulates lateral root development and early symbiotic interaction with *Sinorhizobium meliloti*. *The Plant Cell, 18*, 2680–2693.

Gupta, K. J., Hebelstrup, K. H., Mur, L. A., & Igamberdiev, A. U. (2011). Plant hemoglobins: important players at the crossroads between oxygen and nitric oxide. *FEBS Letters, 585*, 3843–3849.

Gupta, K. J., & Igamberdiev, A. U. (2011). The anoxic plant mitochondrion as a nitrite: NO reductase. *Mitochondrion, 11*, 537–543.

Herold, S., & Puppo, A. (2005). Oxyleghemoglobin scavenges nitrogen monoxide and peroxynitrite: a possible role in functioning nodules? *Journal of Biological Inorganic Chemistry, 10*, 935–945.

Hess, D. T., Matsumoto, A., Kim, S. O., Marshall, H. E., & Stamler, J. S. (2005). Protein S-nitrosylation: purview and parameters. *Nature Reviews, 6*, 150–166.

Hess, D. T., Matsumoto, A., Nudelman, R., & Stamler, J. S. (2001). S-nitrosylation: spectrum and specificity. *Nature Cell Biology, 3*, E46–E49.

Hichri, I., Boscari, A., Castella, C., Rovere, M., Puppo, A., & Brouquisse, R. (2015). Nitric oxide: a multifaceted regulator of the nitrogen-fixing symbiosis. *Journal of Experimental Botany, 66*, 2877–2887.

Hill, R. D. (2012). Non-symbiotic haemoglobins-What's happening beyond nitric oxide scavenging? *AoB Plants, 2012*, pls004.

Horchani, F., Prevot, M., Boscari, A., Evangelisti, E., Meilhoc, E., Bruand, C., et al. (2011). Both plant and bacterial nitrate reductases contribute to nitric oxide production in *Medicago truncatula* nitrogen-fixing nodules. *Plant Physiology, 155*, 1023–1036.

Hyduke, D. R., Jarboe, L. R., Tran, L. M., Chou, K. J., & Liao, J. C. (2007). Integrated network analysis identifies nitric oxide response networks and dihydroxyacid dehydratase as a crucial target in *Escherichia coli*. *Proceedings of the National Academy of Sciences of the United States of America, 104*, 8484–8489.

Igamberdiev, A. U., Ratcliffe, R. G., & Gupta, K. J. (2014). Plant mitochondria: source and target for nitric oxide. *Mitochondrion, 19*(Pt B), 329–333.

Innocenti, G., Pucciariello, C., Le Gleuher, M., Hopkins, J., De Stefano, M., Delledonne, M., et al. (2007). Glutathione synthesis is regulated by nitric oxide in *Medicago truncatula* roots. *Planta, 225*, 1597–1602.

Kanayama, Y., & Yamamoto, Y. (1991). Formation of nitrosylleghemoglobin in nodules of nitrate-treated cowpea and pea-plants. *Plant and Cell Physiology, 32*, 19–24.

Kato, K., Kanahama, K., & Kanayama, Y. (2010). Involvement of nitric oxide in the inhibition of nitrogenase activity by nitrate in Lotus root nodules. *Journal of Plant Physiology, 167*, 238–241.

Keyster, M., Klein, A., Egbichi, I., Jacobs, A., & Ludidi, N. (2011). Nitric oxide increases the enzymatic activity of three ascorbate peroxidase isoforms in soybean root nodules. *Plant Signaling and Behavior, 6*, 956–961.

Kulik, A., Noirot, E., Grandperret, V., Bourque, S., Fromentin, J., Salloignon, P., et al. (2015). Interplays between nitric oxide and reactive oxygen species in cryptogein signalling. *Plant, Cell and Environment, 38*, 331–348.

Leach, J., Keyster, M., Du Plessis, M., & Ludidi, N. (2010). Nitric oxide synthase activity is required for development of functional nodules in soybean. *Journal of Plant Physiology, 167*, 1584–1591.

Lee, H. W., Hitchcock, T. M., Park, S. H., Mi, R., Kraft, J. D., Luo, J., et al. (2010). Involvement of thioredoxin domain-containing 5 in resistance to nitrosative stress. *Free Radical Biology and Medicine, 49*, 872–880.

Maiti, D., Sarkar, T. S., & Ghosh, S. (2012). Detection of S-nitrosothiol and nitrosylated proteins in *Arachis hypogaea* functional nodule: response of the nitrogen fixing symbiont. *PLoS One, 7*, e45526.

Maskall, C. S., Gibson, J. F., & Dart, P. J. (1977). Electron-paramagnetic-resonance studies of leghaemoglobins from soya-bean and cowpea root nodules. Identification of nitrosylleghaemoglobin in crude leghaemoglobin preparations. *Biochemical Journal, 167*, 435–445.

Masson-Boivin, C., Giraud, E., Perret, X., & Batut, J. (2009). Establishing nitrogen-fixing symbiosis with legumes: how many rhizobium recipes? *Trends in Microbiology, 17*, 458–466.

Matamoros, M. A., Baird, L. M., Escuredo, P. R., Dalton, D. A., Minchin, F. R., Iturbe-Ormaetxe, I., et al. (1999). Stress-induced legume root nodule senescence. Physiological, biochemical, and structural alterations. *Plant Physiology, 121*, 97–112.

Matamoros, M. A., Fernández-García, N., Wienkoop, S., Loscos, J., Saiz, A., & Becana, M. (2013). Mitochondria are an early target of oxidative modifications in senescing legume nodules. *New Phytologist, 197*, 873–885.

Matamoros, M. A., Saiz, A., Peñuelas, M., Bustos-Sanmamed, P., Mulet, J. M., Barja, M. V., et al. (2015). Function of glutathione peroxidases in legume root nodules. *Journal of Experimental Botany, 66*, 2979–2990.

Mathieu, C., Moreau, S., Frendo, P., Puppo, A., & Davies, M. J. (1998). Direct detection of radicals in intact soybean nodules: presence of nitric oxide-leghemoglobin complexes. *Free Radical Biology and Medicine, 24*, 1242–1249.

Meakin, G. E., Bueno, E., Jepson, B., Bedmar, E. J., Richardson, D. J., & Delgado, M. J. (2007). The contribution of bacteroidal nitrate and nitrite reduction to the formation of nitrosylleghaemoglobin complexes in soybean root nodules. *Microbiology, 153*, 411–419.

Meilhoc, E., Blanquet, P., Cam, Y., & Bruand, C. (2013). Control of NO level in rhizobium-legume root nodules: not only a plant globin story. *Plant Signaling and Behavior, 8*. http://dx.doi.org/10.4161/psb.25923.

Meilhoc, E., Boscari, A., Bruand, C., Puppo, A., & Brouquisse, R. (2011). Nitric oxide in legume-rhizobium symbiosis. *Plant Science, 181*, 573–581.

Meilhoc, E., Cam, Y., Skapski, A., & Bruand, C. (2010). The response to nitric oxide of the nitrogen-fixing symbiont *Sinorhizobium meliloti*. *Molecular Plant–Microbe Interaction, 23*, 748–759.

Melo, P. M., Silva, L. S., Ribeiro, I., Seabra, A. R., & Carvalho, H. G. (2011). Glutamine synthetase is a molecular target of nitric oxide in root nodules of *Medicago truncatula* and is regulated by tyrosine nitration. *Plant Physiology, 157*, 1505–1517.

Mur, L. A., Mandon, J., Persijn, S., Cristescu, S. M., Moshkov, I. E., Novikova, G. V., et al. (2013). Nitric oxide in plants: an assessment of the current state of knowledge. *AoB Plants, 5*, pls052.

Murakami, E., Nagata, M., Shimoda, Y., Kucho, K., Higashi, S., Abe, M., et al. (2011). Nitric oxide production induced in roots of *Lotus japonicus* by lipopolysaccharide from *Mesorhizobium loti*. *Plant and Cell Physiology, 52*, 610–617.

Nagata, M., Hashimoto, M., Murakami, E., Shimoda, Y., Shimoda-Sasakura, F., Kucho, K., et al. (2009). A possible role of class 1 plant hemoglobin at the early stage of legume-rhizobium symbiosis. *Plant Signaling and Behavior, 4*, 202–204.

Nagata, M., Murakami, E., Shimoda, Y., Shimoda-Sasakura, F., Kucho, K., Suzuki, A., et al. (2008). Expression of a class 1 hemoglobin gene and production of nitric oxide in response to symbiotic and pathogenic bacteria in *Lotus japonicus*. *Molecular Plant–Microbe Interaction, 21*, 1175–1183.

Navascues, J., Perez-Rontome, C., Gay, M., Marcos, M., Yang, F., Walker, F. A., et al. (2012). Leghemoglobin green derivatives with nitrated hemes evidence production of highly reactive nitrogen species during aging of legume nodules. *Proceedings of the National Academy of Sciences of the United States of America, 109*, 2660–2665.

Palmieri, M. C., Sell, S., Huang, X., Scherf, M., Werner, T., Durner, J. R., et al. (2008). Nitric oxide-responsive genes and promoters in *Arabidopsis thaliana*: a bioinformatics approach. *Journal of Experimental Botany, 59*, 177–186.

Perez Guerra, J. C., Coussens, G., De Keyser, A., De Rycke, R., De Bodt, S., Van De Velde, W., et al. (2010). Comparison of developmental and stress-induced nodule senescence in *Medicago truncatula*. *Plant Physiology, 152*, 1574–1584.

Pierre, O., Hopkins, J., Combier, M., Baldacci, F., Engler, G., Brouquisse, R., et al. (2014). Involvement of papain and legumain proteinase in the senescence process of *Medicago truncatula* nodules. *The New Phytologist, 202*, 849–863.

Pii, Y., Crimi, M., Cremonese, G., Spena, A., & Pandolfini, T. (2007). Auxin and nitric oxide control indeterminate nodule formation. *BMC Plant Biology, 7*, 21.

Puppo, A., Pauly, N., Boscari, A., Mandon, K., & Brouquisse, R. (2013). Hydrogen peroxide and nitric oxide: key regulators of the Legume-Rhizobium and mycorrhizal symbioses. *Antioxidants and Redox Signaling, 18*, 2202–2219.

Rhee, K. Y., Erdjument-Bromage, H., Tempst, P., & Nathan, C. F. (2005). S-nitroso proteome of *Mycobacterium tuberculosis*: enzymes of intermediary metabolism and antioxidant defense. *Proceedings of the National Academy of Sciences of the United States of America, 102*, 467–472.

Ribeiro, C. W., Alloing, G., Mandon, K., & Frendo, P. (2014). Redox regulation of differentiation in symbiotic nitrogen fixation. *Biochimica et Biophysica Acta, 1850*, 1469–1478.

Roux, B., Rodde, N., Jardinaud, M. F., Timmers, T., Sauviac, L., Cottret, L., et al. (2014). An integrated analysis of plant and bacterial gene expression in symbiotic root nodules using laser-capture microdissection coupled to RNA sequencing. *Plant Journal, 77*, 817–837.

Sainz, M., Calvo-Begueria, L., Perez-Rontome, C., Wienkoop, S., Abian, J., Staudinger, C., et al. (2015). Leghemoglobin is nitrated in functional legume nodules in a tyrosine residue within the heme cavity by a nitrite/peroxide-dependent mechanism. *Plant Journal, 81*, 723–735.

Sanchez, C., Cabrera, J. J., Gates, A. J., Bedmar, E. J., Richardson, D. J., & Delgado, M. J. (2011). Nitric oxide detoxification in the rhizobia-legume symbiosis. *Biochemical Society Transactions, 39*, 184–188.

Sanchez, C., Gates, A. J., Meakin, G. E., Uchiumi, T., Girard, L., Richardson, D. J., et al. (2010). Production of nitric oxide and nitrosylleghemoglobin complexes in soybean nodules in response to flooding. *Molecular Plant–Microbe Interaction, 23*, 702–711.

Santi, C., Bogusz, D., & Franche, C. (2013). Biological nitrogen fixation in non-legume plants. *Annals of Botany*. http://dx.doi.org/10.1093/aob/mct048.

Sarti, P., Forte, E., Giuffre, A., Mastronicola, D., Magnifico, M. C., & Arese, M. (2012). The chemical interplay between nitric oxide and mitochondrial cytochrome c oxidase: reactions, effectors and pathophysiology. *International Journal of Cell Biology, 2012*, 571067.

Seregelyes, C., Igamberdiev, A. U., Maassen, A., Hennig, J., Dudits, D., & Hill, R. D. (2004). NO-degradation by alfalfa class 1 hemoglobin (Mhb1): a possible link to PR-1a gene expression in Mhb1-overproducing tobacco plants. *FEBS Letters, 571*, 61–66.

Shimoda, Y., Nagata, M., Suzuki, A., Abe, M., Sato, S., Kato, T., et al. (2005). Symbiotic rhizobium and nitric oxide induce gene expression of non-symbiotic hemoglobin in *Lotus japonicus*. *Plant and Cell Physiology, 46*, 99–107.

Shimoda, Y., Shimoda-Sasakura, F., Kucho, K., Kanamori, N., Nagata, M., Suzuki, A., et al. (2009). Overexpression of class 1 plant hemoglobin genes enhances symbiotic nitrogen fixation activity between *Mesorhizobium loti* and *Lotus japonicus*. *Plant Journal, 57*, 254–263.

Sprent, J. I. (2007). Evolving ideas of legume evolution and diversity: a taxonomic perspective on the occurrence of nodulation. *New Phytologist, 174*, 11–25.

Stamler, J. S., Singel, D. J., & Loscalzo, J. (1992). Biochemistry of nitric oxide and its redox-activated forms. *Science, 258*, 1898.

Svistoonoff, S., Hocher, V., & Gherbi, H. (2014). Actinorhizal root nodule symbioses: what is signalling telling on the origins of nodulation? *Current Opinion in Plant Biology, 20*, 11–18.

Tominaga, A., Nagata, M., Futsuki, K., Abe, H., Uchiumi, T., Abe, M., et al. (2009). Enhanced nodulation and nitrogen fixation in the abscisic acid low-sensitive mutant enhanced nitrogen fixation1 of *Lotus japonicus*. *Plant Physiology, 151*, 1965–1976.

Trinchant, J. C., & Rigaud, J. (1982). Nitrite and nitric oxide as inhibitors of nitrogenase from soybean bacteroids. *Applied and Environmental Microbiology, 44*, 1385–1388.

Turner, M., Nizampatnam, N. R., Baron, M., Coppin, S., Damodaran, S., Adhikari, S., et al. (2013). Ectopic expression of miR160 results in auxin hypersensitivity, cytokinin hyposensitivity, and inhibition of symbiotic nodule development in soybean. *Plant Physiology, 162*, 2042–2055.

Van De Velde, W., Guerra, J. C., De Keyser, A., De Rycke, R., Rombauts, S., Maunoury, N., et al. (2006). Aging in legume symbiosis. A molecular view on nodule senescence in *Medicago truncatula*. *Plant Physiology, 141*, 711–720.

Vandelle, E., & Delledonne, M. (2011). Peroxynitrite formation and function in plants. *Plant Science, 181*, 534–539.

Wally, O. S., Mira, M. M., Hill, R. D., & Stasolla, C. (2013). Hemoglobin regulation of plant embryogenesis and plant pathogen interaction. *Plant Signaling and Behavior, 8*.

Waszczak, C., Akter, S., Jacques, S., Huang, J., Messens, J., & Van Breusegem, F. (2015). Oxidative post-translational modifications of cysteine residues in plant signal transduction. *Journal of Experimental Botany, 66*, 2923–2934.

Xue, Y., Liu, Z., Gao, X., Jin, C., Wen, L., Yao, X., et al. (2010). GPS-SNO: computational prediction of protein S-nitrosylation sites with a modified GPS algorithm. *PLoS One, 5*, e11290.

Young, N. D., Debelle, F., Oldroyd, G. E., Geurts, R., Cannon, S. B., Udvardi, M. K., et al. (2011). The Medicago genome provides insight into the evolution of rhizobial symbioses. *Nature, 480*, 520–524.

Yun, B. W., Feechan, A., Yin, M., Saidi, N. B., Le Bihan, T., Yu, M., et al. (2011). S-nitrosylation of NADPH oxidase regulates cell death in plant immunity. *Nature, 478*, 264–268.

CHAPTER ELEVEN

Nitric Oxide Signaling during the Hypersensitive Disease Resistance Response

Elodie Vandelle, Tengfang Ling, Zahra Imanifard, Ruitao Liu, Massimo Delledonne, Diana Bellin[1]
Dipartimento di Biotecnologie, Università degli Studi di Verona, Verona, Italy
[1]Corresponding author: E-mail: diana.bellin@univr.it

Contents

Abstract

Nitric oxide (NO) signaling is known to play a key role in triggering the hypersensitive response (HR) of plants to avirulent pathogens. In this chapter, we have summarized what is currently known about the role of NO in this important biological context. We have discussed NO production and turnover leading to the accumulation of this reactive compound when plants are challenged by pathogens. Unfortunately, enzymatic system for its production and the molecular basis for its accumulation are still largely unknown. Furthermore, we have reviewed the ways by which NO transduces its activity to establish hypersensitive disease resistance response by discussing novel emerging findings about its functions. By considering characterized NO targets, we have shown that NO signaling mainly relies on its reactivity with diverse protein targets. This finally orchestrates the crosstalk of NO with other signaling pathways and modulates defence gene expression, allowing the execution of HR cell death and the triggering of the defence response.

Advances in Botanical Research, Volume 77
ISSN 0065-2296
http://dx.doi.org/10.1016/bs.abr.2015.10.013

1. INTRODUCTION

The ubiquitous endogenous signaling molecule nitric oxide (NO) was first discovered to play a key role in disease resistance in plants (Delledonne, Xia, Dixon, & Lamb, 1998; Durner, Wendehenne, & Klessig, 1998). Since then, NO research in this field has received particular attention.

Immunity in plants relies on a two-branch system, including pathogen-associated molecular pattern (PAMP)-triggered immunity (PTI) and effector-triggered immunity (ETI) (Gohre & Robatzek, 2008; Jones & Dangl, 2006). During PTI, the recognition of PAMPs by membrane-bound pattern recognition receptors activates a first layer of defence mechanisms to inhibit pathogen growth. However, virulent pathogens release effectors inside the plant cell to suppress PTI (Boller & Felix, 2009). Plants have evolved a second layer of immunity (ETI) which relies on resistance proteins that recognize these effectors in a specific manner. Pathogen genes encoding recognized effectors are termed avirulence (*Avr*) genes because the recognition of these effectors by plant resistance proteins results in the avirulence of the microorganism. Indeed, this recognition activates an overwhelming defence response characterizing ETI, the hypersensitive disease resistance response (HR), which typically induces programmed cell death (PCD) at the site of pathogen infection to prevent the pathogen from spreading (Coll, Epple, & Dangl, 2011). The pathogen cannot propagate within the host under these circumstances, resulting in an incompatible interaction. Following resistance-gene-mediated pathogen recognition, a prominent feature of the HR is the rapid production of reactive oxygen species (ROS), known as the oxidative burst, which is primarily characterized by the production of superoxide (O_2^-) and hydrogen peroxide (H_2O_2). The application of NO donors or NO production inhibitors has revealed that NO cooperates with ROS during the induction of HR cell death triggered by incompatible interactions, and functions independently of such intermediates to induce defence-related genes (Delledonne et al., 1998; Durner et al., 1998). Since these initial reports, various pharmacological, biochemical and genetic approaches that modulate NO production have been used to confirm the importance of NO in the HR (Chen, Vandelle, Bellin, & Delledonne, 2014). However, the mechanism of NO signaling is not completely understood and there are several unresolved questions, particularly the source of NO and its mechanism of production.

This chapter discusses the major unresolved issues relating to NO production during incompatible interactions and then summarizes recent research underpinning our current understanding of NO signal transduction and its role in HR cell death and immunity.

2. ORIGINS OF THE NO BURST: STILL SEARCHING FOR AN ANSWER

2.1 NO Production

In animals, NO is synthesized by the calmodulin-dependent enzyme NO synthase (NOS). This enzyme exists as three different isoforms: endothelial eNOS and neuronal nNOS, which are constitutively expressed, and inducible iNOS. All catalyse the NADPH-dependent oxidation of L-arginine to NO and citrulline (Bogdan, 2015).

In plants, the reliable, sensitive and specific detection of NO is challenging, especially during the HR and this has hindered the investigation of NO production. Several established analytical methods have been tested, including laser-based infrared spectroscopy, haemoglobin conversion, the detection of diaminofluorescein and electron paramagnetic resonance. However, these methods either lack specificity or require expensive equipment. Chemiluminescence has recently been used to detect NO in intact *Arabidopsis thaliana* plants and tobacco leaves (*Nicotiana tabacum*) infected with avirulent pathogens (Chen et al., 2014; Gupta et al., 2013). This approach achieved an unequivocal detection of NO emitted by infected plants in real time and will hopefully provide more insight into the origin of NO in plants (Figure 1).

Initial attempts to determine the origin of NO in plants relied on arginine-based NOS inhibitors. These suppress NO production of plant extracts during infections, supporting the existence of an NOS-like enzyme responsible of oxidative NO production in plants (Corpas, Palma, del Rio, & Barroso, 2009; Delledonne et al., 1998; Durner et al., 1998). However, the specificity of such inhibitors is unclear, as it is often the case in pharmacological assays (Rasul et al., 2012). Furthermore, no plant gene or protein homologous to mammalian NOS has been identified thus far, although a structurally similar enzyme was identified in the unicellular green alga *Ostreococcus tauri*, albeit with differences in cofactor requirements and calcium dependence (Correa-Aragunde, Foresi, & Lamattina, 2013; Foresi et al., 2010). NOS-like enzymes therefore appear to be absent in land plants

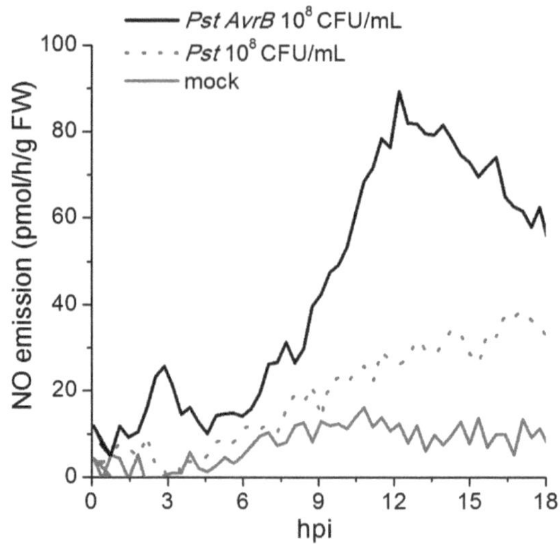

Figure 1 Gas phase NO emission pattern of *Arabidopsis thaliana* (Col) plants in response to infection with virulent (*Pst*) or avirulent (*Pst AvrB*) pathogen revealed by chemiluminescence. *Reprinted with permission from Chen et al. (2014).*

and may have been lost or may have diverged beyond recognition during evolution. In agreement with the first hypothesis, NOS enzymes have not been found in *Physcomitrella patens*, and pharmacological evidence suggests that NO production in this species relies on a reductive pathway (Medina-Andres et al., 2015). In agreement with the second hypothesis, a gene with no significant homology to mammalian NOS was proposed to encode a plant NOS based on genetic evidence (Guo, Okamoto, & Crawford, 2003). However, further studies have shown that this enzyme does not possess true NOS activity, and the protein has been renamed AtNOA1 (nitric oxide associated 1) (Moreau, Lee, Wang, Crane, & Klessig, 2008; Zemojtel et al., 2006). Based on the few differences detected in the NOSoxy domain among the NOS structures described thus far, a focused search for new NOS isoforms differing in the NOSoxy domain could help to identify NOS enzymes in the plant kingdom (Correa-Aragunde et al., 2013).

NO can be produced by oxidation also from polyamines, such as spermine and spermidine, which play a role in drought stress responses and other physiological processes (Tun et al., 2006). Interestingly, candidate genes encoding proteins that may fulfil this role have already been identified (Wimalasekera, Villar, Begum, & Scherer, 2011; Wimalasekera et al., 2015). NO can also be produced by the oxidation of hydroxylamine (Rumer, Gupta, & Kaiser, 2009). However, the relevance of these mechanisms in the HR is unclear.

Several reductive pathways for NO biosynthesis in plants have been also described (Gupta, Fernie, Kaiser, & van Dongen, 2011). These include the xanthine oxidoreductase (XOR) pathway (Corpas et al., 2008), the mitochondrial electron transport chain (ETC) (Planchet, Gupta, Sonoda, & Kaiser, 2005), and the activity of plasma membrane-bound nitrite:NO reductase (Ni:NOR) in roots (Stohr, Strube, Marx, Ullrich, & Rockel, 2001). XOR and ETC work mainly under anoxic conditions and Ni: NOR functions only in the roots, suggesting that none of them are involved in HR. Recently, the class 1 rice (*Oryza sativa*) nonsymbiotic haemoglobin (nsHb1) was shown to generate NO by nitrite reduction, at least *in vitro* (Sturms, DiSpirito, & Hargrove, 2011). Nevertheless, this reaction requires a low oxygen tension, whereas nonsymbiotic haemoglobins in oxygenated tissues act as NO scavengers (Perazzolli et al., 2004).

In contrast to the enzymes discussed above, the cytosolic enzyme nitrate reductase (NR) can generate NO under normal oxygen tension conditions by reducing nitrite to NO in addition to its main function, i.e. the NAD(P) H-dependent reduction of nitrate (NO_3^-) to nitrite (NO_2^-) (Yamasaki, Sakihama, & Takahashi, 1999). The efficiency of this reaction is low and it requires high NO_2^- concentration (Planchet et al., 2005; Rockel, Strube, Rockel, Wildt, & Kaiser, 2002). Therefore, its physiological relevance, especially as a component of plant HR, remains to be demonstrated. NR activity is tightly controlled at several levels and is even regulated in a feedback loop through NO itself during plant development (Frungillo, Skelly, Loake, Spoel, & Salgado, 2014). It is unclear whether this regulation also plays a role in immunity, but feeding plants on NO_3^- rather than ammonia boosts NO production during incompatible interactions, supporting a role for this reductive route in NO biosynthesis (Gupta et al., 2013). The *A. thaliana* double mutant *nia1 nia2* is almost fully deficient for NR activity because these genes encode for the two isoforms of the enzyme in the plant (Wilkinson & Crawford, 1993). This double mutant produced less NO than wild-type plants in response to bacterial infection (Chen et al., 2014; Modolo, Augusto, Almeida, Magalhaes, & Salgado, 2005; Modolo et al., 2006). Importantly, supplementing the *nia1 nia2* mutant with exogenous NO_2^- fully or partially restored the ability to produce NO in response to the pathogen (Chen et al., 2014; Modolo et al., 2005, 2006). In contrast, the mutant phenotype is not abolished by feeding the plants with arginine (Oliveira, Justino, Sodek, & Salgado, 2009). This suggests that NR is mainly required for the synthesis of NO_2^- as a substrate for NO production, and that an additional NR-independent enzyme synthesizes NO from NO_2^-.

As conclusion, although the NO biosynthesis pathways that confer immunity in plants are incompletely understood (Moreau, Lindermayr, Durner, & Klessig, 2010), the combination of genetic, biochemical and pharmacological data suggests the existence of a reductive pathway involving NR itself and/or an unknown NR-independent enzyme for NO biosynthesis, plus an oxidative pathway involving an as yet unidentified enzyme related to mammalian NOS (Figure 2). Redundancy and crosstalk among these different sources would explain our inability to identify mutants solely defined by their inability to produce NO, and this makes it challenging to study NO production and function at a genetic level.

2.2 NO Turnover

NO homeostasis reflects the balance of NO production and NO turnover or conversion into other reactive nitrogen species. The NO burst associated with the HR can therefore be regulated at the level of production but also by mechanisms involved in NO turnover. Given the well-characterized reactivity of NO, competing mechanisms acting during the HR have been largely described and the relevance of some is currently emerging (Figure 2).

NO can be converted into NO_3^- by the NADPH-dependent NO dioxygenase activity of plant haemoglobins (Perazzolli et al., 2004). These proteins may be symbiotic (i.e. localized in the symbiotic root nodules of

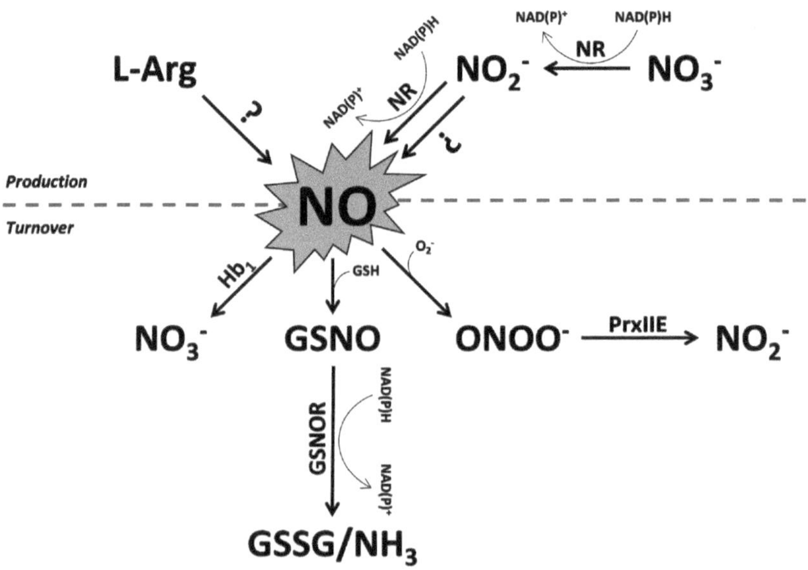

Figure 2 Routes for NO production and turnover regulating the NO burst during the plant hypersensitive disease resistance response.

leguminous plants), but nonsymbiotic and truncated haemoglobins are found in all plants. The characterization of different *A. thaliana* mutants has shown that only the nsHb1 AtHb1 is associated with NO levels in plants (Hebelstrup, Igamberdiev, & Hill, 2007; Perazzolli et al., 2004). Although AtHb1 has a documented role in NO detoxification, the overexpression or silencing of this protein did not affect NO levels or the HR cell death in response to avirulent pathogens in an early study (Perazzolli et al., 2004). More recently, the overexpression of AtHb1 has been shown to compromise NO accumulation in response to avirulent pathogens, associated with a reduction in HR-mediated PCD, whereas AtHb1 silencing enhances the resistance and the modulation of hormones involved in defence. Interestingly, independently of these discrepancies, the AtHb1 expression is also rapidly downregulated in response to infections with the avirulent bacteria *Pseudomonas syringae* pv. *tomato*, suggesting the existence of a mechanism potentiating the NO burst during the HR (Mur et al., 2012). This promising genetic tool to manipulate NO levels has been used to investigate hormone modulation, but not thus far to characterize the role of NO in the HR.

Other mechanisms of NO homeostasis have been reported. For example, NO typically reacts inside plant cells with the antioxidant tripeptide glutathione (GSH) to form *S*-nitrosylated glutathione (GSNO). This compound acts as a natural NO reservoir, which can subsequently either release NO or function directly as a trans-nitrosylating agent. The enzyme GSNO reductase (GSNOR) tightly controls GSNO levels by reducing GSNO to oxidized glutathione (GSSG) and ammonia (NH_3) (Liu et al., 2001). GSNOR activity is suppressed by NO itself through *S*-nitrosylation during nitrogen assimilation (Frugillo et al., 2014; see "Control of Nitrogen Assimilation in Plants through *S*-nitrosothiols" by Frungillo, Spoel, & Salgado, 2016, of this issue) but its role in immunity is unclear. *Arabidopsis thaliana gsnor* knockout mutants accumulate high levels of NO and *S*-nitroso species and have been widely used to study NO/GSNO functions in different biological contexts, including HR. This approach revealed a positive role for *S*-nitrosothiols (SNOs) in HR cell death triggered by pathogens (Yun et al., 2011), although several discrepancies were found when testing plant resistance in this background (Feechan et al., 2005; Holzmeister et al., 2011; Rusterucci, Espunya, Diaz, Chabannes, & Martinez, 2007; Yun et al., 2011). However, this phenomenon is likely to depend on the numerous pleiotropic phenotypes of *gsnor* mutants (Kwon et al., 2012; Xu, Guerra, Lee, & Vierling, 2013).

NO can also react promptly with O_2^- in a diffusion-limited reaction leading to the production of peroxynitrite ($ONOO^-$), a potent oxidizing and nitrating species which, nevertheless, is not cytotoxic in plants (Delledonne, Zeier, Marocco, & Lamb, 2001; Vandelle & Delledonne, 2011). The nonenzymatic biosynthesis of $ONOO^-$ is tightly controlled by the (enzymatic) formation of its precursors. The availability of O_2^- can modulate the NO burst (and vice versa) integrating NO/H_2O_2 signaling during the HR according to the so-called balance model (Delledonne et al., 2001).

3. NO SIGNAL TRANSDUCTION DURING THE HR

NO signal transduction involves a highly amplified and integrated signaling system. This mainly relies on its unique chemical features and reactivity, directly or indirectly affecting a large number of different protein targets and ultimately triggering immunity (Bellin, Asai, Delledonne, & Yoshioka, 2013; Besson-Bard et al., 2008; Leitner, Vandelle, Gaupels, Bellin, & Delledonne, 2009).

NO can modify protein functions directly by reacting with protein-associated transition metals. The most biologically relevant process is haem nitrosylation. For example, this process modulates the activity of the soluble guanylate cyclase in animals to form an iron—nitrosyl complex. The resulting conformational change triggers the conversion of guanosine triphosphate to cyclic guanosine-monophosphate (cGMP), thus transducing the biological activity of NO (Denninger & Marletta, 1999). Despite preliminary reports of pathogen-induced (Meier et al., 2009) and NO-induced cGMP accumulation in plants (Durner et al., 1998; Isner & Maathuis, 2011), there is no evidence yet for a parallel NO-dependent soluble guanylate cyclase that mediates NO signaling. The link between NO and cGMP accumulation/cGMP signaling would deserve to be further explored in plants (Leiter et al., 2009).

Changes in cellular reduction—oxidation (redox) status typically occur when cells are infected with pathogens. Redox-related protein modifications occur in this condition, among which the *S*-nitrosylation allowing the transduction of NO signals. *S*-nitrosylation is a posttranslational modification mediated either by nitrosylating NO derivatives or GSNO. It involves the covalent and reversible binding of NO to the sulfhydryl group of reactive cysteine residues to generate SNOs. Biologically relevant signal transduction relies on target specificity. Accordingly, only a small subset of cysteine residues in proteins possess sulfhydryl groups with particularly low pKa making them susceptible to redox-based posttranslational modifications, including *S*-nitrosylation (Spadaro et al., 2010). Specific redox sensitive target proteins modified in this manner can act as molecular switches responding to changes in cellular redox status caused by pathogens (Spoel & Loake, 2011; Yu, Yun, Spoel, & Loake, 2012; Yu, Lamattina, Spoel, & Loake, 2014). The transient nature of signaling also requires that *S*-nitrosylation is reversible. The indirect regulation of protein *S*-nitrosylation is achieved by GSNOR, which regulates the level of GSNO (Feechan et al., 2005). Importantly, the oxidoreductase TRXh5 was recently shown to possess a potent direct denitrosylation activity, providing selective reversibility for protein-SNO signaling during the establishment of plant immunity (Kneeshaw, Gelineau, Tada, Loake, & Spoel, 2014). TRXh5 can discriminate between different protein-SNO substrates derived from free NO or GSNO and its selective activity contributes, via the specific denitrosylation of the transcriptional coactivator nonexpresser of pathogenesis-related genes 1 (NPR1) and its subsequent nuclear relocalization, to salicylic acid-dependent disease resistance (Kneeshaw et al., 2014). Several groups have performed a computational identification of cysteine residues targeted by *S*-nitrosylation based on consensus motifs (Chaki, Kovacs, Spannagl, & Lindermayr, 2014; Kovacs & Lindermayr, 2013), resulting in a large number of potential targets. By applying such tools or through experimental approaches, a large number of *S*-nitrosylated putative target proteins has been identified (Chaki et al., 2014; Lindermayr, Saalbach, & Durner, 2005; Romero-Puertas et al., 2008). However, the specific functional impact of *S*-nitrosylation during the establishment of plant HR was only explored in a small number of candidates, revealing that *S*-nitrosylation mainly affects the formation of disulfide bonds or protein oxidation, induces changes in protein conformation and localization and impacts biological activity directly or by influencing the binding of cofactors. These mechanisms are discussed in more detail below.

Interestingly, NO modulates its own signaling activity under stress conditions through the *S*-nitrosylation and hence inactivation of peroxiredoxin II E (PrxIIE), an enzyme that detoxifies $ONOO^-$, thus causing the accumulation of $ONOO^-$ during the HR (Romero-Puertas et al., 2007) (Figure 3). This has been confirmed by using the $ONOO^-$ specific dye HK-green2 (Gaupels, Spiazzi-Vandelle, Yang, & Delledonne, 2011). As specified above, although $ONOO^-$ promotes PCD in animals, it does not appear to fulfil a similar role in plants (Delledonne et al., 2001) and is instead emerging as a potential signaling molecule that acts by selectively nitrating tyrosine

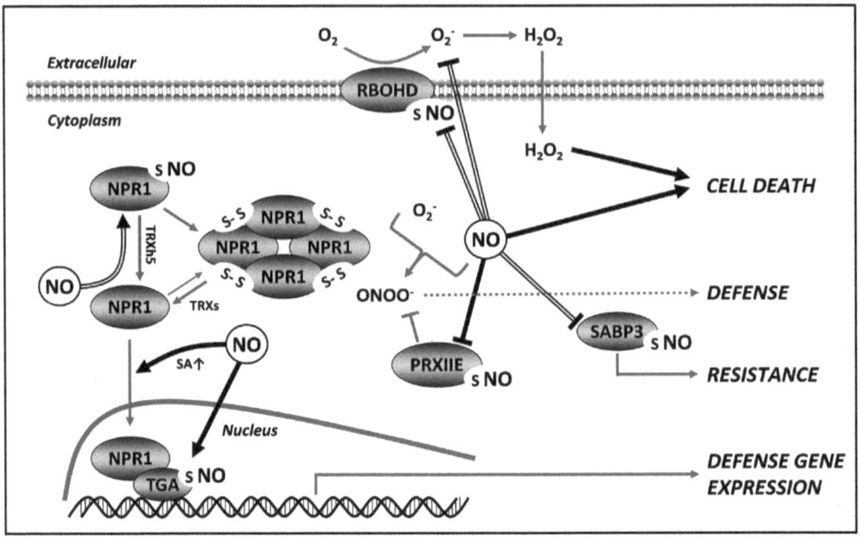

Figure 3 NO functions on plant defence during the hypersensitive disease resistance response. Positive (green (black in print versions) arrows) and negative/feedback control (red (double line in print versions) arrows) are shown. Only plant proteins targeted by S-nitrosylation characterized in the specific context of the HR are included.

residues (Vandelle & Delledonne, 2011). This posttranslational modification involves the addition of a nitro-group to the *ortho*-position of the aromatic ring of tyrosine residues, forming 3-nitrotyrosine (Radi, 2004). There is preliminary evidence that protein nitration in plants can achieve selective activity inhibition and can trigger selective proteasome-mediated degradation of targets which would be involved in signaling, although this has not been demonstrated in the context of defence responses thus far (Álvarez et al., 2011; Melo, Silva, Ribeiro, Seabra, & Carvalho, 2011; Castillo et al., 2015). The accumulation of $ONOO^-$ during the HR causes a parallel increase in the abundance of nitrated plant proteins (Romero-Puertas et al., 2007) which include proteins involved in photosynthesis, ATP synthesis, the Calvin cycle, glycolysis and nitrate assimilation (Cecconi et al., 2009). No clear functional role for protein nitration has been elucidated in the context of the HR, but some important immunity-related candidate proteins can be nitrated *in vitro* and the potential role in defence and immunity for this modification is discussed in more detail below (Begara-Morales et al., 2014, 2015; Chaki et al., 2013; Holzmeister et al., 2015).

NO extensively crosstalks with other signaling pathways. For example, NO can interact with ROS signaling, hormone signaling, downstream mitogen-activated protein kinase cascades, second messengers such as

Ca^{2+} and cGMP (Courtois et al., 2008; Jeandroz et al., 2013; Trapet et al., 2015; Yoshioka, Mase, Yoshioka, Kobayashi, & Asai, 2011) and new emerging interplayers (Choi et al., 2014; Gao et al., 2014; Joudoi et al., 2013; Liu et al., 2013; Mandal et al., 2012). We will focus on the relevance of some of such crosstalks for the HR in the following sections. Ultimately, NO causes the modulation of gene expression during the HR, and the identification of transcription factors regulated by *S*-nitrosylation helps to clarify the links in this chain. Related to this issue, the targeted proteomic analysis of nuclear *S*-nitrosylated proteins is expected to enhance our understanding of NO signal transduction also in plant defence (Chaki et al., 2015; Mengel, Chaki, Shekariesfahlan, & Lindermayr, 2013; see "S-Nitrosylation of Nuclear Proteins: New Pathways in Regulation of Gene Expression" by Kovacs, Ageeva, König, & Lindermayr, 2016, of this issue).

4. THE ROLE OF NO IN THE HR CELL DEATH

4.1 Crosstalk with ROS

The HR during incompatible plant–pathogen interactions involves the collapse of plant tissue due to PCD at the infection site, restricting pathogen growth by blocking its access to nutrients and thus preventing the spread of the disease. This form of PCD is similar in many ways to apoptosis and autophagy in animal cells, although some features are unique to plants (Mur, Kenton, Lloyd, Ougham, & Prats, 2008). Whereas PCD in animals can be triggered by $ONOO^-$, the equivalent process in plants during the HR requires a balance between NO and H_2O_2 levels (Delledonne et al., 2001; de Pinto, Tommasi, & De Gara, 2002; see "Nitric Oxide and Reactive Oxygen Species in PCD Signaling" by Locato, Paradiso, Sabetta, De Gara, & de Pinto, 2016, of this issue). HR-mediated PCD in soybean (*Glycine max*) cells infected with avirulent *P. syringae* pv. *glycinea* was impaired by high levels of exogenous NO and could only be restored by providing high exogenous H_2O_2.

The oxidative burst associated with the HR mainly arises from the activity of membrane-bound RBOH proteins. These are homologous to the gp91 subunit of the mammalian enzyme NADPH oxidase and produce O_2^- in the apoplast through the transfer of electrons from NADPH to molecular oxygen (Mittler et al., 2011; Torres, Dangl, & Jones, 2002). H_2O_2 is then produced by the dismutation of O_2^- catalysed by superoxide dismutase (SOD). H_2O_2 levels are tightly controlled by the activity of these enzymes as well as antioxidant enzymes such as catalase (CAT) and ascorbate peroxidase (APX).

During the HR, NO biosynthesis is concurrent with the oxidative burst. A large, intricate and still unresolved crosstalk network is established among these reactive species and NO largely controls both ROS synthesis and homeostasis. Although NO promotes the oxidative burst induced by pathogen elicitors (Rasul et al., 2012), it also inhibits the oxidative burst triggered by avirulent pathogens or the elicitor cryptogein which induces PCD. Indeed, NO scavenging by 2-(4-carboxyphenyl)-4,4,5,5-tetramethylimidazoline-1-oxyl-3-oxide (cPTIO) enhances the oxidative burst (Kulik et al., 2015; Tada et al., 2004). Radical chemistry may facilitate this negative regulation. During the pathogen-triggered HR, NO can react rapidly with O_2^- to form $ONOO^-$ and thus work directly as an antioxidant. Through its interaction with O_2^- and direct competition with SOD activity, NO could therefore inhibit the oxidative burst, prevent the accumulation of H_2O_2 and avoid PCD. The negative regulation of ROS by NO is also supported by the observed direct inhibition of NADPH oxidase by *S*-nitrosylation, which reduces the affinity of the enzyme for its cofactor FAD (Yun et al., 2011). *Arabidopsis thaliana* AtRBOHD was shown to be *S*-nitrosylated during the HR triggered by the avirulent *P. syringae* pv. *tomato* DC3000 AvrB. *In silico*, structural modelling indicated that the *S*-nitrosylation of Cys890 would disrupt the side chain position of Phe921, which is required to bind FAD, thus explaining the loss of activity (Yun et al., 2011). The NO-dependent regulation of AtRBOHD would help to prevent the generation of excess ROS leading to the induction of PCD (Figure 3).

To add further layers of complexity to the relationship between NO and the oxidative burst, NO has recently been shown to modify and inhibit three of the seven *A. thaliana* SODs in vitro by tyrosine nitration, namely MSD1, CSD3 and FSD3 (Holzmeister et al., 2015). The relevance of this phenomenon during the HR is currently unclear, but it could help to fine-tune the accumulation of different ROS to regulate the induction of PCD. Accordingly NO also targets a number of diverse ROS-detoxifying enzymes by *S*-nitrosylation or nitration according to *in vitro* studies, including APX, monodehydroascorbate reductase, CAT, PrxIIE and PrxIIF, all of which being involved in H_2O_2 detoxification (Begara-Morales et al., 2014, 2015; Camejo et al., 2015; Lin et al., 2012; Ortega-Galisteo et al., 2012; de Pinto et al., 2013; Romero-Puertas et al., 2007; Yang et al., 2015). During the PCD triggered by H_2O_2 or heat stress in tobacco cell suspension cultures, *S*-nitrosylation reduced the activity of cytosolic APX *in vivo* (de Pinto et al., 2013) although controversial finding arise in different conditions triggering oxidative stress (Yang et al., 2015).

Finally, it is also apparent that ROS can in turn modulate NO production (Bright, Desikan, Hancock, Weir, & Neill, 2006; Srivastava, Gonugunta, Puli, & Raghavendra, 2009; Wang, Ries, Wu, Yang, & Crawford, 2010). This has been shown to affect the induction and execution of PCD in some biological contexts and should therefore be also explored in more detail (Kulik et al., 2015; Lin et al., 2012).

4.2 NO and Programmed Cell Death Execution

Although many experimental strategies have been applied to screen for the components of NO-mediated PCD during the HR, only a few have been identified and their mechanisms remain largely unknown. The comparative transcriptomic analysis of NO-treated and untreated plants has yielded a number of candidates (Grun, Lindermayr, Sell, & Durner, 2006; Huang et al., 2002; Kulik et al., 2015; Lin et al., 2012; Parani et al., 2004; Polverari et al., 2003). More focused approaches using specific mutants, e.g. catalase-deficient tobacco plants (CAT1AS) that allow the modulation of H_2O_2 levels *in planta* according to light exposure, have been developed as an experimental system for the independent modulation of NO and H_2O_2 levels. These approaches have led to the identification of genes whose expression is specifically modulated by the balanced action of these molecules, suggesting that they are involved in PCD although no direct association has been confirmed (Zago et al., 2006). More recently, cryptogein elicitor treatment combined with NO scavengers or mutants with impaired ROS production has facilitated the identification of genes modulated by the elicitor during the induction of PCD via an NO/ROS-mediated process. This approach revealed that both NO and ROS act in concert particularly on genes encoding proteins with ubiquitin ligase activity (Kulik et al., 2015).

Proteomic approaches have also been used to identify PCD components during the HR. *S*-nitrosylated proteins were studied either during a pathogen-triggered HR or in more focused experiments in which PCD was triggered by ROS and NO. PCD is induced in *noe1* rice mutants by strong light or in tobacco cell suspension cultures by treatment with H_2O_2 or cryptogein, and each of these conditions was demonstrated to involve the combined action of NO and ROS (Astier et al., 2012; Lin et al., 2012; Romero-Puertas et al., 2008; Vannini et al., 2012). Many redox-related proteins involved in ROS homeostasis were identified in these experiments, as well as additional candidates which homologs mediate apoptosis in animal cells. For example, the glycolytic enzyme glyceraldehyde 3-phosphate dehydrogenase (GAPDH) was found to be *S*-nitrosylated in these studies. In mammals, the *S*-nitrosylation of this protein is known to promote interactions with the E3 ubiquitin ligase Siah1, resulting in nuclear translocation (Mannick, 2007). In the nucleus, SNO-GAPDH stabilizes Siah1 and facilitates the ubiquitination and degradation of nuclear proteins, thus triggering PCD (Foster, Forrester, & Stamler, 2009; Hara et al., 2005; Tristan, Shahani, Sedlak, & Sawa, 2011). The nuclear relocalization of *S*-nitrosylated GAPDH was also observed in plants, although it is not clear whether *S*-nitrosylation is directly responsible for nuclear uptake (Holtgrefe et al., 2008; Vescovi et al., 2013). Furthermore, cryptogein treatment induced the *S*-nitrosylation of the chaperone-like CDC48 AAA+ ATPase, thus triggering a conformational change that abolished its activity *in vitro* (Astier et al., 2012). In animals and yeast, CDC48 targets ubiquitinylated proteins for degradation by the proteasome and its inhibition can trigger PCD (Deichsel, Mouysset, & Hoppe, 2009).

Apoptosis in animals involves caspases, a family of proteases whose activity is controlled by *S*-nitrosylation. Plants do not appear to express animal-like caspases. However, related proteins known as metacaspases are involved in HR-mediated PCD and these proteins have therefore been tested for *S*-nitrosylation. Accordingly, the *A. thaliana* metacaspase 9 MC9 precursor was found to be inhibited by *S*-nitrosylation like animal caspases, but the activity of the mature protein does not appear to be modulated by NO in plants (Belenghi et al., 2007).

5. NO AND IMMUNITY IN PLANTS

In addition to its role in the HR cell death, NO also contributes to plant defence responses by activating directly or indirectly (through crosstalk with other pathways) defence response genes. Early reports described the effect of NO on defence genes such as phenylalanine ammonia-lyase and pathogenesis-related protein 1 (*PR1*) (Delledonne et al., 1998; Durner et al., 1998). A key regulator of immunity in plants is the transcriptional coactivator NPR1. The corresponding gene was identified because the mutant failed to trigger the induction of PR genes following infection and was more susceptible to pathogens (Cao, Bowling, Gordon, & Dong, 1994). NPR1 is a central player in several immunity-related signaling pathways (including ETI) and a key component of salicylic acid-mediated signal transduction (Pajerowska-Mukhtar, Emerine, & Mukhtar, 2013). Two studies have reported evidence that salicylic acid binds to NPR1 directly and indirectly, leading to the activation of defence genes (Fu et al., 2012; Wu et al., 2012). Importantly, NPR1 undergoes oxidation and *S*-nitrosylation that both control its activity (Mou, Fan, & Dong, 2003; Tada et al., 2008; Wu et al., 2012). In unchallenged cells, NPR1 is present as an oligomer with intermolecular redox-sensitive disulfide bridges, and the complex is sequestered in the cytoplasm. Redox changes induced by pathogens and the accumulation of salicylic acid cause a reduction and monomerization of the protein and, consequently, the monomers are translocated to the nucleus, inducing the induction of specific resistance genes (Mou et al., 2003). The equilibrium between NPR1 active monomers and inactive oligomers is a key regulatory component controlling immunity. The *S*-nitrosylation of NPR1 at the predicted oligomerization interface favours the formation of disulfide bonds that promote oligomerization whereas, as mentioned above, the active monomeric form is released by a TRXh5-dependent reaction and accumulates in the nucleus (Tada et al., 2008). Oligomerization triggered by *S*-nitrosylation may be required to maintain NPR1 oligomer/monomer homeostasis, thereby facilitating the steady supply of monomeric protein to support salicylic acid-dependent gene expression (Figure 3). However NO also promotes NPR1-dependent defence responses by facilitating the translocation of NPR1 into the nucleus (Lindermayr, Sell, Muller, Leister, & Durner, 2010) (Figure 3). This additional regulatory mechanism has recently been shown to involve the NO-dependent regulation of GSH biosynthesis and accumulation, which increases salicylic acid levels and thus activates NPR1-dependent defence responses (Kovacs, Durner, & Lindermayr, 2015). Indeed, the activation of *PR* genes by GSNO requires both NPR1 and salicylic acid. However, the nuclear translocation of NPR1 was much slower when driven by GSNO instead of salicylic acid. This finding suggests that this process was not the direct consequence of NPR1 *S*-nitrosylation but, instead, was dependent on a signaling pathway involving GSH biosynthesis, possibly triggered by initially NO-induced GSH oxidation acting upstream of salicylic acid accumulation. This reveals additional crosstalks among NO, GSH and salicylic acid pathways in the establishment of immunity in plants.

In the nucleus, NPR1 acts as a transcriptional coactivator by interacting with the TGACG motif binding factor (TGA) family of basic domain/leucine zipper (bZIP) transcription factors, regulating their ability to bind response elements in the promoter of *PR1* and other defence genes (Despres et al., 2003). TGA proteins can be also oxidized and *S*-nitrosylated. The oxidized form carries disulfide bonds that block interactions with NPR1.

The GSNO-mediated *S*-nitrosylation of cysteine residues protects TGA proteins from oxidative modification, thus promoting DNA binding and stability when NPR1 is translocated into the nucleus (Lindermayr et al., 2010) (Figure 3). The DNA-binding activity of the transcription factor AtMYB30, which regulates defence genes during the establishment of immunity and the induction of PCD, is also inhibited *in vitro* by *S*-nitrosylation, which modifies its secondary structure and thermal stability (Tavares et al., 2014). Therefore the modulation of transcriptional regulators by *S*-nitrosylation could represent a general mechanism for the control of gene expression by NO.

The *A. thaliana* salicylic acid binding protein 3 (AtSABP3) is *S*-nitrosylated *in vivo* during the later stages of an immune response, revealing further crosstalks between the NO and salicylic acid signaling pathways (Wang et al., 2009). Indeed, SABP3 shows high affinity for salicylic acid and also possesses carbonic anhydrase (CA) activity (Slaymaker et al., 2002). The latter may be required for lipid biosynthesis. Indeed, lipids with roles in salicylic acid signaling may represent an integral component of innate immunity (Hoang & Chapman, 2002; Kachroo, Shanklin, Shah, Whittle, & Klessig, 2001). Even if CA role in defence is still unclear, SABP3 is a positive regulator of immunity. *S*-nitrosylation suppresses both its carbonic anhydrase activity and its ability to bind salicylic acid and could therefore initiate a negative feedback loop that modulates plant defence responses during the HR (Wang et al., 2009) (Figure 3). These findings reinforce the connection between NO and salicylic acid and the general role of crosstalks in the establishment of immunity.

6. CONCLUSIONS

In this chapter, we have summarized what is currently known about the role of NO during HR triggered by an avirulent pathogen. We have discussed the production of NO when plants are challenged by pathogens, the mechanisms for NO signal transduction to establish the HR and considered in detail those rare cases in which the specific function of NO has been at least partly elucidated. The emerging general view is that NO transduces its function in plant HR primarily by reacting directly with a range of different target proteins, rather than through a unique receptor. However, such view could just arise from our still fragmentary knowledge and we cannot exclude the possibility that classical receptor-based signaling or more unifying sensing mechanisms, as found in other biological context (Gibbs et al., 2014), may also be involved. According to available data on its specific function, NO appears to have a dual role in the promotion of plant defence during HR (Gross, Durner, & Gaupels, 2013). It can act positively, inducing defence gene expression, signaling events and PCD by its pro-oxidant activity, but it can also generate feedback controls on the activity of enzymes involved in PCD or in resistance and acts itself as an antioxidant in this context (Figure 3). The outcome of these counteracting effects probably depends on spatiotemporal regulation to fine-tune the responses in different contexts to ensure its overall positive effect on defence during HR. Further research will help to yield a more unifying and cohesive view. Still the greatest challenge remains the unclear enzymatic origin of NO during plant–pathogen interactions.

ACKNOWLEDGEMENT

We thank the Future in Research Program (FIRB 2010 – RBFR10S1LJ_001) funded by the Italian Ministry of Education, University and Research for financial support to E.V.

REFERENCES

Álvarez, C., Lozano-Juste, J., Romero, L. C., García, I., Gotor, C., & León, J. (2011). Inhibition of *Arabidopsis* O-acetylserine(thiol)lyase A1 by tyrosine nitration. *Journal of Biological Chemistry, 286*(1), 578–586. http://dx.doi.org/10.1074/jbc.M110.147678.

Astier, J., Besson-Bard, A., Lamotte, O., Bertoldo, J., Bourque, S., Terenzi, H., et al. (2012). Nitric oxide inhibits the ATPase activity of the chaperone-like AAA plus ATPase CDC48, a target for S-nitrosylation in cryptogein signalling in tobacco cells. *Biochemical Journal, 447*, 249–260. http://dx.doi.org/10.1042/bj20120257.

Begara-Morales, J. C., Sanchez-Calvo, B., Chaki, M., Mata-Perez, C., Valderrama, R., Padilla, M. N., et al. (2015). Differential molecular response of monodehydroascorbate reductase and glutathione reductase by nitration and S-nitrosylation. *Journal of Experimental Botany, 66*(19), 5983–5996. http://dx.doi.org/10.1093/jxb/erv306.

Begara-Morales, J. C., Sanchez-Calvo, B., Luque, F., Leyva-Perez, M. O., Leterrier, M., Corpas, F. J., et al. (2014). Differential transcriptomic analysis by RNA-Seq of GSNO-responsive genes between *Arabidopsis* roots and leaves. *Plant and Cell Physiology, 55*(6), 1080–1095. http://dx.doi.org/10.1093/pcp/pcu044.

Belenghi, B., Romero-Puertas, M. C., Vercammen, D., Brackenier, A., Inze, D., Delledonne, M., et al. (2007). Metacaspase activity of *Arabidopsis thaliana* is regulated by S-nitrosylation of a critical cysteine residue. *Journal of Biological Chemistry, 282*(2), 1352–1358. http://dx.doi.org/10.1074/jbc.M608931200. M608931200 [pii].

Bellin, D., Asai, S., Delledonne, M., & Yoshioka, H. (2013). Nitric oxide as a mediator for defense responses. *Molecular Plant–Microbe Interactions, 26*(3), 271–277. http://dx.doi.org/10.1094/Mpmi-09-12-0214-Cr.

Besson-Bard, A., Courtois, C., Gauthier, A., Dahan, J., Dobrowolska, G., Jeandroz, S., et al. (2008). Nitric oxide in plants: production and cross-talk with Ca^{2+} signaling. *Molecular Plant, 1*(2), 218–228. http://dx.doi.org/10.1093/mp/ssm016.

Bogdan, C. (2015). Nitric oxide synthase in innate and adaptive immunity: an update. *Trends in Immunology, 36*(3), 161–178. http://dx.doi.org/10.1016/j.it.2015.01.003.

Boller, T., & Felix, G. (2009). A renaissance of elicitors: perception of microbe-associated molecular patterns and danger signals by pattern-recognition receptors. *Annual Review of*

Plant Biology, 60(1), 379–406. http://dx.doi.org/10.1146/annurev.arplant.57.032905.105346.
Bright, J., Desikan, R., Hancock, J. T., Weir, I. S., & Neill, S. J. (2006). ABA-induced NO generation and stomatal closure in *Arabidopsis* are dependent on H_2O_2 synthesis. *Plant Journal, 45*(1), 113–122. http://dx.doi.org/10.1111/j.1365-313X.2005.02615.x. TPJ2615 [pii].
Camejo, D., Ortiz-Espin, A., Lazaro, J. J., Romero-Puertas, M. C., Lazaro-Payo, A., Sevilla, F., et al. (2015). Functional and structural changes in plant mitochondrial PrxII F caused by NO. *Journal of Proteomics, 119*, 112–125. http://dx.doi.org/10.1016/j.jprot.2015.01.022.
Cao, H., Bowling, S. A., Gordon, A. S., & Dong, X. (1994). Characterization of an *Arabidopsis* mutant that is nonresponsive to inducers of systemic acquired resistance. *Plant Cell, 6*(11), 1583–1592. http://dx.doi.org/10.1105/tpc.6.11.1583.
Castillo, M. C., Lozano-Juste, J., Gonzalez-Guzman, M., Rodriguez, L., Rodriguez, P. L., & Leon, J. (2015). Inactivation of PYR/PYL/RCAR ABA receptors by tyrosine nitration may enable rapid inhibition of ABA signaling by nitric oxide in plants. *Science Signaling, 8*(392), ra89. http://dx.doi.org/10.1126/scisignal.aaa7981.
Cecconi, D., Orzetti, S., Vandelle, E., Rinalducci, S., Zolla, L., & Delledonne, M. (2009). Protein nitration during defense response in *Arabidopsis thaliana*. *Electrophoresis, 30*(14), 2460–2468. http://dx.doi.org/10.1002/elps.200800826.
Chaki, M., Carreras, A., Lopez-Jaramillo, J., Begara-Morales, J. C., Sanchez-Calvo, B., Valderrama, R., et al. (2013). Tyrosine nitration provokes inhibition of sunflower carbonic anhydrase (beta-CA) activity under high temperature stress. *Nitric Oxide: Biology and Chemistry/Official Journal of the Nitric Oxide Society, 29*, 30–33. http://dx.doi.org/10.1016/j.niox.2012.12.003.
Chaki, M., Kovacs, I., Spannagl, M., & Lindermayr, C. (2014). Computational prediction of candidate proteins for s-nitrosylation in *Arabidopsis thaliana*. *PLoS One, 9*(10). http://dx.doi.org/10.1371/journal.pone.0110232.
Chaki, M., Shekariesfahlan, A., Ageeva, A., Mengel, A., von Toerne, C., Durner, J., et al. (2015). Identification of nuclear target proteins for S-nitrosylation in pathogen-treated *Arabidopsis thaliana* cell cultures. *Plant Science, 238*, 115–126. http://dx.doi.org/10.1016/j.plantsci.2015.06.011.
Chen, J., Vandelle, E., Bellin, D., & Delledonne, M. (2014). Detection and function of nitric oxide during the hypersensitive response in *Arabidopsis thaliana*: where there's a will there's a way. *Nitric Oxide-Biology and Chemistry, 43*, 81–88. http://dx.doi.org/10.1016/j.niox.2014.06.008.
Choi, J. S., Islam, M. N., Ali, M. Y., Kim, Y. M., Park, H. J., Sohn, H. S., et al. (2014). The effects of C-glycosylation of luteolin on its antioxidant, anti-Alzheimer's disease, anti-diabetic, and anti-inflammatory activities. *Archives of Pharmacal Research, 37*(10), 1354–1363. http://dx.doi.org/10.1007/s12272-014-0351-3.
Coll, N. S., Epple, P., & Dangl, J. L. (2011). Programmed cell death in the plant immune system. *Cell Death and Differentiation, 18*(8), 1247–1256. http://dx.doi.org/10.1038/cdd.2011.37.
Corpas, F. J., Chaki, M., Fernandez-Ocana, A., Valderrama, R., Palma, J. M., Carreras, A., et al. (2008). Metabolism of reactive nitrogen species in pea plants under abiotic stress conditions. *Plant and Cell Physiology, 49*(11), 1711–1722. http://dx.doi.org/10.1093/pcp/pcn144.
Corpas, F. J., Palma, J. M., del Rio, L. A., & Barroso, J. B. (2009). Evidence supporting the existence of L-arginine-dependent nitric oxide synthase activity in plants. *New Phytologist, 184*(1), 9–14. http://dx.doi.org/10.1111/j.1469-8137.2009.02989.x. NPH2989 [pii].
Correa-Aragunde, N., Foresi, N., & Lamattina, L. (July 4, 2013). Structure diversity of nitric oxide synthases (NOS): the emergence of new forms in photosynthetic organisms. *Frontiers in Plant Science, 4*, 232. http://dx.doi.org/10.3389/fpls.2013.00232.

Courtois, C., Besson, A., Dahan, J., Bourque, S., Dobrowolska, G., Pugin, A., et al. (2008). Nitric oxide signalling in plants: interplays with Ca^{2+} and protein kinases. *Journal of Experimental Botany, 59*(2), 155–163. http://dx.doi.org/10.1093/Jxb/Erm197.

Deichsel, A., Mouysset, J., & Hoppe, T. (2009). The ubiquitin-selective chaperone CDC-48/p97, a new player in DNA replication. *Cell Cycle, 8*(2), 185–190.

Delledonne, M., Xia, Y., Dixon, R. A., & Lamb, C. (1998). Nitric oxide functions as a signal in plant disease resistance. *Nature London, 394*(6693), 585–588.

Delledonne, M., Zeier, J., Marocco, A., & Lamb, C. (2001). Signal interactions between nitric oxide and reactive oxygen intermediates in the plant hypersensitive disease resistance response. *Proceedings of the National Academy of Sciences of the United States of America, 98*(23), 13454–13459.

Denninger, J. W., & Marletta, M. A. (1999). Guanylate cyclase and the NO/cGMP signaling pathway. *Biochimica et Biophysica Acta-Bioenergetics, 1411*(2–3), 334–350.

Despres, C., Chubak, C., Rochon, A., Clark, R., Bethune, T., Desveaux, D., et al. (2003). The *Arabidopsis* NPR1 disease resistance protein is a novel cofactor that confers redox regulation of DNA binding activity to the basic domain/leucine zipper transcription factor TGA1. *Plant Cell, 15*(9), 2181–2191.

Durner, J., Wendehenne, D., & Klessig, D. F. (1998). Defense gene induction in tobacco by nitric oxide, cyclic GMP, and cyclic ADP-ribose. *Proceedings of the National Academy of Sciences of the United States of America, 95*(17), 10328–10333.

Feechan, A., Kwon, E., Yun, B.-W., Wang, Y., Pallas, J. A., & Loake, G. J. (2005). A central role for S-nitrosothiols in plant disease resistance. *Proceedings of the National Academy of Science of the United States of America, 102*(22), 8054–8059.

Foresi, N., Correa-Aragunde, N., Parisi, G., Calo, G., Salerno, G., & Lamattina, L. (2010). Characterization of a nitric oxide synthase from the plant kingdom: NO generation from the green alga *Ostreococcus tauri* is light irradiance and growth phase dependent. *Plant Cell, 22*(11), 3816–3830. http://dx.doi.org/10.1105/tpc.109.073510.

Foster, M. W., Forrester, M. T., & Stamler, J. S. (2009). A protein microarray-based analysis of S-nitrosylation. *Proceedings of the National Academy of Science of the United States of America, 106*(45), 18948–18953. http://dx.doi.org/10.1073/pnas.0900729106.

Frungillo, L., Skelly, M. J., Loake, G. J., Spoel, S. H., & Salgado, I. (2014). S-nitrosothiols regulate nitric oxide production and storage in plants through the nitrogen assimilation pathway. *Nature Communications, 5*. http://dx.doi.org/10.1038/ncomms6401.

Frungillo, L., Spoel, S. H., & Salgado, I. (2016). Control of nitrogen assimilation in plants through *S*-nitrosothiols. *Nitric Oxide and Signaling in Plants,* 77, 55–78.

Fu, Z. Q., Yan, S., Saleh, A., Wang, W., Ruble, J., Oka, N., et al. (2012). NPR3 and NPR4 are receptors for the immune signal salicylic acid in plants. *Nature, 486*(7402), 228–232. http://dx.doi.org/10.1038/nature11162.

Gao, Q. M., Yu, K. S., Xia, Y., Shine, M. B., Wang, C. X., Navarre, D., et al. (2014). Mono- and digalactosyldiacylglycerol lipids function nonredundantly to regulate systemic acquired resistance in plants. *Cell Reports, 9*(5), 1681–1691. http://dx.doi.org/10.1016/j.celrep.2014.10.069.

Gaupels, F., Spiazzi-Vandelle, E., Yang, D., & Delledonne, M. (2011). Detection of peroxynitrite accumulation in *Arabidopsis thaliana* during the hypersensitive defense response. *Nitric Oxide, 25*(2), 222–228.

Gibbs, D. J., Md Isa, N., Movahedi, M., Lozano-Juste, J., Mendiondo, G. M., Berckhan, S., et al. (2014). Nitric oxide sensing in plants is mediated by proteolytic control of group VII ERF transcription factors. *Molecular Cell, 53*(3), 369–379. http://dx.doi.org/10.1016/j.molcel.2013.12.020.

Gohre, V., & Robatzek, S. (2008). Breaking the barriers: microbial effector molecules subvert plant immunity. *Annual Review of Phytopathology, 52*(46), 189–215. http://dx.doi.org/10.1146/annurev.phyto.46.120407.110050.

Gross, F., Durner, J., & Gaupels, F. (2013). Nitric oxide, antioxidants and prooxidants in plant defence responses. *Frontiers in Plant Science, 4*. http://dx.doi.org/10.3389/Fpls.2013.00419.
Grun, S., Lindermayr, C., Sell, S., & Durner, J. (2006). Nitric oxide and gene regulation in plants. *Journal of Experimental Botany, 57*(3), 507–516. http://dx.doi.org/10.1093/jxb/erj053. erj053 [pii].
Guo, F. Q., Okamoto, M., & Crawford, N. M. (2003). Identification of a plant nitric oxide synthase gene involved in hormonal signaling. *Science, 302*(5642), 100–103. http://dx.doi.org/10.1126/science.1086770302/5642/100.
Gupta, K. J., Brotman, Y., Segu, S., Zeier, T., Zeier, J., Persijn, S. T., et al. (2013). The form of nitrogen nutrition affects resistance against *Pseudomonas syringae* pv. phaseolicola in tobacco. *Journal of Experimental Botany, 64*(2), 553–568. http://dx.doi.org/10.1093/jxb/ers348.
Gupta, K. J., Fernie, A. R., Kaiser, W. M., & van Dongen, J. T. (2011). On the origins of nitric oxide. *Trends in Plant Science, 16*(3), 160–168. http://dx.doi.org/10.1016/j.tplants.2010.11.007. S1360-1385(10)00253-0 [pii].
Hara, M. R., Agrawal, N., Kim, S. F., Cascio, M. B., Fujimuro, M., Ozeki, Y., et al. (2005). S-nitrosylated GAPDH initiates apoptotic cell death by nuclear translocation following Siah1 binding. *Nature Cell Biology, 7*(7), 665–674. http://dx.doi.org/10.1038/ncb1268.
Hebelstrup, K. H., Igamberdiev, A. U., & Hill, R. D. (2007). Metabolic effects of hemoglobin gene expression in plants. *Gene, 398*(1–2), 86–93. http://dx.doi.org/10.1016/j.gene.2007.01.039.
Hoang, C. V., & Chapman, K. D. (2002). Biochemical and molecular inhibition of plastidial carbonic anhydrase reduces the incorporation of acetate into lipids in cotton embryos and tobacco cell suspensions and leaves. *Plant Physiology, 128*(4), 1417–1427. http://dx.doi.org/10.1104/pp.010879.
Holtgrefe, S., Gohlke, J., Starmann, J., Druce, S., Klocke, S., Altmann, B., et al. (2008). Regulation of plant cytosolic glyceraldehyde 3-phosphate dehydrogenase isoforms by thiol modifications. *Physiologia Plantarum, 133*(2), 211–228. http://dx.doi.org/10.1111/j.1399-3054.2008.01066.x.
Holzmeister, C., Frohlich, A., Sarioglu, H., Bauer, N., Durner, J., & Lindermayr, C. (2011). Proteomic analysis of defense response of wildtype *Arabidopsis thaliana* and plants with impaired NO- homeostasis. *Proteomics, 11*(9), 1664–1683. http://dx.doi.org/10.1002/pmic.201000652.
Holzmeister, C., Gaupels, F., Geerlof, A., Sarioglu, H., Sattler, M., Durner, J., et al. (2015). Differential inhibition of *Arabidopsis* superoxide dismutases by peroxynitrite-mediated tyrosine nitration. *Journal of Experimental Botany, 66*(3), 989–999. http://dx.doi.org/10.1093/jxb/eru458.
Huang, X., Kiefer, E., von Rad, U., Ernst, D., Foissner, I., & Durner, J. (2002). Nitric oxide burst and nitric oxide-dependent gene induction in plants. *Trends in Plant Science, 40*, 625–631.
Isner, J.-C., & Maathuis, F. J. M. (2011). Measurement of cellular cGMP in plant cells and tissues using the endogenous fluorescent reporter FlincG. *Plant Journal, 65*(2), 329–334. http://dx.doi.org/10.1111/j.1365-313X.2010.04418.x.
Jeandroz, S., Lamotte, O., Astier, J., Rasul, S., Trapet, P., Besson-Bard, A., et al. (2013). There's more to the picture than meets the eye: nitric oxide cross talk with Ca^{2+} signaling. *Plant Physiology, 163*(2), 459–470. http://dx.doi.org/10.1104/pp.113.220624.
Jones, J. D. G., & Dangl, J. L. (2006). The plant immune system. *Nature, 444*(7117), 323–329. http://dx.doi.org/10.1038/Nature05286.
Joudoi, T., Shichiri, Y., Kamizono, N., Akaike, T., Sawa, T., Yoshitake, J., et al. (2013). Nitrated cyclic GMP modulates guard cell signaling in *Arabidopsis*. *Plant Cell, 25*(2), 558–571. http://dx.doi.org/10.1105/tpc.112.105049.
Kachroo, P., Shanklin, J., Shah, J., Whittle, E. J., & Klessig, D. F. (2001). A fatty acid desaturase modulates the activation of defense signaling pathways in plants. *Proceedings of the*

National Academy of Science of the United States of America, 98(16), 9448–9453. http://dx.doi.org/10.1073/pnas.151258398.

Kneeshaw, S., Gelineau, S., Tada, Y., Loake, G. J., & Spoel, S. H. (2014). Selective protein denitrosylation activity of thioredoxin-h5 modulates plant immunity. *Molecular Cell, 56*(1), 153–162. http://dx.doi.org/10.1016/j.molcel.2014.08.003.

Kovacs, I., Ageeva, A., König, E. E., & Lindermayr, C. (2016). S-Nitrosylation of nuclear proteins: New pathways in regulation of gene expression. *Nitric Oxide and Signaling in Plants, 77*, 15–39.

Kovacs, I., Durner, J., & Lindermayr, C. (2015). Crosstalk between nitric oxide and glutathione is required for NONEXPRESSOR OF PATHOGENESIS-RELATED GENES 1 (NPR1)-dependent defense signaling in *Arabidopsis thaliana*. *New Phytologist*. http://dx.doi.org/10.1111/nph.13502.

Kovacs, I., & Lindermayr, C. (2013). Nitric oxide-based protein modification: formation and site-specificity of protein S-nitrosylation. *Frontiers in Plant Science, 4*. http://dx.doi.org/10.3389/fpls.2013.00137.

Kulik, A., Noirot, E., Grandperret, V., Bourque, S., Fromentin, J., Salloignon, P., et al. (2015). Interplays between nitric oxide and reactive oxygen species in cryptogein signalling. *Plant Cell and Environment, 38*(2), 331–348. http://dx.doi.org/10.1111/pce.12295.

Kwon, E., Feechan, A., Yun, B.-W., Hwang, B.-H., Pallas, J. A., Kang, J.-G., et al. (2012). AtGSNOR1 function is required for multiple developmental programs in *Arabidopsis*. *Planta, 236*(3), 887–900. http://dx.doi.org/10.1007/s00425-012-1697-8.

Leitner, M., Vandelle, E., Gaupels, F., Bellin, D., & Delledonne, M. (2009). NO signals in the haze nitric oxide signalling in plant defence. *Current Opinion in Plant Biology, 12*(4), 451–458. http://dx.doi.org/10.1016/j.pbi.2009.05.012.

Lin, A., Wang, Y., Tang, J., Xue, P., Li, C., Liu, L., et al. (2012). Nitric oxide and protein s-nitrosylation are integral to hydrogen peroxide-induced leaf cell death in rice. *Plant Physiology, 158*(1), 451–464. http://dx.doi.org/10.1104/pp.111.184531. pp.111.184531 [pii].

Lindermayr, C., Saalbach, G., & Durner, J. (2005). Proteomic identification of S-nitrosylated proteins in *Arabidopsis*. *Plant Physiology, 137*(3), 921–930. http://dx.doi.org/10.1104/pp.104.058719. pp.104.058719 [pii].

Lindermayr, C., Sell, S., Muller, B., Leister, D., & Durner, J. (2010). Redox regulation of the NPR1-TGA1 system of *Arabidopsis thaliana* by nitric oxide. *Plant Cell, 22*(8), 2894–2907. http://dx.doi.org/10.1105/tpc.109.066464. tpc.109.066464 [pii].

Liu, L. M., Hausladen, A., Zeng, M., Que, L., Heitman, J., & Stamler, J. S. (2001). A metabolic enzyme for S-nitrosothiol conserved from bacteria to humans. *Nature, 410*(6827), 490–494.

Liu, W.-Z., Kong, D.-D., Gu, X.-X., Gao, H.-B., Wang, J.-Z., Xia, M., et al. (2013). Cytokinins can act as suppressors of nitric oxide in *Arabidopsis*. *Proceedings of the National Academy of Sciences of the United States of America, 110*(4), 1548–1553. http://dx.doi.org/10.1073/pnas.1213235110.

Locato, V., Paradiso, A., Sabetta, W., De Gara, L., & de Pinto, M. C. (2016). Nitric oxide and reactive oxygen species in pcd signaling. *Nitric Oxide and Signaling in Plants,* 77, 165–192.

Mandal, M. K., Chandra-Shekara, A. C., Jeong, R. D., Yu, K. S., Zhu, S. F., Chanda, B., et al. (2012). Oleic acid-dependent modulation of NITRIC OXIDE ASSOCIATED1 protein levels regulates nitric oxide-mediated defense signaling in *Arabidopsis*. *Plant Cell, 24*(4), 1654–1674. http://dx.doi.org/10.1105/tpc.112.096768.

Mannick, J. B. (2007). Regulation of apoptosis by protein S-nitrosylation. *Amino Acids, 32*(4), 523–526. http://dx.doi.org/10.1007/s00726-006-0427-6.

Medina-Andres, R., Solano-Peralta, A., Pablo Saucedo-Vazquez, J., Napsucialy-Mendivil, S., Arturo Pimentel-Cabrera, J., Elena Sosa-Torres, M., et al. (2015). The nitric oxide production in the moss *Physcomitrella patens* is mediated by nitrate reductase. *PLoS One, 10*(3). http://dx.doi.org/10.1371/journal.pone.0119400.

Meier, S., Madeo, L., Ederli, L., Donaldson, L., Pasqualini, S., & Gehring, C. (2009). Deciphering cGMP signatures and cGMP-dependent pathways in plant defence. *Plant Signal Behav, 4*(4), 307–309.

Melo, P. M., Silva, L. S., Ribeiro, I., Seabra, A. R., & Carvalho, H. G. (2011). Glutamine synthetase is a molecular target of nitric oxide in root nodules of *Medicago truncatula* and is regulated by tyrosine nitration. *Plant Physiology, 157*(3), 1505–1517. http://dx.doi.org/10.1104/pp.111.186056. pp.111.186056 [pii].

Mengel, A., Chaki, M., Shekariesfahlan, A., & Lindermayr, C. (2013). Effect of nitric oxide on gene transcription – S-nitrosylation of nuclear proteins. *Frontiers in Plant Science, 4*. http://dx.doi.org/10.3389/fpls.2013.00293.

Mittler, R., Vanderauwera, S., Suzuki, N., Miller, G., Tognetti, V. B., Vandepoele, K., et al. (2011). ROS signaling: the new wave? *Trends in Plant Science, 16*(6), 300–309. http://dx.doi.org/10.1016/j.tplants.2011.03.007.

Modolo, L. V., Augusto, O., Almeida, I. M. G., Magalhaes, J. R., & Salgado, I. (2005). Nitrite as the major source of nitric oxide production by *Arabidopsis thaliana* in response to *Pseudomonas syringae*. *FEBS Letters, 579*(17), 3814–3820. http://dx.doi.org/10.1016/j.febslet.2005.05.078.

Modolo, L. V., Augusto, O., Almeida, I. M. G., Pinto-Maglio, C. A. F., Oliveira, H. C., Seligman, K., et al. (2006). Decreased arginine and nitrite levels in nitrate reductase-deficient *Arabidopsis thaliana* plants impair nitric oxide synthesis and the hypersensitive response to *Pseudomonas syringae*. *Plant Science, 171*(1), 34–40.

Moreau, M., Lee, G. I., Wang, Y., Crane, B. R., & Klessig, D. F. (2008). AtNOS/A1 is a functional *Arabidopsis thaliana* cGTPase and not a nitric oxide synthase. *Journal of Biological Chemistry*. http://dx.doi.org/10.1074/jbc.M804838200.

Moreau, M., Lindermayr, C., Durner, J., & Klessig, D. F. (2010). NO synthesis and signaling in plants—where do we stand? *Physiologia Plantarum, 138*(4), 372–383. http://dx.doi.org/10.1111/j.1399-3054.2009.01308.x.

Mou, Z., Fan, W., & Dong, X. (2003). Inducers of plant systemic acquired resistance regulate NPR1 function through redox changes. *Cell, 113*(7), 935–944. S009286740300429X [pii].

Mur, L. A. J., Kenton, P., Lloyd, A. J., Ougham, H., & Prats, E. (2008). The hypersensitive response; the centenary is upon us but how much do we know? *Journal of Experimental Botany, 59*(3), 501–520.

Mur, L. A. J., Sivakumaran, A., Mandon, J., Cristescu, S. M., Harren, F. J. M., & Hebelstrup, K. H. (2012). Haemoglobin modulates salicylate and jasmonate/ethylene-mediated resistance mechanisms against pathogens. *Journal of Experimental Botany, 63*(12), 4375–4387. http://dx.doi.org/10.1093/jxb/ers116.

Oliveira, H. C., Justino, G. C., Sodek, L., & Salgado, I. (2009). Amino acid recovery does not prevent susceptibility to *Pseudomonas syringae* in nitrate reductase double-deficient *Arabidopsis thaliana* plants. *Plant Science, 176*(1), 105–111.

Ortega-Galisteo, A. P., Rodríguez-Serrano, M., Pazmiño, D. M., Gupta, D. K., Sandalio, L. M., & Romero-Puertas, M. C. (2012). S-nitrosylated proteins in pea (*Pisum sativum* L.) leaf peroxisomes: changes under abiotic stress. *Journal of Experimental Botany*. http://dx.doi.org/10.1093/jxb/err414.

Pajerowska-Mukhtar, K. M., Emerine, D. K., & Mukhtar, M. S. (2013). Tell me more: roles of NPRs in plant immunity. *Trends in Plant Science, 18*(7), 402–411. http://dx.doi.org/10.1016/j.tplants.2013.04.004.

Parani, M., Rudrabhatla, S., Myers, R., Weirich, H., Smith, B., Leaman, D. W., et al. (2004). Microarray analysis of nitric oxide responsive transcripts in *Arabidopsis*. *Plant Biotechnology Journal, 2*(4), 359–366.

Perazzolli, M., Dominici, P., Romero-Puertas, M. C., Zago, E., Zeier, J., Sonoda, M., et al. (2004). *Arabidopsis* nonsymbiotic hemoglobin AHb1 modulates nitric oxide bioactivity. *Plant Cell, 16*(10), 2785–2794.

de Pinto, M. C., Locato, V., Sgobba, A., del Carmen Romero-Puertas, M., Gadaleta, C., Delledonne, M., et al. (2013). S-nitrosylation of ascorbate peroxidase is part of programmed cell death signaling in tobacco bright yellow-2 cells. *Plant Physiology, 163*(4), 1766–1775. http://dx.doi.org/10.1104/pp.113.222703.

de Pinto, M. C., Tommasi, F., & De Gara, L. (2002). Changes in the antioxidant systems as part of the signaling pathway responsible for the programmed cell death activated by nitric oxide and reactive oxygen species in tobacco bright-yellow 2 cells. *Plant Physiology, 130*(2), 698–708.

Planchet, E., Gupta, K. J., Sonoda, M., & Kaiser, W. M. (2005). Nitric oxide (NO) emission from tobacco leaves and cell suspensions: rate limiting factors and evidence for the involvement of mitochondrial electron transport. *Plant Journal, 41*, 732–743.

Polverari, A., Molesini, B., Pezzotti, M., Buonaurio, R., Marte, M., & Delledonne, M. (2003). Nitric oxide-mediated transcriptional changes in *Arabidopsis thaliana*. *Molecular Plant–Microbe Interactions, 16*(12), 1094–1105. http://dx.doi.org/10.1094/MPMI.2003.16.12.1094.

Radi, R. (2004). Nitric oxide, oxidants, and protein tyrosine nitration. *Proceedings of the National Academy of Science of the United States of America, 101*(12), 4003–4008. http://dx.doi.org/10.1073/pnas.03074461010307446101.

Rasul, S., Dubreuil-Maurizi, C., Lamotte, O., Koen, E., Poinssot, B., Alcaraz, G., et al. (2012). Nitric oxide production mediates oligogalacturonide-triggered immunity and resistance to *Botrytis cinerea* in *Arabidopsis thaliana*. *Plant Cell and Environment, 35*(8), 1483–1499. http://dx.doi.org/10.1111/j.1365-3040.2012.02505.x.

Rockel, P., Strube, F., Rockel, A., Wildt, J., & Kaiser, W. M. (2002). Regulation of nitric oxide (NO) production by plant nitrate reductase *in vivo* and *in vitro*. *Journal of Experimental Botany, 53*, 103–110.

Romero-Puertas, M. C., Campostrini, N., Matte, A., Righetti, P. G., Perazzolli, M., Zolla, L., et al. (2008). Proteomic analysis of S-nitrosylated proteins in *Arabidopsis thaliana* undergoing hypersensitive response. *Proteomics, 8*(7), 1459–1469. http://dx.doi.org/10.1002/pmic.200700536.

Romero-Puertas, M. C., Laxa, M., Matte, A., Zaninotto, F., Finkemeier, I., Jones, A. M., et al. (2007). S-nitrosylation of peroxiredoxin II E promotes peroxynitrite-mediated tyrosine nitration. *Plant Cell, 19*(12), 4120–4130. http://dx.doi.org/10.1105/tpc.107.055061. tpc.107.055061 [pii].

Rumer, S., Gupta, K. J., & Kaiser, W. M. (2009). Plant cells oxidize hydroxylamines to NO. *Journal of Experimental Botany, 60*(7), 2065–2072. http://dx.doi.org/10.1093/jxb/erp077.

Rusterucci, C., Espunya, M. C., Diaz, M., Chabannes, M., & Martinez, M. C. (2007). S-nitrosoglutathione reductase affords protection against pathogens in *Arabidopsis*, both locally and systemically. *Plant Physiology, 143*(3), 1282–1292. http://dx.doi.org/10.1104/pp.106.091686.

Slaymaker, D. H., Navarre, D. A., Clark, D., del Pozo, O., Martin, G. B., & Klessig, D. F. (2002). The tobacco salicylic acid-binding protein 3 (SABP3) is the chloroplast carbonic anhydrase, which exhibits antioxidant activity and plays a role in the hypersensitive defense response. *Proceedings of the National Academy of Sciences of the United States of America, 99*(18), 11640–11645. http://dx.doi.org/10.1073/pnas.182427699.

Spadaro, D., Yun, B. W., Spoel, S. H., Chu, C., Wang, Y. Q., & Loake, G. J. (2010). The redox switch: dynamic regulation of protein function by cysteine modifications. *Physiologia Plantarum, 138*(4), 360–371. http://dx.doi.org/10.1111/j.1399-3054.2009.01307.x. PPL1307 [pii].

Spoel, S. H., & Loake, G. J. (2011). Redox-based protein modifications: the missing link in plant immune signalling. *Current Opinion in Plant Biology, 14*(4), 358–364. http://dx.doi.org/10.1016/j.pbi.2011.03.007.

Srivastava, N., Gonugunta, V., Puli, M., & Raghavendra, A. (2009). Nitric oxide production occurs downstream of reactive oxygen species in guard cells during stomatal closure induced by chitosan in abaxial epidermis of *Pisum sativum*. *Planta, 229*(4), 757–765.

Stohr, C., Strube, F., Marx, G., Ullrich, W. R., & Rockel, P. (2001). A plasma membrane-bound enzyme of tobacco roots catalyses the formation of nitric oxide from nitrite. *Planta, 212*(5–6), 835–841.

Sturms, R., DiSpirito, A. A., & Hargrove, M. S. (2011). Plant and cyanobacterial hemoglobins reduce nitrite to nitric oxide under anoxic conditions. *Biochemistry, 50*(19), 3873–3878. http://dx.doi.org/10.1021/bi2004312.

Tada, Y., Mori, T., Shinogi, T., Yao, N., Takahashi, S., Betsuyaku, S., et al. (2004). Nitric oxide and reactive oxygen species do not elicit hypersensitive cell death but induce apoptosis in the adjacent cells during the defense response of oat. *Molecular Plant–Microbe Interactions, 17*(3), 245–253.

Tada, Y., Spoel, S. H., Pajerowska-Mukhtar, K., Mou, Z., Song, J., Wang, C., et al. (2008). Plant immunity requires conformational changes [corrected] of NPR1 via S-nitrosylation and thioredoxins. *Science, 321*(5891), 952–956. http://dx.doi.org/10.1126/science.1156970, 1156970 [pii].

Tavares, C. P., Vernal, J., Delena, R. A., Lamattina, L., Cassia, R., & Terenzi, H. (2014). S-nitrosylation influences the structure and DNA binding activity of AtMYB30 transcription factor from *Arabidopsis thaliana*. *Biochimica et Biophysica Acta-Proteins and Proteomics, 1844*(4), 810–817. http://dx.doi.org/10.1016/j.bbapap.2014.02.015.

Torres, M. A., Dangl, J. L., & Jones, J. D. (2002). *Arabidopsis* gp91phox homologues *AtrbohD* and *AtrbohF* are required for accumulation of reactive oxygen intermediates in the plant defense response. *Proceedings of the National Academy of Science of the United States of America, 99*(1), 517–522.

Trapet, P., Kulik, A., Lamotte, O., Jeandroz, S., Bourque, S., Nicolas-Francès, V., et al. (2015). NO signaling in plant immunity: a tale of messengers. *Phytochemistry, 112*(0), 72–79. http://dx.doi.org/10.1016/j.phytochem.2014.03.015.

Tristan, C., Shahani, N., Sedlak, T. W., & Sawa, A. (2011). The diverse functions of GAPDH: views from different subcellular compartments. *Cell Signal, 23*(2), 317–323. http://dx.doi.org/10.1016/j.cellsig.2010.08.003.

Tun, N. N., Santa-Catarina, C., Begum, T., Silveira, V., Handro, W., Floh, E. I. S., et al. (2006). Polyamines induce rapid biosynthesis of nitric oxide (NO) in *Arabidopsis thaliana* seedlings. *Plant Cell Physiology, 47*(3), 346–354.

Vandelle, E., & Delledonne, M. (2011). Peroxynitrite formation and function in plants. *Plant Science, 181*(5), 534–539. http://dx.doi.org/10.1016/j.plantsci.2011.05.002. S0168-9452(11)00132-4 [pii].

Vannini, C., Marsoni, M., Cantara, C., De Pinto, M. C., Locato, V., De Gara, L., et al. (2012). The soluble proteome of tobacco bright yellow-2 cells undergoing H(2)O(2)-induced programmed cell death. *Journal of Experimental Botany, 63*(8), 3137–3155. http://dx.doi.org/10.1093/jxb/ers031.

Vescovi, M., Zaffagnini, M., Festa, M., Trost, P., Lo Schiavo, F., & Costa, A. (2013). Nuclear accumulation of cytosolic glyceraldehyde-3-phosphate dehydrogenase in cadmium-stressed *Arabidopsis* roots. *Plant Physiology, 162*(1), 333–346. http://dx.doi.org/10.1104/pp.113.215194.

Wang, Y. Q., Feechan, A., Yun, B. W., Shafiei, R., Hofmann, A., Taylor, P., et al. (2009). S-nitrosylation of AtSABP3 antagonizes the expression of plant immunity. *Journal of Biological Chemistry, 284*(4), 2131–2137. http://dx.doi.org/10.1074/jbc.M806782200. M806782200 [pii].

Wang, Y., Ries, A., Wu, K., Yang, A., & Crawford, N. M. (2010). The *Arabidopsis* prohibitin gene PHB3 functions in nitric oxide-mediated responses and in hydrogen peroxide-induced nitric oxide accumulation. *Plant Cell*. http://dx.doi.org/10.1105/tpc.109.072066.

Wilkinson, J. Q., & Crawford, N. M. (1993). Identification and characterization of a chlorate-resistant mutant of *Arabidopsis thaliana* with mutations in both nitrate reductase structural genes NIA1 and NIA2. *Molecular and General Genetics, 239*(1–2), 289–297.

Wimalasekera, R., Schaarschmidt, F., Angelini, R., Cona, A., Tavladoraki, P., & Scherer, G. F. (2015). POLYAMINE OXIDASE2 of *Arabidopsis* contributes to ABA mediated plant developmental processes. *Plant Physiology and Biochemistry, 96,* 231–240. http://dx.doi.org/10.1016/j.plaphy.2015.08.003.

Wimalasekera, R., Villar, C., Begum, T., & Scherer, G. F. (2011). COPPER AMINE OXIDASE1 (CuAO1) of *Arabidopsis thaliana* contributes to abscisic acid- and polyamine-induced nitric oxide biosynthesis and abscisic acid signal transduction. *Molecular Plant, 4*(4), 663–678. http://dx.doi.org/10.1093/mp/ssr023. ssr023 [pii].

Wu, Y., Zhang, D., Chu, J. Y., Boyle, P., Wang, Y., Brindle, I. D., et al. (2012). The *Arabidopsis* NPR1 protein is a receptor for the plant defense hormone salicylic acid. *Cell Reports, 1*(6), 639–647. http://dx.doi.org/10.1016/j.celrep.2012.05.008.

Xu, S., Guerra, D., Lee, U., & Vierling, E. (2013). S-nitrosoglutathione reductases are low-copy number, cysteine-rich proteins in plants that control multiple developmental and defense responses in *Arabidopsis*. *Frontiers in Plant Science, 4*. http://dx.doi.org/10.3389/fpls.2013.00430.

Yamasaki, H., Sakihama, Y., & Takahashi, S. (1999). An alternative pathway for nitric oxide production in plants: new features of an old enzyme. *Trends in Plant Science, 4*(4), 128–129.

Yang, H., Mu, J., Chen, L., Feng, J., Hu, J., Li, L., et al. (2015). S-Nitrosylation positively regulates ascorbate peroxidase activity during plant stress responses. *Plant Physiology, 167*(4), 1604–1615. http://dx.doi.org/10.1104/pp.114.255216.

Yoshioka, H., Mase, K., Yoshioka, M., Kobayashi, M., & Asai, S. (2011). Regulatory mechanisms of nitric oxide and reactive oxygen species generation and their role in plant immunity. *Nitric Oxide-Biology and Chemistry, 25*(2), 216–221. http://dx.doi.org/10.1016/j.niox.2010.12.008.

Yu, M., Lamattina, L., Spoel, S. H., & Loake, G. J. (2014). Nitric oxide function in plant biology: a redox cue in deconvolution. *New Phytologist, 202*(4), 1142–1156. http://dx.doi.org/10.1111/nph.12739.

Yu, M., Yun, B.-W., Spoel, S. H., & Loake, G. J. (2012). A sleigh ride through the SNO: regulation of plant immune function by protein S-nitrosylation. *Current Opinion in Plant Biology, 15*(4), 424–430. http://dx.doi.org/10.1016/j.pbi.2012.03.005.

Yun, B. W., Feechan, A., Yin, M., Saidi, N. B., Le Bihan, T., Yu, M., et al. (2011). S-nitrosylation of NADPH oxidase regulates cell death in plant immunity. *Nature, 478*(7368), 264–268. http://dx.doi.org/10.1038/nature10427. nature10427 [pii].

Zago, E., Morsa, S., Dat, J. F., Alard, P., Ferrarini, A., Inze, D., et al. (2006). Nitric oxide- and hydrogen peroxide-responsive gene regulation during cell death induction in tobacco. *Plant Physiology, 141*(2), 404–411. http://dx.doi.org/10.1104/pp.106.078444. pp.106.078444 [pii].

Zemojtel, T., Fröhlich, A., Palmieri, M. C., Kolanczyk, M., Mikula, I., Wyrwicz, L. S., et al. (2006). Plant nitric oxide synthase: a never-ending story? *Trends in Plant Science, 11*(11), 524–525.

CHAPTER TWELVE

Nitric Oxide-Mediated Chemical Signaling during Systemic Acquired Resistance

Pradeep Kachroo[1], Gah-Hyun Lim, Aardra Kachroo
Department of Plant Pathology, University of Kentucky, Lexington, KY, USA
[1]Corresponding author: Email: pk62@uky.edu

Contents

Abstract

Reactive nitrogen (nitric oxide, NO) and oxygen species are important free radicals that play a vital role in a number of physiological responses, including plant defence. More recently they have been shown to play an important role in systemic acquired resistance (SAR), which is induced in response to primary infection and confers broad-spectrum disease resistance to secondary infections. NO functions upstream of reactive oxygen species (ROS) and other SAR signals (azelaic acid (AzA) and glycerol-3-phosphate (G3P)) and is regulated by galactolipid digalactosyl diacylglycerol as well as fatty acid (oleic acid). Oleic acid and other C18 unsaturated fatty acids containing a double bond at carbon 9 also serve as a precursor to AzA, which functions upstream of G3P. The NO—ROS branch of SAR pathway functions in parallel to salicylic acid, and these branches likely undergo crosstalk to regulate the optimal induction of signals. This review summarizes the recent advances in the role of free radicals in SAR and discusses its relationship to other SAR inducers.

Plants are constantly challenged by microbial pathogens, and this is further compounded by their static nature and the absence of an active circulatory system. However, plants utilize several novel mechanisms to defend themselves from pathogen infections. These include physical barriers like the cuticle and cell wall, which can also perform a signaling role (Xia et al., 2009, 2010, 2012), and induction of defence against pathogens, which in turn is associated with accumulation of antimicrobial compounds. The typical host defence responses induced against pathogens include nonhost

Advances in Botanical Research, Volume 77
ISSN 0065-2296
http://dx.doi.org/10.1016/bs.abr.2015.10.010

resistance, PTI (pathogen-associated molecular patterns (PAMP)-triggered immunity), and ETI (effector-triggered immunity). PTI is induced upon recognition of pathogen elicitors by extracellular pattern-recognition receptors in the plant (Shabab et al., 2008; Shan et al., 2008; Shang et al., 2006; Thordal-Christensen, 2003; Zhou & Chai, 2008; Zipfel & Rathjen, 2008). In comparison, ETI is induced in response to specialized pathogen elicitors called avirulence (avr) factors, which are recognized by the host encoded resistance (R) proteins. ETI is highly species-specific and results in a much stronger and durable resistance that in general can prevent the spread of the pathogen (Flor, 1971; Jones & Dangl, 2006).

Induction of local defences is associated with elicitation of systemic responses that confers broad-spectrum resistance throughout the plant. Induced systemic response (ISR) is triggered in response to root colonization by beneficial bacteria (Van Wees, Van der Ent, & Pieterse, 2008), whereas systemic acquired resistance (SAR) is activated in response to pathogen infection (Gao, Kachroo, & Kachroo, 2014; Kachroo & Robin, 2013; Shah & Zeier, 2013; Spoel & Dong, 2012; Wendehenne, Gao, Kachroo, & Kachroo, 2014). The signaling pathways required for ISR and SAR appear to be distinct except for a common requirement of NPR1 (non-expressor for PR-1). NPR1 is a positive regulator of salicylic acid(SA)-signaling pathway but is required downstream of ethylene response in the ISR pathway. Thus, the requirement for NPR1 in SA-mediated signaling leading to *PR-1* expression and SAR does not overlap with the role of NPR1 in ISR. This is further supported by the observation that SAR, but not ISR, is associated with changes in gene expression in the distal tissues (Pieterse et al., 2000; Verhagen et al., 2004). However, ISR does augment changes in gene expression profiles after secondary infections, suggesting that it primes the distal tissues against secondary infection. Likewise, induction of SAR has been associated with priming or a faster response to secondary infections (Jaskiewicz, Conrath, & Peterhänsel, 2011; Pieterse, Poelman, van Wees, & Dicke, 2013). SAR can be transferred to the immediate next generation of progeny via modifications in the chromatin structure (Luna, Bruce, Roberts, Flors, & Ton, 2012; Slaughter et al., 2012), suggesting the heritable nature of SAR signaling.

1. SALICYLIC ACID METABOLISM IN RELATION TO SAR

SA is one of the critical components of SAR (Figure 1). The biosynthesis of SA occurs via the shikimic acid pathway, in which chorismic acid is

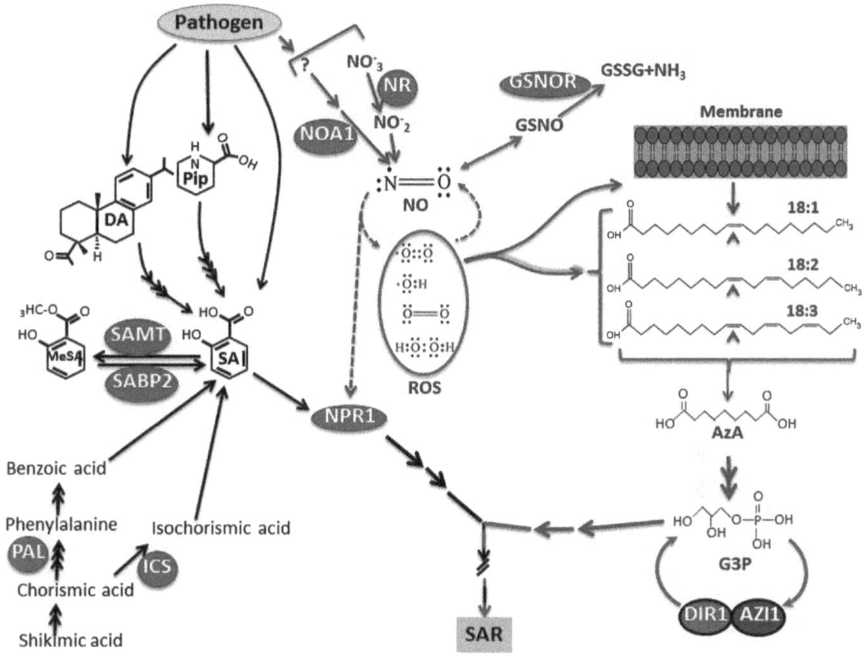

Figure 1 A simplified scheme showing components of the SAR signaling pathway. Inoculation of avirulent pathogen leads to accumulation of salicylic acid (SA) and nitric oxide (NO). NO acts upstream of reactive oxygen species (ROS), which include the superoxide radical, singlet oxygen, hydroxyl radical and hydrogen peroxide. These acts in an additive manner to catalyse oxidation of C18 unsaturated fatty acids (FA) that contain double bond on carbon 9 (shown by arrowhead). NO and ROS operate in a feedback loop, since mutants defective in ROS biosynthesis do not accumulate NO after pathogen inoculation. Hydrolysis of Δ9 double bond on C18 fatty acids generates AzA, which triggers biosynthesis of G3P via upregulation of genes encoding G3P biosynthetic enzymes, glycerol kinase (GK) and G3P dehydrogenase (G3Pdh). G3P and the lipid transfer proteins DIR1 and AZI1 operate in a feedback loop. DIR1 and AZI1 interact with self and each other. Cellular NO levels are regulated via their storage as GSNO (S-nitrosoglutathione), which can be reduced to glutathione disulfide (GSSG) and NH_3 by GSNOR (S-nitrosoglutathione reductase). The SA and NO/ROS pathways crosstalk at several levels and one of these steps includes nitrosylation of NPR1, a positive regulator of the SA pathway. Critical enzymes and proteins are shown in red. NOA1, nitric oxide associated 1; NR, nitrate reductase; AzA, azelaic acid; G3P, glycerol-3-phosphate; PAL, phenylalanine ammonia lyase; ICS, isochorismate synthase; SABP2, SA binding protein 2; SAMT, SA methyltransferase; DA, dehydroabietinal; Pip, pipecolic acid; NPR1, non-expressor for PR-1. (See colour plate)

converted to SA either via phenylalanine ammonia lyase (PAL), which converts phenylalanine to cinnamic acid, and isochorismate synthase (ICS), which catalyses conversion of chorismic acid to isochorismic acid (Strawn et al., 2007; Wildermuth, Dewdney, Wu, & Ausubel, 2001). Initial evidence supporting a role for SA in SAR came from analysis of transgenic plants expressing the bacterial gene encoding salicylate hydroxylase, an enzyme that catalyses conversion of SA to catechol. More recent work has supported a role for SA in SAR by utilizing genetic mutants that are impaired in ICS or PAL activities. Intriguingly, even though ICS contributes to over 98% of pathogen-induced SA, PAL isoforms are still required for normal SAR in Arabidopsis. Since the PAL-derived branch for SA biosynthesis also contributes to the synthesis of many other metabolites including the antimicrobial phenylpropanoids, it is possible that the compromised SAR in *pal* plants is related to their deficiency in other metabolites rather than SA.

Pathogen infection results in SA accumulation in both the local (infected) and distal (uninfected) tissues, although the level of induction in the distal tissues is rather nominal in comparison to the increase in local tissues. This, together with the finding that radioisotope-labelled SA was transported from infected to distal leaves, suggests a role for transported SA in systemic signaling. Notably however, no significant differences in SA levels were observed between SAR competent wild-type and SAR-compromised SA-deficient scions, which were grafted onto pathogen-infected SA-deficient rootstocks. This suggests that increase in SA accumulation may not be a prerequisite for the normal induction of SAR (Vernooij et al., 1994). Furthermore, SA-deficient rootstocks of plants expressing SA hydroxylase or plants suppressed in *PAL* expression were capable of activating SAR in the leaves of wild-type scions. These data argue against a role for transported SA in SAR (Pallas, Paiva, Lamb, & Dixon, 1996; Vernooij et al., 1994).

Endogenous SA is metabolized to several derivatives, including the glucose conjugate SA 2-O-β-D-glucose (SAG), the SA glucose ester (SGE), the methylated ester methyl SA (MeSA) and the hydroxylated form of gentisic acid (Dean & Delaney, 2008; Enyedi, Yalpani, Silverman, & Raskin, 1992; Edwards, 1994; Lee & Raskin, 1998, 1999; Song et al., 2008). MeSA is a well-characterized volatile organic compound that is biologically inactive and functions when converted back to SA. The conversion of MeSA to SA is mediated by SABP2, an esterase with SA-binding activity. SA inhibits SABP2 resulting in the accumulation of MeSA (Kumar & Klessig, 2003). Conversion of SA to MeSA is mediated by SA methyltransferases, which can utilize either SA or benzoic acid as substrates (Chen et al.,

2003; Effmert et al., 2005; Koo et al., 2007). Following SAR induction, MeSA accumulates in the inoculated leaves, some of which translocates to the distal tissues where it is reconverted to SA via SABP2 (Park, Kaiyomo, Kumar, Mosher, & Klessig, 2007). Consequently, conversion of SA to MeSA and its reversible reaction are important for normal SAR (Liu, Dahl, & Klessig, 2011). MeSA is required in the distal tissues between 48 and 72 h after primary infection to confer SAR (Park et al., 2009). In contrast, the SAR signal(s) translocate to distal tissues within 4–6 h of primary infection (Chanda et al., 2011; Chaturvedi et al., 2012), thus suggesting that MeSA is one of the late SAR signals.

2. FREE RADICALS AND THEIR ROLE IN SAR

Pharmacological studies had previously suggested a role for the free radicals NO and ROS in SAR (Alvarez et al., 1998; Song & Goodman, 2001). Recent work conclusively demonstrated the genetic requirements for NO and ROS in SAR using mutants impaired in NO and ROS biosynthesis (Wang et al., 2014; Figure 1). Notably, the requirement for NO and ROS is concentration dependent such that either too little or too much accumulation of NO or ROS can impair SAR. This corelates with the observation that NO modulates gene expression in a concentration-dependent manner and exogenous application of 0.1 mM NO donor has different effects on Arabidopsis transcriptome than application of 1 mM NO donor (Parani et al., 2004). Likewise, localized application of higher concentrations of NO or ROS is ineffective in inducing SAR. Moreover, mutations in *GSNOR1* (*S-Nitrosoglutathione Reductase*) or *NOX1 (NO overproducing)*, which elevate NO and/or S-nitrosothiols levels constitutively, compromise SAR (Wang et al., 2014). In contrast, plants expressing antisense *GSNOR1* show enhanced SAR, likely due to a partial reduction in GSNOR1 activity, which increases NO levels to an extent that is ideal for stimulating SAR (Espunya, De Michele, Gómez-Cadenas, & Martínez, 2012; Rustérucci, Espunya, Díaz, Chabannes, & Martínez, 2007).

Efforts to identify the mammalian equivalent of NO synthase activity in plants have been unsuccessful, although NO synthase-like activity has been documented by several independent groups (Durner, Wendehenne, & Klessig, 1998; Guo, Okamoto, & Crawford, 2003). Recent efforts have focused on two groups of proteins, which directly or indirectly appear to be involved in NO biosynthesis and/or accumulation. These include the

NO associated protein 1 (AtNOA1) and the nitrate reductases (NR; NIA1 and NIA2). NOA1 encodes a GTPase and is present in a subcompartment within the chloroplasts (Gas et al., 2008; Mandal et al., 2012; Moreau, Lee, Wang, Crane, & Klessig, 2008). How NOA1-encoded GTPase functions in NO biosynthesis and/or accumulation remains unknown at present. One possibility is that the GTPase activity on NOA1 regulates the synthesis of enzyme(s) required for NO biosynthesis. Alternatively, NOA1 might serve as an important catalytic component of a larger complex that facilitates NO production in plants. In contrast to NOA1, the NIA1 and NIA2 enzymes are extraplastidal and generate NO via nitrate (Crawford, 2006; Mandal et al., 2012). Earlier work has also suggested at least two pathways for NO production in isolated soybean chloroplasts, one of which was dependent on nitrite (Jasid, Simontacchi, Bartoli, & Puntarulo, 2006). This is consistent with the observation that *noa1 nia1* or *noa1 nia2* double mutants are fully compromised in NO accumulation as well as SAR. This suggests that the NOA1 and NIA activities are partially redundant for NO generation/accumulation and that loss of either *NIA* isoform in the *noa1* background is sufficient to compromise NO levels and thereby SAR. NIA2 is thought to be the major isoform that accounts for ~90% of the total NR activity. However, the relative contribution of NIA isoforms to NO generation and NR activity appears to be variable. For instance, NIA1 is the major contributor of abscisic acid-induced NO (Neil et al., 2008) and NR activity in response to cytokinin treatment (Yu, Sukumaran, & Márton, 1998), but NIA2 is required for SA-induced NO production (Hao et al., 2010). The differential regulation of the *NIA1* and *NIA2* genes (Chen, Acedo, Dewdney, Goodman, & Conkling, 1991; Yu et al., 1998) and proteins (Park, Song, & Seo, 2011; Wang, Du, Li, Ren, & Song, 2010) suggests that these proteins likely play distinct roles in nitrogen and NO metabolism, depending upon the signaling response.

Both NO and ROS function as important signaling molecules in a number of systemic responses including systemic acquired acclimation, which is triggered in response to abiotic stresses (Mittler & Blumwald, 2015). ROS biosynthesis is catalysed by respiratory burst oxidase homologues (RBOH) that are homologues of mammalian gp91^{phox} protein, one of the several components present in the NADPH oxidase complex. RBOH are plasma membrane localizing proteins that have FAD- and NADPH-binding sites, six transmembrane domains and a cytosolic N-terminal domain containing calcium binding elongation-factor (EF)-hand motifs. Several studies have shown a role for Ca^{2+}-dependent and -independent processes in regulation of ROS and RBOH protein members (Kadota et al., 2004, 2014; Li et al., 2014; Ogaasawara et al., 2008; Segonzac et al., 2011; Zhang et al., 2009). The Arabidopsis genome encodes 10 *RBOH* but only two of these (*RBOHD* and *RBOHF*) are expressed throughout the plant and thus far known to participate in defence (Sagi & Fluhr, 2006; Torres, Jones, & Dangl, 2005; Wang et al., 2014; Wendehenne et al., 2014). Other isoforms are involved in pollen tube growth, xylem differentiation, seed ripening or root hair formation (Barcelo, 2005; Bioisson-Dernier et al., 2013; Carol et al., 2005; Foreman et al., 2003; Kaya et al., 2014; Lassig, Gutermuth, Bey, Konrad, & Romeis, 2014; Takeda et al., 2008), suggesting that RBOH isoforms function in a cell- and tissue-specific manner. Both RBOHD and RBOHF proteins catalyse the formation of superoxide radical (Torres et al., 2005; Wang et al., 2014), which has a short half-life and is converted to the more stable and membrane diffusible species H_2O_2. Interestingly, although RBOHD is thought to play a major role in plant defence (Torres et al., 2005), mutations in either RBOHD or RBOHF can compromise SAR. The RBOHF isoform is also required for lignified casparian strip formation (Lee, Rubio, Alassimone, & Geldner, 2013) and shoot Na^+ homoeostasis of Arabidopsis plants grown in saline soils (Jiang et al., 2012). These observations suggest that RBOH isoforms can have varied roles in a cell-, tissue- and organ-specific manner. A recent study has shown that parasitic nematodes stimulate RBOHD and RBOHF isoforms to produce ROS that promote infection by restricting the spread of cell death signals (Siddique et al., 2014). This further emphasizes a role for ROS levels in modulating host responses to pathogens and worms.

3. RELATIONSHIP AMONG FREE RADICALS AND OTHER SAR SIGNALS AND LIPIDS

SAR involves several other chemical signals besides SA, NO and ROS, including pipecolic acid (a nonprotein amino acid derivative of lysine, Pip; Návarová, Bernsdorff, Döring, & Zeier, 2012), dehydroabietinal (a diterpenoid, DA; Chaturvedi et al., 2012), azelaic acid (C_9 dicarboxylic acid, AzA) (Jung, Tschaplinkski, Wang, Glazebrook, & Greenberg, 2009) and glycerol-3-phosphate (phosphorylated sugar derivative, G3P) (Chanda et al., 2011; Gao, Yu et al., 2014; Mandal et al., 2011; Wang et al., 2014; Yu et al., 2013). Based on their interaction with each other, these signals can be arranged into two branches of the SAR pathway; one driven by SA and the other driven by NO and ROS. Pip and DA likely feed into the SA branch since both of these chemicals can stimulate SA biosynthesis in the absence of pathogen infection (Chaturvedi et al., 2012; Návarová et al., 2012). However, the relationship between Pip, DA and the NO/ROS driven branch has not been investigated as yet. AzA and G3P function downstream of NO/ROS where AzA induces G3P synthesis by inducing the transcription of G3P biosynthetic genes (Yu et al., 2013). The NO↔ROS→AzA→G3P pathway appears to be distinct from the SA pathway since exogenous SA cannot restore SAR in mutants defective in NO, ROS or G3P biosynthesis. Conversely, NO/ROS cannot confer SAR on mutants defective in SA synthesis or signaling (Wang et al., 2014). Consistent with these results, NO- and SA-deficient mutants accumulate normal levels of SA and NO, respectively. The parallel signaling of the NO↔ROS→AzA→G3P and SA pathways provides multiple points of coregulation, which may be essential for optimal regulatory control of such a key signaling mechanism. For instance, NO S-nitrosylates NPR1 and TGA1, two key components of SA signaling (Lindermayr, Sell, Müller, Leister, & Durner, 2010; Tada et al., 2008), while EDS1 (enhances disease susceptibility 1), another component of the SA pathway, was shown to regulate the biosynthesis of ONA (9-oxononanoic acid (see below))/AzA (Wittek et al., 2014).

NO application was also unable to confer SAR in the *rbohD* and *rbohF* mutants, suggesting that ROS acts downstream of NO. This conclusion was further supported by the fact that ROS was able to confer SAR in the NO deficient *noa1 nia2* plants. However, the *rbohD* and *rbohF* mutants showed reduced accumulation of NO, suggesting that the superoxide radical and NO operate in a feedback loop. The *noa1 nia2* and *rbohD* or *rbohF* mutants were unable to induce wild-type levels of G3P and exogenous application of G3P restored SAR in these mutants (Wang et al., 2014). This suggests that both NO and ROS act upstream of G3P in the SAR pathway.

AzA confers SAR by inducing biosynthesis of G3P accumulation (Yu et al., 2013). Consequently, plants defective in the G3P synthesizing enzymes are insensitive to AzA. G3P is made via the G3P dehydrogenase (G3Pdh)-mediated reduction of dihydroxyacetone phosphate. There are several isoforms of G3Pdh in Arabidopsis, though interestingly only one of the plastidal isoforms, encoded by the *GLY1/SFD1* gene, contributes to a subpool of G3P that is utilized in the biosynthesis of plastidal lipids (Chanda et al., 2011). Consequently, mutations in *G3Pdh* isoforms other than *GLY1* do not have any effect on the plastidal lipid profile but at least two other isoforms besides *GLY1* are required for SAR (Chanda et al., 2011). Likewise, glycerol kinase (GK) also contributes to pathogen-induced G3P levels and SAR. These observations suggest that the pathogen-induced G3P pool is derived from the combined actions of G3Pdh and GK enzymes, which are present in different cellular compartments.

AzA is derived upon the oxidative cleavage of C18 unsaturated FAs that contain a double bond at carbon 9. ROS also appear to be required for the oxidation of C18 unsaturated fatty acids and different species of ROS appear to function additively in SAR (Wang et al., 2014). Thus, oleic acid (18:1), linoleic acid (18:2) and linolenic acid (18:3), which all contain a double bond at carbon 9, can serve as precursors of 9-oxononanoic acid ONA (monocarboxylic acid), which is the immediate precursor of AzA. Exogenous application of either 18:1 or 18:2 can trigger AzA biosynthesis and confers robust SAR equivalent to that induced by AzA or G3P (Yu et al., 2013). The fatty acids that serve as precursors for AzA are in turn derived from the major membrane galactolipids, monogalactoside diacylglycerol (MGDG) and digalactoside diacylglycerol (DGDG). MGDG and DGDG are almost exclusively present on the chloroplastic membranes and constitute ~80% of total plant lipids (Gao, Yu et al., 2014; Yu et al., 2013; Zoeller et al., 2012). AzA likely complexes with other lipids since a very small portion of exogenous AzA stays in its free form (Gao, Yu et al., 2014; Yu et al., 2013; Zoeller et al., 2012).

Besides serving as a precursor for AzA, the DGDG galactolipid also functions at an upstream step in SAR where it is required for pathogen-induced SA and NO biosynthesis (Gao, Yu et al., 2014). Interestingly, petiole exudates from pathogen-infected *dgd1* plants were able to confer SAR in wild-type plants, suggesting that SAR signal(s) operating upstream of SA and NO branch point are present in the *dgd1* plants and can induce SA and NO levels in plants containing the DGDG lipid. The *dgd1* plants were unable to induce *ICS1* after pathogen infection, suggesting that SAR involves DGDG-dependent retrograde signaling between the chloroplast and nucleus (Gao, Yu et al., 2014). Furthermore, replacement of the terminal galactose sugar in DGDG with glucose was unable to restore pathogen-induced NO or SA accumulation or SAR, even though it restored the morphological and photosynthesis defects in *dgd1* plants. This highlights the importance of the terminal galactose sugar moiety in SAR signaling.

NO levels in plants are also regulated by the fatty acid 18:1 (Mandal et al., 2012), which in addition to serving as a precursor for AzA (see above), directly binds the AtNOA1 protein. 18:1 binding then results in the degradation of AtNOA1 in a protease-dependent mechanism (Mandal et al., 2012). Conversely, reduction in 18:1 levels increases NOA1 levels, by inducing NO synthesis and triggering the upregulation of NO-responsive nuclear genes, thereby activating disease resistance. Interestingly, 18:1 also inhibits NO synthase activity in humans (Davda et al., 1995), suggesting that plants and humans might use conserved mechanisms to regulate NO levels. In animals, NO-mediated nitration of C18 fatty acids can transform the fatty acids into important signaling molecules (Baker et al., 2005; Jain et al., 2008). It would be worthwhile to investigate if plants contain nitrated fatty acids and whether these participate in defence signaling.

18:1 is synthesized by the plastidal enzyme stearoyl-ACP-desaturase (SACPD; SSI2/FAB2), which regulates the ratio of saturated to monounsaturated fatty acids in plants (Kachroo & Lapchyk, 2003; Kachroo, Kachroo, Lapchyk, Hildebrand, & Klessig, 2003; Kachroo, Shanklin, Shah, Whittle, & Klessig, 2001). The *noa1 nia1/2* double mutant abolishes constitutive NO biosynthesis in *ssi2* (suppressor of SA insensitivity) plants and restores defence signaling in these plants (Mandal et al., 2012; Wang et al., 2014). Like *ssi2*, mutations in *NOX1* or *GSNOR1* also result in the accumulation of increased NO (El-Shehety et al., 2015; Wang et al., 2014). However, unlike *ssi2* plants, *gsnor1* and *nox1* plants do not show constitutive defence activation (Mandal et al., 2012; Xu, Guerra, Lee, & Vierling, 2013). Furthermore, unlike *ssi2* plants, pathogen-infected *gsnor1* plants contain significantly reduced SA (Feechan et al., 2005). Thus, constitutive NO accumulation can result in very different phenotypes, likely based on the precise NO levels in different subcellular locations and within specific genetic backgrounds.

4. FATTY ACID FLUX AND SAR

During fatty acid biosynthesis in the plastids, the 18:1 fatty acid is either deposited on the G3P backbone or exported out of plastids (Kachroo & Kachroo, 2009). The acylation reaction, catalysed by the *ACT1*-encoded plastid-localized G3P acyltransferase, serves as the initial step for biosynthesis of plastidal lipids. Consequently, a mutation in *ACT1* severely compromises plastidal lipids results in the accumulation of the G3P and 18:1 substrates. This correlates with the normal SAR and increased basal resistance to necrotrophic fungal pathogens in the *act1* mutant (Chanda et al., 2008,

2011). Conversely, overexpression of *ACT1* results in the depletion of the 18:1 pool, particularly when glycerol is supplied exogenous (glycerol is rapidly converted to G3P which is then acylated by 18:1, Kachroo, Lapchyk, et al., 2003; Kachroo et al., 2004; Kachroo, Kachroo et al., 2003; Kachroo et al., 2007; Kachroo, Venugopal, Navarre, Lapchyk, & Kachroo, 2005; Venugopal et al., 2009). The ACT1-catalysed reaction prefers the 18:1-acyl carrier protein (ACP) 4 conjugate as a substrate (Xia et al., 2009). Accordingly, *acp4* mutant plants exhibit several identical phenotypes as *act1* plants. Thus, a mutation in *ACP4* also results in an increase in the 18:1 pool. A mutation in *acp4* also reduces the overall fatty acid flux, causing a defect in cuticle biogenesis, which role is important in SAR (Xia et al., 2009). Thus, many cuticle-defective mutants are also compromised SAR (Xia et al., 2009, 2010, 2012). Notably, a cuticular defect is associated with increased ROS biosynthesis (L'Haridon et al., 2011) and could be responsible for the compromised SAR in at least some of the cuticle defective mutants.

ACKNOWLEDGEMENTS

Work in our laboratories is supported by grants from National Science Foundation (IOS#0749731, #1457121).

REFERENCES

Alvarez, M. E., Pennell, R. I., Meijer, P.-J., Ishikawa, A., Dixon, R. A., & Lamb, C. (1998). Reactive oxygen intermediates mediate a systemic signal network in the establishment of plant immunity. *Cell, 92*, 773–784.

Baker, P. R., Lin, Y., Schopfer, F. J., Woodcock, S. R., Groeger, A. L., Batthyany, C., et al. (2005). Fatty acid transduction of nitric oxide signaling: multiple nitrated unsaturated fatty acid derivatives exist in human blood and urine and serve as endogenous peroxisome proliferator-activated receptor ligands. *Journal of Biological Chemistry, 280*, 42464–42475.

Barcelo, A. R. (2005). Xylem parenchyma cells deliver the H_2O_2 necessary for lignification in differentiating xylem vessels. *Planta, 220*, 747–756.

Boisson-Dernier, A., Lituiev, D. S., Nestorova, A., Franck, C. M., Thirugnanarajah, S., & Grossniklaus, U. (2013). ANXUR receptor-like kinases coordinate cell wall integrity with growth at the pollen tube tip via NADPH oxidases. *PLoS Biology, 11*, e1001719.

Carol, R. J., Takeda, S., Linstead, P., Durrant, M. C., Kakesova, H., Derbyshire, P., et al. (2005). A RhoGDP dissociation inhibitor spatially regulates growth in root hair cells. *Nature, 438*, 1013–1016.

Chanda, B., Venugopal, S. C., Kulshrestha, K., Navarre, D., Downie, B., Kachroo, A., et al. (2008). Glycerol-3-phosphate levels are associated with basal resistance to the hemibiotrophic fungus *Colletotrichim higginsianum* in Arabidopsis. *Plant Physiology, 147*, 2017–2029.

Chanda, B., Xia, Y., Mandal, M. K., Sekine, K. T., Gao, Q. M., Selote, D., et al. (2011). Glycerol-3-phosphate is a critical mobile inducer of systemic immunity in plants. *Nature Genetics, 43*, 421–427.

Chaturvedi, R., Venables, B., Petros, R. A., Nalam, V., Li, M., Wang, X., et al. (2012). An abietane diterpenoid is a potent activator of systemic acquired resistance. *Plant Journal, 71*, 161–172.

Chen, C.-L., Acedo, G. N., Dewdney, J., Goodman, H. M., & Conkling, M. A. (1991). Differential expression of the two *Arabidopsis* nitrate reductase genes. *Plant Physiology, 96*, 275–279.

Chen, F., D'Auria, J. C., Tholl, D., Ross, J. R., Gershenzon, J., Noel, J. P., et al. (2003). An *Arabidopsis thaliana* gene for methylsalicylate biosynthesis, identified by a biochemical genomics approach, has a role in defense. *Plant Journal, 36*, 577–588.

Crawford, N. M. (2006). Mechanism of nitric oxide synthesis in plants. *Journal of Experimental Botany, 57*, 471–478.

Davda, R. K., Stepniakowski, K. T., Lu, G., Ullian, E., Goodfriend, T. L., & Egan, B. M. (1995). Oleic acid inhibits endothelial nitric oxide synthase by a protein kinase C-independent mechanism. *Hypertension, 26*, 764–770.

Dean, J. V., & Delaney, S. P. (2008). Metabolism of salicylic acid in wild-type, ugt74f1 and ugt74f2 glucosyltransferase mutants of *Arabidopsis thaliana. Physiologia Plantarum, 132*, 417–425.

Durner, J., Wendehenne, D., & Klessig, D. F. (1998). Defense gene induction in tobacco by nitric oxide, cyclic GMP, and cyclic ADP-ribose. *Proceedings of the National Academy of Sciences of the United States of America, 95*, 10328–10333.

Edwards, R. (1994). Conjugation and metabolism of salicylic acid in tobacco. *Journal of Plant Physiology, 143*, 609–614.

Effmert, U., Saschenbrecker, S., Ross, J., Negre, F., Fraser, C. M., Noel, J. P., et al. (2005). Floral benzenoid carboxyl methyltransferases: from *in vitro* to *in planta* function. *Phytochemistry, 66*, 1211–1230.

El-Shetehy, M., Wang, C., Shine, M. B., Yu, K., Navarre, D., Wendehenne, D., et al. (2015). Nitric oxide and reactive oxygen species are required for systemic acquired resistance in plants. *Plant Signaling and Behavior*. http://dx.doi.org/10.1080/15592324.2014.998544.

Enyedi, A. J., Yalpani, N., Silverman, P., & Raskin, I. (1992). Localization, conjugation, and function of salicylic acid in tobacco during the hypersensitive reaction to tobacco mosaic virus. *Proceedings of the National Academy of Sciences of the United States of America, 89*, 2480–2484.

Espunya, M. C., De Michele, R., Gómez-Cadenas, A., & Martínez, M. C. (2012). S-Nitrosoglutathione is a component of wound- and salicylic acid-induced systemic responses in *Arabidopsis thaliana. Journal of Experimental Botany, 63*, 3219–3227.

Feechan, A., Kwon, E., Yun, B.-W., Wang, Y., Pallas, J. A., & Loake, G. J. (2005). A central role for S-nitrosothiols in plant disease resistance. *Proceedings of the National Academy of Sciences of the United States of America, 102*, 8054–8059.

Flor, H. H. (1971). Current status of the gene-for-gene concept. *Annual Review of Phytopathology, 9*, 275–296.

Foreman, J., Demidchik, V., Bothwell, J. H. F., Mylona, P., Miedema, H., Torres, M. A., et al. (2003). Reactive oxygen species produced by NADPH oxidase regulate plant cell growth. *Nature, 422*, 442–446.

Gao, Q.-M., Kachroo, A., & Kachroo, P. (2014). Chemical inducers of systemic immunity in plants. *Journal of Experimental Botany, 65*, 1849–1855.

Gao, Q. M., Yu, K., Xia, Y., Shine, M. B. S., Wang, C., Navarre, D., et al. (2014). Mono- and di-galactosyldiacylglycerol lipids function nonredundantly to regulate systemic acquired resistance in plants. *Cell Reports, 9*, 1681–1691.

Gas, E., Flores-Pérez, U., Sauret-Gueto, S., Gas, E., Jarvis, P., & Rodríguez-Concepcióna, M. (2008). A mutant impaired in the production of plastome-encoded proteins uncovers a mechanism for the homeostasis of isoprenoid biosynthetic enzymes in Arabidopsis plastids. *Plant Cell, 20*, 1303–1315.

Guo, F. Q., Okamoto, M., & Crawford, N. M. (2003). Identification of a plant nitric oxide synthase gene involved in hormonal signaling. *Science, 302*, 100–103.
Hao, F., Zhao, S., Dong, H., Zhang, H., Sun, L., & Miao, C. (2010). Nia1 and Nia2 are involved in exogenous salicylic acid-induced nitric oxide generation and stomatal closure in Arabidopsis. *Journal of Integrative Plant Biology, 52*, 298–307.
Jain, K., Siddam, A., Marathi, A., Roy, U., Falck, J. R., & Balazy, M. (2008). Mechanism of oleic acid nitration by •NO_2. *Free Radical Biology and Medicine, 45*, 269–283.
Jasid, S., Simontacchi, M., Bartoli, C. G., & Puntarulo, S. (2006). Chloroplasts as a nitric oxide cellular source. Effect of reactive nitrogen species on chloroplastic lipids and proteins. *Plant Physiology, 142*, 1246–1255.
Jaskiewicz, M., Conrath, U., & Peterhänsel, C. (2011). Chromatin modification acts as a memory for systemic acquired resistance in the plant stress response. *EMBO Reports, 12*, 50–55.
Jiang, C., Belfield, E. J., Mithani, A., Visscher, A., Ragoussis, J., Mott, R., et al. (2012). ROS-mediated vascular homeostatic control of root-to-shoot soil Na delivery in Arabidopsis. *EMBO Journal, 31*, 4359–4370.
Jones, J. D. G., & Dangl, J. L. (2006). The plant immune system. *Nature, 444*, 323–329.
Jung, H. W., Tschaplinkski, T. J., Wang, L., Glazebrook, J., & Greenberg, J. T. (2009). Priming in systemic plant immunity. *Science, 324*, 89–91.
Kachroo, A., & Kachroo, P. (2009). Fatty acid-derived signals in plant defense. *Annual Review of Phytopathology, 47*, 153–176.
Kachroo, A., Lapchyk, L., Fukushigae, H., Hildebrand, D., Klessig, D., & Kachroo, P. (2003). Plastidal fatty acid signaling modulates salicylic acid- and jasmonic acid-mediated signaling in the Arabidopsis *ssi2* mutant. *Plant Cell, 15*, 2952–2965.
Kachroo, A., & Robin, G. P. (2013). Systemic signaling during plant defense. *Current Opinion in Plant Biology, 16*, 527–533.
Kachroo, A., Shanklin, J., Lapchyk, L., Whittle, E., Hildebrand, D., & Kachroo, P. (2007). The Arabidopsis stearoyl-acyl carrier protein-desaturase family and the contribution of leaf isoforms to oleic acid synthesis. *Plant Molecular Biology, 63*, 257–271.
Kachroo, A., Venugopal, S. C., Lapchy, L., Falcone, D., Hildebrand, D., & Kachroo, P. (2004). Oleic acid levels regulated by glycerolipid metabolism modulate defense gene expression in Arabidopsis. *Proceedings of the National Academy of Sciences of the United States of America, 101*, 5152–5157.
Kachroo, P., Kachroo, A., Lapchyk, L., Hildebrand, D., & Klessig, D. F. (2003). Restoration of defective cross talk in *ssi2* mutants: role of salicylic acid, jasmonic acid, and fatty acids in SSI2-mediated signaling. *Molecular Plant–Microbe Interactions, 16*, 1022–1029.
Kachroo, P., Shanklin, J., Shah, J., Whittle, E., & Klessig, D. (2001). A fatty acid desaturase modulates the activation of defense signaling pathways in plants. *Proceedings of the National Academy of Sciences of the United States of America, 98*, 9448–94453.
Kachroo, P., Venugopal, S. C., Navarre, D. A., Lapchyk, L., & Kachroo, A. (2005). Role of salicylic acid and fatty acid desaturation pathways in ssi2-mediated signaling. *Plant Physiology, 139*, 1717–1735.
Kadota, Y., Goh, T., Tomatsu, H., Tamauchi, R., Higashi, K., Muto, S., et al. (2004). Cryptogein-induced initial events in tobacco BY-2 cells: pharmacological characterization of molecular relationship among cytosolic Ca^{2+} transients, anion efflux and production of reactive oxygen species. *Plant and Cell Physiology, 45*, 160–170.
Kadota, Y., Sklenar, J., Derbyshire, P., Stransfeld, L., Asai, S., Ntoukakis, V., et al. (2014). Direct regulation of the NADPH oxidase RBOHD by the PRR-associated kinase BIK1 during plant immunity. *Molecular Cell, 54*, 43–55.
Kaya, H., Nakajima, R., Iwano, M., Kanaoka, M. M., Kimura, S., Takeda, S., et al. (2014). Ca^{2+}-activated reactive oxygen species production by Arabidopsis RbohH and RbohJ is essential for proper pollen tube tip growth. *Plant Cell, 26*, 1069–1080.

Koo, Y. J., Kim, M. A., Kim, E. H., Song, J. T., Jung, C., Moon, J.-K., et al. (2007). Overexpression of salicylic acid carboxyl methyltransferase reduces salicylic acid-mediated pathogen resistance in *Arabidopsis thaliana*. *Plant Molecular Biology, 64*, 1–15.

Kumar, D., & Klessig, D. F. (2003). High-affinity salicylic acid–binding protein 2 is required for plant innate immunity and has salicylic acid stimulated lipase activity. *Proceedings of the National Academy of Sciences of the United States of America, 100*, 16101–16106.

L'Haridon, F., Besson-Bard, A., Binda, M., Serrano, M., Abou-Mansour, E., Balet, F., et al. (2011). A permeable cuticle is associated with the release of reactive oxygen species and induction of innate immunity. *PLoS Pathogens, 7*, e1002148.

Lassig, R., Gutermuth, T., Bey, T. D., Konrad, K. R., & Romeis, T. (2014). Pollen tube NAD(P)H oxidases act as a speed control to dampen growth rate oscillations during polarized cell growth. *Plant Journal, 78*, 94–106.

Lee, H.-I., & Raskin, I. (1998). Glucosylation of salicylic acid in *Nicotiana tabacum* cv. Xanthi-nc. *Phytopathology, 88*, 692–697.

Lee, H.-I., & Raskin, I. (1999). Purification, cloning, and expression of a pathogen inducible UDP-glucose:salicylic acid glucosuyltransferase from tobacco. *Journal of Biological Chemistry, 274*, 36637–36642.

Lee, Y., Rubio, M. C., Alassimone, J., & Geldner, N. (2013). A mechanism for localized lignin deposition in the endodermis. *Cell, 153*, 402–412.

Li, L., Li, M., Yu, L., Zhou, Z., Liang, X., Liu, Z., et al. (2014). The FLS2-associated kinase BIK1 directly phosphorylates the NADPH oxidase RbohD to control plant immunity. *Cell Host and Microbe, 15*, 329–338.

Lindermayr, C., Sell, S., Müller, B., Leister, D., & Durner, J. (2010). Redox regulation of the NPR1-TGA1 system of *Arabidopsis thaliana* by nitric oxide. *Plant Cell, 22*, 2894–2907.

Liu, P.-P., Dahl, C. C., & Klessig, D. F. (2011). The extent to which methyl salicylate is required for systemic acquired resistance is dependent on exposure to light after infection. *Plant Physiology, 157*, 2216–2226.

Luna, E., Bruce, T. J., Roberts, M. R., Flors, V., & Ton, J. (2012). Next-generation systemic acquired resistance. *Plant Physiology, 158*, 844–853.

Mandal, M., Chanda, B., Xia, Y., Yu, K., Sekine, K., Gao, Q.-M., et al. (2011). Glycerol-3-phosphate and systemic immunity. *Plant Signaling and Behavior, 6*, 1871–1874.

Mandal, M. K., Chandra-Shekara, A. C., Jeong, R. D., Yu, K., Zhu, S., Chanda, B., et al. (2012). Oleic acid-dependent modulation of NITRIC OXIDE ASSOCIATED1 protein levels regulates nitric oxide-mediated defense signaling in Arabidopsis. *Plant Cell, 24*, 1654–1674.

Mittler, R., & Blumwald, E. (2015). The roles of ROS and ABA in systemic acquired acclimation. *Plant Cell, 27*, 64–70.

Moreau, M., Lee, G. I., Wang, Y., Crane, B. R., & Klessig, D. F. (2008). AtNOS/AtNOA1 is a functional *Arabidopsis thaliana* cGTPase and not a nitric-oxide synthase. *Journal of Biological Chemistry, 283*, 32957–32967.

Návarová, H., Bernsdorff, F., Döring, A. C., & Zeier, J. (2012). Pipecolic acid, an endogenous mediator of defense amplification and priming, is a critical regulator of inducible plant immunity. *Plant Cell, 24*, 5123–5141.

Neill, S., Barros, R., Bright, J., Desikan, R., Hancock, J., Harrison, J., et al. (2008). Nitric oxide, stomatal closure, and abiotic stress. *Journal of Experimental Botany, 59*, 165–176.

Ogasawara, Y., Kaya, H., Hiraoka, G., Yumoto, F., Kimura, S., Kadota, Y., et al. (2008). Synergistic activation of the Arabidopsis NADPH oxidase AtrbohD by Ca^{2+} and phosphorylation. *Journal of Biological Chemistry, 283*, 8885–8892.

Pallas, J. A., Paiva, N. L., Lamb, C., & Dixon, R. A. (1996). Tobacco plants epigenetically suppressed in phenylalanine ammonia-lyase expression do not develop systemic acquired resistance in response to infection by tobacco mosaic virus. *Plant Journal, 10*, 281–293.

Parani, M., Rudrabhatla, S., Myers, R., Weirich, H., Smith, B., Leaman, D. W., et al. (2004). Microarray analysis of nitric oxide responsive transcripts in Arabidopsis. *Plant Biotechnology Journal, 2*, 359–366.

Park, B. S., Song, J. T., & Seo, H. S. (2011). *Arabidopsis* nitrate reductase activity is stimulated by the E3 SUMO ligase AtSIZ1. *Nature Communications, 2*, 400.

Park, S. W., Kaiyomo, E., Kumar, D., Mosher, S. L., & Klessig, D. F. (2007). Methyl salicylate is a critical mobile signal for plant systemic acquired resistance. *Science, 318*, 113–116.

Park, S.-W., Liu, P.-P., Forouhar, F., Vlot, A. C., Tong, L., Tietjen, K., et al. (2009). Use of a synthetic salicylic acid analog to investigate the roles of methyl salicylate and its esterases in plant disease resistance. *Journal of Biological Chemistry, 284*, 7307–7317.

Pieterse, C. M. J., van Pelt, J. A., Ton, J., Parchmann, S., Mueller, M. J., Buchala, A. J., et al. (2000). Rhizobacteria-mediated induced systemic resistance (ISR) in Arabidopsis requires sensitivity to jasmonate and ethylene but is not accompanied by an increase in their production. *Physiological and Molecular Plant Pathology, 57*, 123–134.

Pieterse, C. M. J., Poelman, E. H., van Wees, A. C. M., & Dicke, M. (November 21, 2013). Induced plant responses to microbes and insects. *Frontiers in Plant Science, 4*, 3.

Rustérucci, C., Espunya, M. C., Díaz, M., Chabannes, M., & Martínez, M. C. (2007). S-nitrosoglutathione reductase affords protection against pathogens in Arabidopsis, both locally and systemically. *Plant Physiology, 143*, 1282–1292.

Sagi, M., & Fluhr, R. (2006). Production of reactive oxygen species by plant NADPH oxidases. *Plant Physiology, 141*, 336–340.

Segonzac, C., Feike, D., Gimenez-Ibanez, S., Hann, D. R., Zipfel, C., & Rathjen, J. P. (2011). Hierarchy and roles of pathogen-associated molecular pattern-induced responses in *Nicotiana benthamiana*. *Plant Physiology, 156*, 687–699.

Shabab, M., Shindo, T., Gu, C., Kaschani, F., Pansuriya, T., Chintha, R., et al. (2008). Fungal effector protein AVR2 targets diversifying defense-related cys proteases of tomato. *Plant Cell, 20*, 1169–1183.

Shah, J., & Zeier, J. (2013). Long-distance communication and signal amplification in systemic acquired resistance. *Frontiers in Plant Science, 4*, 30.

Shan, L., He, P., Li, J., Heese, A., Peck, S. C., Nürnberger, T., et al. (2008). Bacterial effectors target the common signaling partner BAK1 to disrupt multiple MAMP receptor-signaling complexes and impede plant immunity. *Cell Host and Microbe, 4*, 17–27.

Shang, Y., Li, X., Cui, H., He, P., Thilmony, R., Chintamanani, S., et al. (2006). RAR1, a central player in plant immunity, is targeted by *Pseudomonas syringae* effector AvrB. *Proceedings of the National Academy of Sciences of the United States of America, 103*, 19200–19205.

Siddique, S., Matera, C., Radakovic, Z. S., Hasan, M. S., Gutbrod, P., Rozanska, E., et al. (2014). Parasitic worms stimulate host NADPH oxidases to produce reactive oxygen species that limit plant cell death and promote infection. *Science Signaling, 8*, ra33.

Slaughter, A., Daniel, X., Flors, V., Luna, E., Hohn, B., & Mauch-Mani, B. (2012). Descendants of primed Arabidopsis plants exhibit resistance to biotic stress. *Plant Physiology, 158*, 835–843.

Song, F., & Goodman, R. M. (2001). Activity of nitric oxide is dependent on, but is partially required for function of, salicylic acid in the signaling pathway in tobacco systemic acquired resistance. *Molecular Plant–Microbe Interactions, 14*, 1458–1462.

Song, J. T., Koo, Y. J., Seo, H. K., Kim, M. C., Choi, Y. D., & Kim, J. H. (2008). Overexpression of AtSGT1, an Arabidopsis salicylic acid glucosyltransferase, leads to increased susceptibility to *Pseudomonas syringae*. *Phytochemistry, 69*, 1128–1134.

Spoel, S., & Dong, X. (2012). How do plants achieve immunity? Defence without specialized immune cells. *Nature Reviews Immunology, 12*, 89–100.

Strawn, M. A., Marr, S. K., Inoue, K., Inada, N., Zubieta, C., & Wildermuth, M. C. (2007). Arabidopsis isochorismate synthase functional in pathogen-induced salicylate biosynthesis exhibits properties consistent with a role in diverse stress responses. *Journal of Biological Chemistry, 282*, 5919–5933.

Tada, Y., Spoel, S. H., Pajerowska-Mukhtar, K., Mou, Z., Song, J., Wang, C., et al. (2008). Plant immunity requires conformational changes of NPR1 via S-nitrosylation and thioredoxins. *Science, 321*, 952–956.

Takeda, S., Gapper, C., Kaya, H., Bell, E., Kuchitsu, K., & Dolan, L. (2008). Local positive feedback regulation determines cell shape in root hair cells. *Science, 319*, 1241–1244.

Thordal-Christensen, H. (2003). Fresh insights into processes of nonhost resistance. *Current Opinion in Plant Biology, 6*, 351–357.

Torres, M. A., Jones, J. D., & Dangl, J. L. (2005). Pathogen-induced, NADPH oxidase-derived reactive oxygen intermediates suppress spread of cell death in *Arabidopsis thaliana. Nature Genetics, 37*, 1130–1134.

Van Wees, S. C., Van der Ent, S., & Pieterse, C. M. (2008). Plant immune responses triggered by beneficial microbes. *Current Opinion in Plant Biology, 11*, 443–448.

Venugopal, S. C., Jeong, R. D., Mandal, M., Zhu, S., Chandra-Shekara, A. C., Xia, Y., et al. (2009). Enhanced disease susceptibility and salicylic acid act redundantly to regulate resistance gene-mediated signaling. *PLoS Genetics, 5*, e1000545.

Verhagen, B. W. M., Glazebrook, J., Zhu, T., Chang, H.-S., Van Loon, L. C., & Pieterse, C. M. J. (2004). The transcriptome of rhizobacteria-induced systemic resistance in Arabidopsis. *Molecular Plant–Microbe Interactions, 17*, 895–908.

Vernooij, B., Friedrich, L., Morse, A., Reist, R., Kolditz-Jawhar, R., Ward, E., et al. (1994). Salicylic acid is not the translocated signal responsible for inducing systemic acquired resistance but is required in signal transduction. *Plant Cell, 6*, 959–965.

Wang, C., El-Shetehy, M., Shine, M. B., Yu, K., Navarre, D., Wendehenne, D., et al. (2014). Free radicals mediate systemic acquired resistance. *Cell Reports, 7*, 348–355.

Wang, P., Du, Y., Li, Y., Ren, D., & Song, C.-P. (2010). Hydrogen peroxide-mediated activation of MAP kinase 6 modulates nitric oxide biosynthesis and signal transduction in *Arabidopsis. Plant Cell, 22*, 2981–2998.

Wendehenne, D., Gao, Q. M., Kachroo, A., & Kachroo, P. (2014). Free radical-mediated systemic immunity in plants. *Current Opinion in Plant Biology, 20*, 127–134.

Wildermuth, M. C., Dewdney, J., Wu, G., & Ausubel, F. M. (2001). Isochorismate synthase is required to synthesize salicylic acid for plant defence. *Nature, 414*, 562–565.

Wittek, F., Hoffmann, T., Kanawati, B., Bichlmeier, M., Knappe, C., Wenig, M., et al. (2014). Arabidopsis ENHANCED DISEASE SUSCEPTIBILITY1 promotes systemic acquired resistance via azelaic acid and its precursor 9-oxo nonanoic acid. *Journal of Experimental Botany, 65*, 5919–5931.

Xia, Y., Gao, Q. M., Yu, K., Lapchyk, L., Navarre, D., Hildebrand, D., et al. (2009). An intact cuticle in distal tissues is essential for the induction of systemic acquired resistance in plants. *Cell Host and Microbe, 5*, 151–165.

Xia, Y., Yu, K., Gao, Q. M., Wilson, E. V., Navarre, D., Kachroo, P., et al. (2012). Acyl CoA binding proteins are required for cuticle formation and plant responses to microbes. *Frontiers in Plant Science, 3*, 224–242.

Xia, Y., Yu, K., Navarre, D., Seebold, K., Kachroo, A., & Kachroo, P. (2010). The glabra1 mutation affects cuticle formation and plant responses to microbes. *Plant Physiology, 154*, 833–846.

Xu, S., Guerra, D., Lee, U., & Vierling, E. (2013). S-nitrosoglutathione reductases are low-copy number, cysteine-rich proteins in plants that control multiple developmental and defense responses in Arabidopsis. *Frontiers in Plant Science, 4*, 30.

Yu, K., Soares, J. M., Mandal, M. K., Wang, C., Chanda, B., Gifford, A. N., et al. (2013). A feedback regulatory loop between G3P and lipid transfer proteins DIR1 and AZI1 mediates azelaic-acid-induced systemic immunity. *Cell Reports, 3*, 1266–1278.

Yu, X., Sukumaran, S., & Márton, L. (1998). Differential expression of the Arabidopsis *Nia1* and *Nia2* genes. *Plant Physiology, 116*, 1091–1096.

Zhang, Y., Zhu, H., Zhang, Q., Li, M., Yan, M., Wang, R., et al. (2009). Phospholipase dalpha1 and phosphatidic acid regulate NADPH oxidase activity and production of reactive oxygen species in ABA-mediated stomatal closure in Arabidopsis. *Plant Cell, 21*, 2357–2377.

Zhou, J. M., & Chai, J. (2008). Plant pathogenic bacterial type III effectors subdue host responses. *Current Opinion in Microbiology, 11*, 179–185.

Zipfel, C., & Rathjen, J. P. (2008). Plant Immunity: AvrPto targets the frontline. *Current Biology., 18*, 218–220.

Zoeller, M., Stingl, N., Krischke, M., Fekete, A., Waller, S.-B., & Mueller, M. J. (2012). Lipid profiling of the Arabidopsis hypersensitive response reveals specific lipid peroxidation and fragmentation processes: biogenesis of pimelic and azelaic acid. *Plant Physiology, 160*, 365–378.

CHAPTER THIRTEEN

The Role of Nitric Oxide in Development and Pathogenesis of Biotrophic Phytopathogens – Downy and Powdery Mildews

Michaela Sedlářová*, Lucie Kubienová§, Zuzana Drábková Trojanová*, Lenka Luhová§, Aleš Lebeda*, Marek Petřivalský§,[1]

*Department of Botany, Faculty of Science, Palacký University in Olomouc, Olomouc, Czech Republic
§Department of Biochemistry, Faculty of Science, Palacký University in Olomouc, Olomouc, Czech Republic
[1]Corresponding author: E-mail: marek.petrivalsky@upol.cz

Contents

Abstract

Nitric oxide (NO) and reactive nitrogen species are involved in multiple signaling and regulatory pathways across all organisms. This review provides a survey of current knowledge on the role of NO in the development and growth of plant pathogens, highlighting biotrophic parasites depending on an intimate relationship with their hosts. In plant–pathogen interactions, the major focus has been dedicated to the sources, metabolism and functions of NO as a key component of plant signaling and immunity. On the side of pathogens, much less information has been gathered about the sources and fate of NO. Recent findings suggest that the germination, oriented growth and active penetration of filamentous microorganisms, such as oomycetes and fungi, to the host tissues involve endogenous production of NO and reactive oxygen species (ROS). Methodological approaches available so far constraint more detailed studies of NO metabolism in downy and powdery mildews, as these obligate

Advances in Botanical Research, Volume 77
ISSN 0065-2296
http://dx.doi.org/10.1016/bs.abr.2015.10.002

biotrophs cannot be detached from their host plants. We summarize data obtained using two model pathosystems, lettuce – downy mildew and tomato – powdery mildew, which confirm the NO production by biotrophic parasites aiming to penetrate host tissues but omit their priming. The need for a fine balancing of NO and ROS levels during their development and pathogenesis is proposed and discussed in relation to differences in biological and pathophysiological characteristics of both groups of pathogens.

List of Abbreviations

cGMP	Cyclic guanosine monophosphate
GSNO	*S*-Nitrosoglutathione
hpi	Hours post inoculation
HR	Hypersensitive response
MAPK	Mitogen-activated protein kinase
NOS	Nitric oxide synthase
NR	Nitrate reductase
PR	Pathogenesis-related (protein)
RNS	Reactive nitrogen species
ROS	Reactive oxygen species
SAR	Systemic acquired resistance
SOD	Superoxide dismutase

1. NITRIC OXIDE IN PLANT RESPONSES TO PATHOGEN ATTACK

Insights into the molecular mechanisms of plant–microbe interactions revealed that nitric oxide (NO) belongs to the key components of pathogen recognition as well as activation of local and systemic defence mechanisms (Asai, Mase, & Yoshioka, 2010; Leitner, Vandelle, Gaupels, Bellin, & Delledonne, 2009; Trapet et al., 2015; Wendehenne, Durner, & Klessig, 2004; Zaninotto, La Camera, Polverari, & Delledonne, 2006). Moreover, NO is also a well-established player in many forms of plant symbiotic interactions (reviewed in Hichri et al. (2015)). NO interacts with signaling pathways of reactive oxygen species (ROS) (Bolwell & Daudi, 2009; Delledonne, Zeier, Marocco, & Lamb, 2001; Lehmann, Serrano, L'Haridon, Tjamos, & Metraux, 2015; Scheler, Durner, & Astier, 2013; Wang, Loake, & Chu, 2013) and phytohormones (recently summarized e.g. by Mur, Prats, Pierre, Hall, & Hebelstrup (2013)) and its biological activity is linked to cellular redox changes and post-translational protein modifications mediated by NO-derived reactive nitrogen species (RNS) (Arasimowicz-Jelonek & Floryszak-Wieczorek, 2014; Astier & Lindermayr,

2012; Mur et al., 2012; Mur, Prats, Pierre, Hall, & Hebelstrup, 2013). Increased levels of plant RNS under stress conditions, such as pathogen infection, are believed to be the result of temporarily and spatially orchestrated events of NO and ROS bursts, induced by activations of corresponding enzyme systems localized in the plasma membrane or within plant cells. RNS subsequently cause protein post-translational modifications by *S*-nitrosylation of cysteine residues or nitration of tyrosine and tryptophan residues (Martínez-Ruiz, Cadenas, & Lamas, 2011). NO-dependent protein modifications, reversible due to action of specific enzymes, are utilized as signaling switches that regulate protein functions (reviewed by Astier and Lindenmayer (2012)). Cellular redox changes modify cysteine thiols to form *S*-nitrosothiols and disulfide bridges. *S*-nitrosylation seems to be the major NO-dependent post-translational protein alteration, which can inhibit bacterial effectors activity in challenged plant cells and thus disarm pathogen virulence machinery (Yu, Lamattina, Spoel, & Loake, 2014). The regulation of cellular protein *S*-nitrosylation in plant development and responses to biotic stress stimuli is modulated by the activity of *S*-nitrosoglutathione (GSNO) reductase (Xu, Guerra, Lee, & Vierling, 2013). In some plant tissues, local cellular level can be controlled also by nonsymbiotic plant haemoglobins which effectively oxidize NO to NO_3^-.

Plant-derived NO is required for the initiation of local defence mechanisms, e.g. hypersensitive response (HR), modulation of gene expression through transcription factors and synthesis of pathogenesis-related (PR) proteins (Lin et al., 2012; Mur, Kenton, Lloyd, Ougham, & Prats, 2008; Wendehenne et al., 2004; Zaninotto et al., 2006; Zeier et al., 2004). Modes of NO action during HR includes the host cell death induced by H_2O_2 and executed by ROS, restricting pathogen invasion in race-specific interactions (Delledonne, Polverari, & Murgia, 2003; Lin et al., 2011). Detailed review of the crosstalks between NO and ROS in plant HR and mechanisms of senescence, i.e. programmed cell death, has been recently brought by Wang et al. (2013).

In plant cells challenged by pathogens, NO can react with superoxide anion (O_2^-) in a diffusion-limited reaction to form peroxynitrite ($ONOO^-$). This strong oxidant and nitrating species seems to exert its effects in plant cell signaling through protein nitration, as observed in case of mitogen-activated protein kinase (MAPK) signaling cascade in *Arabidopsis thaliana* (Vandelle & Delledonne, 2011). Superoxide dismutase (SOD), an enzyme scavenging O_2^-, is downregulated by peroxynitrite and thus leads to specific feedback strengthening this signal in plant organelles.

Mitochondrial, peroxisomal and chloroplastic SOD3 but no other SODs were found as being strongly inhibited by peroxynitrite (Holzmeister et al., 2015). Furthermore, NO is able to amplify the oxidative status of cells by inhibiting H_2O_2 degrading enzymes, i.e. catalase and ascorbate peroxidase, through *S*-nitrosylation (Trapet et al., 2015).

In potato infected by *Phytophthora infestans*, the causing agent of late blight, peroxynitrite-mediated protein nitration was found to be linked to primary metabolism, defence responses and redox balance mainly in the compatible interaction during the early stages of pathogenesis. Hemibiotrophic strategy of *P. infestans* (behaves as biotroph in the initial phases of interaction and later switches to necrotrophy) is tightly linked with manipulation of the redox balance in infected cells, to avoid HR in early stages of infection but to facilitate later necrosis to feed from dead tissues. The group of Arasimowicz-Jelonek suggested a role for peroxynitrite as the main mediator of potato defence to oomycete pathogen through precise control of redox imbalance that might unlock expression of genes coding for PR proteins in the resistant plants (Arasimowicz-Jelonek & Floryszak-Wieczorek, 2014). However, the precise timing and components of signaling pathways remain to be investigated. On the other hand, tyrosine nitration of proteins has been found more intensive in susceptible than in resistant plants, indicating a higher level of nitrosative stress induced by pathogen challenge (Chaki et al., 2009; Saito, Yamamoto-Katou, Yoshioka, Doke, & Kawakita, 2006).

Transduction of NO systemic signal to plant organs distant from infected tissue is connected with galactolipids. It has been reported that mainly digalactosyldiacylglycerol contributes to biosynthesis of NO and salicylic acid leading to systemic acquired resistance (SAR) (El-Shetehy et al., 2015).

2. SOURCES OF NO IN PHYTOPATHOGENS

In animals, NO is produced mainly in the arginine-dependent oxidative pathway catalyzed by the enzymatic activity of nitric oxide synthases (NOS) (Martínez-Ruiz et al., 2011). During the past two decades, a closer attention has been paid to NO sources and functions also in other eukaryotes. In addition to nonenzymatic reduction of nitrite in acidic conditions, several sources of NO are supposed to operate in plants: NOS-like enzyme, nitrate reductase (NR), nitrite:NO reductase and polyamine oxidases (Arasimowicz & Floryszak-Wieczorek, 2007; Corpas, Hayashi, Mano, Nishimura, & Barroso, 2009; del Río, Corpas, & Barroso, 2004; Stöhr & Ullrich, 2002; Wilson, Neill, & Hancock, 2008; Yamasaki,

2000; Yamasaki & Cohen, 2006; Zemojtel et al., 2006). Unlike in higher plants, NOS occurrence has been demonstrated only in algae (Foresi et al., 2010).

NO production was observed in cells of fungi and fungi-like organisms (oomycetes, slime moulds) associated with plants, including pathogens (Conrath, Amoroso, Kohle, & Sultemeyer, 2004; Piterková et al., 2011; Prats, Carver, & Mur, 2008; Samalova et al., 2013), mycobionts of lichens (Catalá et al., 2010) and saprotrophs from distinct taxonomical groups, e.g. zygomycete *Phycomyces* (Maier, Hecker, Rockel, & Ninnemann, 2001), ascomycetes *Morchella* and *Ptychoverpa* (Sedlářová, personal communication) and wood decay basidiomycete *Phanerochaete* (Servent, Ducrocq, Henry, Guissani, & Lenfant, 1991). Beside plant pathogens, experiments performed with human fungal pathogens also imply the involvement of NO and RNS in their development. The opportunistic pathogen *Aspergillus fumigatus* was found to be sensitive to the cytotoxic action of acidified nitrite solution while, interestingly, the germination of fungal spores was highly induced by the same concentration of GSNO (Kunert, 1995). However, a short-term exposure to gaseous NO applied exogenously inhibited the growth of *Aspergillus niger, Monilinia fructicola* and *Penicillium italicum* (Lazar, Wills, Ho, Harris, & Spohr, 2008). Independently of their nutritional strategy, all fungi apparently require NO accumulation for the prompt activation of their metabolism, either during spore germination (Sedlářová et al., 2011; Wang & Higgins, 2005), penetration of parasites (Samalova et al., 2013), formation of sporangiophores (Maier et al., 2001) and complex fruiting bodies (Song, Jeong, & Choi, 2000) or in the course of lichen rehydration (Catalá et al., 2010). Nevertheless, until now the mechanisms of NO synthesis in these organisms have not been elucidated. Similarly to plants, both oxidative and reductive pathways are likely to be involved. A limited number of reports on fungal metabolism of NO include identification of NOS enzyme in the mushroom *Flammulina velutipes* (Song et al., 2000) and localization of nitrite-dependent NO production in cultures of phytopathogenic *Pythium*, *Botrytis* and *Fusarium* spp. (Conrath et al., 2004). Application of chemical compounds targeting enzymes potentially involved in NO production by downy and powdery mildews brought some contradictory results, showing that both NOS and NR might be involved in NO production in variable extent (Petřivalský et al., 2007; Piterková et al., 2011). L-arginine-dependent enzyme activity was identified in the conidia of *Oidium neolycopersici*, the cause of tomato powdery mildew (Piterková et al., 2009).

It has been repeatedly suggested that measurements of NO production and levels in plant systems should be performed using proper methodological approaches (Mur, Mandon, Cristescu, Harren, & Prats, 2011; Mur et al., 2012; Planchet, Sonoda, Zeier, & Kaiser, 2006; Rümer, Krischke, Fekete, Mueller, & Kaiser, 2012). As summarized in this review, similarly to the plant NO field, the major part of results reporting production of NO in plant pathogens has been obtained by histochemical staining using diaminofluorescein derivatives as presumable NO-sensitive fluorescent probes followed by a confocal microscopy imaging. The specificity and validity of this method has been questioned for plant samples (Rümer et al., 2012), which obviously raised also doubts on its applicability for NO detection in microbial cells. Altogether, the accumulated evidence on the production of NO in phytopathogens shows some parallels with other organisms but the diversity of NO sources and its metabolism in individual groups of phytopathogens needs to be experimentally addressed.

3. NO IN THE PATHOGENESIS OF FUNGAL AND HEMIBIOTROPHIC PHYTOPATHOGENS

The production of NO detected in fungi led to the assumption that it might also serve as one of the pathogen virulence determinants. Compatibility or incompatibility together with the life strategy of a phytopathogen, i.e. biotrophy or necrotrophy, determines the fate of infected plant organs, for which precise spatiotemporal tuning of relative NO concentration seems to be indispensable (Asai et al., 2010; Glazebrook, 2005; Samalova et al., 2013; Scheler et al., 2013; Sedlářová et al., 2011).

Localized production of NO and ROS is a typical feature of fungal pathogenesis, associated with the process of active penetration of fungal structures into the host cells (Prats et al., 2008; Prats, Mur, Sanderson, & Carver, 2005; Sedlářová, Luhová, Petřivalský, & Lebeda, 2007; Sedlářová et al., 2011; Tománková, Luhová, Petřivalský, Peč, & Lebeda, 2006; Yoshioka, Asai, Yoshioka, & Kobayashi, 2009). Experimental evidence suggests that phytopathogens utilize NO as an endogenous signaling component to regulate their mycelial growth and sporulation (Samalova et al., 2013). Recently, a dose-dependent increase in NO production was detected in spores of the tomato pathogen *Fusarium solani* f. sp. *eumartii* treated with chitosan (Terrile, Mansilla, Albertengo, Rodríguez, & Casalongué, 2015). It was concluded that the increased NO production was involved in the activation of NO-mediated cell death leading to the observed dose-dependent decrease in spore cell viability, underlying the observed fungicide effect of chitosan on *Fusarium*-infected tomato plants.

The heterogeneity of plant responses to biotic stress, on local and systemic levels, raised question concerning the origin of NO and RNS. The possibility of an NO diffusion from fungal to plant cells and vice versa through perihaustorial membrane should be also considered; however, direct evidence for crosstalk between the fungus and the plant has been reported only for the necrotrophic *Botrytis cinerea* (Turrion-Gomez & Benito, 2011). NO plays a key role in the plant basal resistance against *B. cinerea*, in contrast to ROS, which are negative regulators of *Nicotiana benthamiana* resistance (Asai et al., 2010). Authors proposed the existence of a feedback mechanism, where an NO concentration above or below certain threshold can increase or reduce the plant defence against this necrotrophic fungal pathogen. Similar hypothesis was formulated for the hemibiotrophic oomycete *P. infestans* where peroxynitrite-induced redox imbalance might regulate expression of PR proteins but also other defence mechanisms (Arasimowicz-Jelonek & Floryszak-Wieczorek, 2014).

To avoid the nitrosative stress triggered by infected host tissues, fungi can employ inducible NO-detoxifying enzymes flavohaemoglobins that are known to confer protection in *B. cinerea* (Turrion-Gomez, Eslava, & Benito, 2010) and *Aspergillus* spp. In *Aspergillus oryzae*, flavohaemoglobins promote oxidative damages by hydrogen peroxide, corroborating a close link between ROS and RNS metabolism and signaling known for plants (Gessler, Averyanov, & Belozerskaya, 2007; Zhou et al., 2010).

NO is also involved in the growth and development of hemibiotrophic pathogens which display the highest flexibility among plant parasites. Indeed, hemibiotrophs are able to switch their metabolism and life strategy from biotrophy to necrotrophy and saprotrophy within days. Such extreme plasticity during their relatively short life cycle is allowed by progressive expression of many genes coding various effectors (Brunner, Torriani, Croll, Stukenbrock, & McDonald, 2013). Only one study has been reported on NO production during early developmental stages, i.e. spore germination and appressoria formation, in the rice blast ascomycete *Magnaporthe oryzae* (Samalova et al., 2013). The importance of regulation of endogenous level of NO and its metabolites is further supported by a recent finding showing that GSNO reductase is required for conidiation and contributes to virulence of *M. oryzae*, i.e. increased penetration rate and biotrophic phases (Zhang et al. 2015). Nevertheless, a complex study concerning changes in

NO metabolism during the whole life cycle of hemibiotroph pathogens remains one of the future challenges.

4. NO IN THE PATHOGENESIS OF DOWNY MILDEWS

Downy and powdery mildews, together with rusts and smuts which are omitted in this review, belong to plant biotrophic pathogens of great economic significance and represent a hot issue in actual crop protection worldwide. Basic differences in biological and physiological characteristics of both groups are summarized in Table 1. A high degree of host specificity has evolved within obligate biotrophs (Panstruga, 2003; Micali, Göllner, Humphry, Consonni, & Panstruga, 2008). Utilization of biotrophs as model organisms in physiological studies brings some methodological constraints, as they cannot be grown on agar media in vitro or only for a limited time, such as in the case of conidia germination of powdery mildews. Thus, it is challenging to analyze NO metabolism in biotrophic pathogens apart from host plants, namely species adapted to grow intercellularly within plant tissues such as downy mildews, rusts or smuts (Mur et al., 2012). The fact that they need to maintain host tissues alive implies that NO and ROS concentration needs to be balanced to avoid the onset of HR in host cells (Heller & Tudzynski, 2011; Mur, Carver, & Prats, 2006; Mur et al., 2008; Tada et al., 2004).

Table 1 Comparison of Downy and Powdery Mildews, Obligate Biotrophic Parasites which Infect Photosynthetically Active Organs (Mostly Leaves) of Plants but Differ in their Biological Characteristics

	Downy Mildews	Powdery Mildews
Taxonomy		
Kingdom	Chromalveolata	Fungi
Phyllum	Heterokontophyta	Ascomycota
Class	Oomycetes	Leotiomycetes
Order	Peronosporales	Erysiphales
Growth	• Inside tissues, intercellular	• On the surface of organs
Haustoria (feeding organs)	• Invade mesophyll cells	• Invade epidermal cells
Material of cell wall	• Cellulose	• Chitin
Optimal season	• Humid weather – spring, early summer	• Dry and hot weather – late summer, early autumn

Biotrophic oomycetes (Peronosporales, Oomycetes, Chromalveolata), causative agents of downy mildews, are distinct from the true fungi in many features, but share some similarities as mycelial growth, feeding structures established in host cells (haustoria) and strategies to avoid host recognition (Figure 1). *Bremia lactucae* Regel infects leaves and causes downy

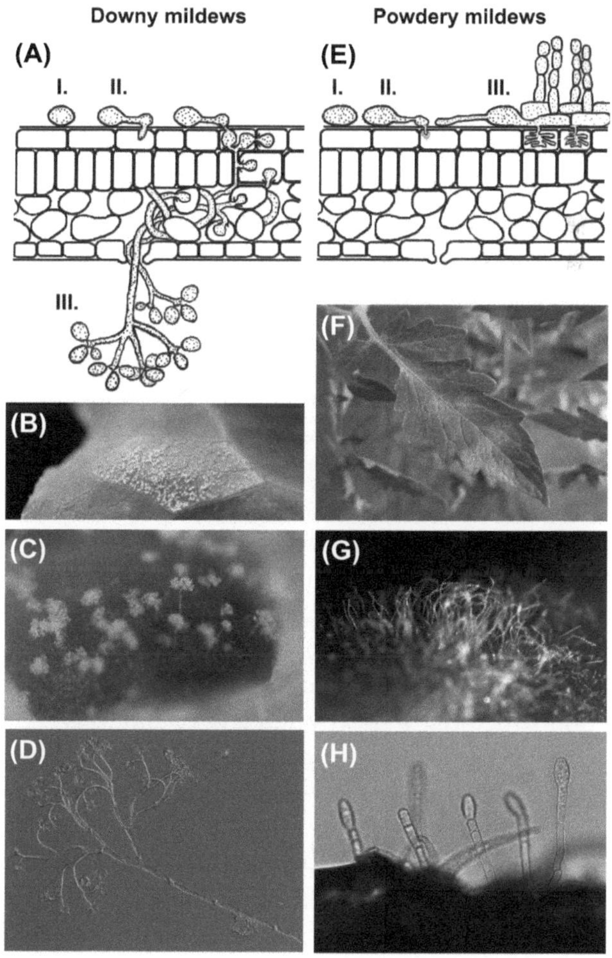

Figure 1 Comparison of downy (A–D) and powdery mildews (E–H). Developmental phases (A, E) represent: (I) spore deposition, (II) germination followed by penetration, (III) development of mycelia (A – endotrophic, E – ectotrophic) and formation of asexual spores. Typical symptoms on leaves include sporulation on abaxial side, often limited by veins, for downy mildews (B, C) and sporulation on adaxial side by powdery mildews (F, G). Details of conidiophore morphology are demonstrated for *Bremia lactucae* (D) and *Oidium neolycopersici* (H). *Drawing and photos: M. Sedlářová.* (See colour plate)

mildew in lettuce, with a high economic impact on lettuce crop yields worldwide. Due to high genotypic and phenotypic variability, the *Lactuca* spp. *B. lactucae* pathosystem has been used to study plant biotrophic oomycete interactions from the field conditions to molecular level (Lebeda & Pink, 1998; Lebeda, Pink & Mieslerová, 2001; Lebeda, Pink, & Astley, 2002; Lebeda, Sedlářová, Lynn, & Pink, 2006; Lebeda, Sedlářová, Petřivalský, & Prokopová, 2008).

In histological characterization of lettuce and wild *Lactuca* spp. responses to *B. lactucae*, the initiation of HR was shown to correlate with pathogen race developmental stages (Lebeda & Pink, 1998; Lebeda et al., 2006; Sedlářová, Lebeda, & Pink, 2001) influenced by ROS levels and antioxidative enzyme activities (Lebeda et al., 2008; Sedlářová et al., 2007). During the pathogen penetration of epidermal cells several hours post inoculation, local concentrations of NO and ROS strongly increase, probably released by both pathogen and host cells. In susceptible interaction, *B. lactucae* establishes an infection by forming the hyphae between plant cells and haustoria in adjacent mesophyll cells. Initiation of HR that confers race-specific resistance is associated with a high accumulation of NO and H_2O_2 early after the inoculation, though its timing and intensity varies among the specific interactions (Lebeda et al., 2008; Sedlářová et al., 2007). Activities of antioxidant enzymes, mainly of peroxidase, were found to be induced in plant tissues neighbouring the micro-HR, i.e. HR localized to a single penetrated cell and sometimes in several surrounding cells. Peroxidase activity and production of NO and H_2O_2 were detected in *B. lactucae* infection structures during initial stages of pathogenesis (Sedlářová et al., 2007, 2011). Similarly, a weak signal for NO was also detected in *Plasmopara halstedii* infection structures, a downy mildew of sunflower (Trojanová & Sedlářová, unpublished). In another study using NO modulators infiltrated into lettuce leaf discs, structures of *B. lactucae* localized outside of plant leaves (e.g. germ tubes and appressoria) were found less affected by NO donor and NO scavenger compared to primary and secondary vesicles formed within penetrated epidermal cells and intercellular hyphae (Petřivalský et al., 2007). This implies a possibility of plant—pathogen crosstalk similar to that previously suggested in the context of *B. cinerea* pathogenesis (Turrion-Gomez & Benito, 2011). NO released from the NO donor sodium nitroprusside delayed pathogen development in *Lactuca sativa* and *Lactuca saligna* but not in *Lactuca virosa* where it is slower anyway (Sedlářová, Lebeda, et al., 2001; Sedlářová et al., 2011). In contrast, the application of the NO scavenger 2-phenyl-4,4,5,5-tetramethylimidazoline-1-oxyl-3-oxide

(PTIO) resulted in an increased frequency of primary vesicles, but not of other infection structures. The ROS scavenger rutin reduced significantly pathogen invasion in all genotypes. In experiments with translaminar application of NO modulators, responses of *Lactuca* spp. genotypes were also affected, mainly in *L. virosa* where its characteristic extensive HR was decreased following treatment with NO donor, NO scavenger as well as ROS scavenger (Sedlářová et al., 2011). Our results indicated that higher concentration of antioxidants in tissue retarded downy mildew and that the fluctuation of the balance between ROS and NO may interfere with the onset of HR as previously suggested (Delledonne et al., 1998; Delledonne et al., 2003). This could indicate that changes in redox state affect plant responses to biotrophs in a process similar to their interactions with necrotrophs (Asai et al., 2010). In infected cells of resistant *Lactuca* spp., increased NO production was detected before visible HR symptoms (Sedlářová et al., 2011). This early NO burst was preceded by H_2O_2 accumulation (Sedlářová et al., 2007). The downstream targets of NO action include tyrosine nitration and *S*-nitrosylation of thiol groups of specific protein cysteine residues (Lin et al., 2012; Mengel, Chaki, Shekariesfahlan, & Lindermayr, 2013). Protein nitration was reported to be increased in susceptible but not in the resistant cultivar of sunflower infected by downy mildew (*P. halstedii*) (Chaki et al., 2009) similarly to potato interaction with hemibiotrophic *P. infestans* (Arasimowicz-Jelonek & Floryszak-Wieczorek, 2014). Post-translational protein modifications were proposed as a marker of nitrosative stress in the sunflower—downy mildew interaction (Chaki et al., 2009). Recently, cytoskeleton reorganization was proposed to be triggered by nitrosative stress (Lipka & Müller, 2014). A cyclic guanosine monophosphate (cGMP)-dependent protein kinase pathway regulates cytoskeleton dynamics and motility. Tyrosine nitration competes with or even blocks protein tyrosine phosphorylation which is crucial for both tubulin and actin functionality (Yemets, Krasylenko, Lytvyn, Sheremet, & Blume, 2011). Reorganization of cytoskeleton found in penetrated epidermal cells or in mesophyll cells with haustoria of *B. lactucae* (Sedlářová, Binarová, & Lebeda, 2001) might be regulated by such local nitrosative stress. Reorganization of cytoskeleton, microtubules and actin filaments, is a well-known phenomenon in plant cell—microbe interactions, involved in the intracellular transfer of organelles (Takemoto & Hardham, 2014). In what extent NO-dependent protein modifications affect signaling pathways and metabolism in pathogen cells, however, still remains to be elucidated in details.

Both partners in plant–fungus interactions have been shown to produce NO (Mur et al., 2006; Piterková et al., 2011, 2009; Prats et al., 2008) and similar relationship can be expected in oomycete pathogenesis. NO was localized in *B. lactucae* and *P. halstedii* by confocal microscopy namely in the tips of germ tubes, appressoria and haustoria (Sedlářová et al., 2011; Trojanová & Sedlářová, unpublished). The source of NO in downy mildews has not been identified so far. Functional NR has been demonstrated in *P. infestans* (Pieterse, van't Klooster, van den Berg-Velthuis, & Govers, 1995) but there are no reports on the potential involvement of NR in NO production in oomycetes. In our leaf disc tests, the growth of *B. lactucae* was not affected by NOS or NR inhibitors (Petřivalský et al., 2007). Unfortunately, downy mildews as biotrophic endoparasites cannot be cultivated in vitro or detached off infected tissues, which makes difficult attempts to identify NO sources in pathogen structures.

5. NO IN THE PATHOGENESIS OF POWDERY MILDEWS

In vivo imaging using fluorescein-based NO probes and confocal microscopy suggested the presence of NO in cytoplasm of several biotrophic phytopathogens, mainly in powdery mildews (Figure 2) whose study is easier than with downy mildews due to their epiphytic mycelium (Piterková et al., 2011, 2009; Prats et al., 2008, 2005). However, similar to plant models, a caution is required in the interpretation of these results, as the specificity of used fluorescent probe for the detection of NO has been recently questioned (Rűmmer et al. 2012).

Oidium neolycopersici (Kiss et al., 2001) causes powdery mildew epidemics on glasshouse tomatoes (Lebeda et al., 2014). While the majority of tomato (*Solanum lycopersicum*) cultivars show high susceptibility to the powdery mildew, many sources of potential resistance have been identified among wild *Solanum* spp. (Lebeda & Mieslerová, 2002; Mieslerová, Lebeda, & Chetelat, 2000). Resistant *Solanum* spp. utilizes HR, papillae formation or both mechanisms, to prevent their cells from *O. neolycopersici* ingress (Huang, Groot, Meijer-Dekens, Niks, & Lindhout, 1998; Mieslerová, Lebeda, & Kennedy, 2004). It is known that HR does not abolish mycelium growth completely but it usually suppresses pathogen reproduction. Several types of HR, differing in extent and affectivity, have been recently distinguished among *Solanum* spp.—*O. neolycopersici* interactions (Seifi et al., 2014).

Interactions of *O. neolycopersici* with susceptible *S. lycopersicum* (cv. Amateur), moderately resistant *Solanum chmielewskii* (LA 2663) and highly

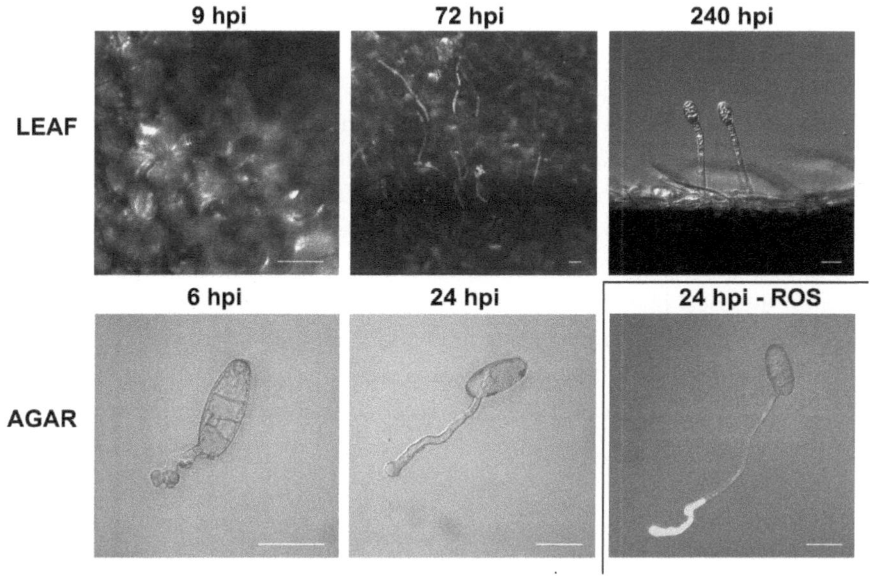

Figure 2 Nitric oxide (NO) and powdery mildews development. The presence of NO is observed during the germination (9 hpi) and formation of haustoria and mycelium (72 hpi) when grown on host plants or detached leaves. A weak signal for NO was also detected during the conidiogenesis (240 hpi). On the contrary, NO was not observed in *Oidium neolycopersici* conidia germinated on agar medium enriched with tomato leaf extract during 24 hpi, while a high ROS signal was localized in the tips of germ tubes. Bar represents 20 μm. *Photos: M. Sedlářová.* (See colour plate)

resistant *Solanum habrochaites* (LA 2128) have been studied from many perspectives (reviewed in Lebeda et al. (2014)). Initial work, focused on timing of pathogen development and host responses demonstrated that moderately resistant *S. chmielewskii* developed more intense HR than highly resistant *S. habrochaites*. Next, oxidative processes were analyzed and the degree of resistance was found to correlate with the intensity of hydrogen peroxide production and peroxidase activity while the extent of HR corresponded with CAT activity (Mlíčková, Luhová, Lebeda, Mieslerová, & Peč, 2004; Tománková et al., 2006). Later experiments focused on NO production disclosed NOS-like enzyme activity in pathogen structures as well as in host leaves and confirmed the necessity of NO for activation of both local and systemic tomato resistance (Piterková et al., 2011, 2009). Early burst of NO was recorded in penetrated cells of moderately resistant *S. chmielewskii* prior to visible signs of HR (Piterková et al., 2011), similarly to other plant–powdery mildew interactions (reviewed by Mur et al. (2006)). On the other hand, early increase of NO production in the susceptible *S. lycopersicum* (4–8 hours post inoculation (hpi)) seems to be effectively scavenged, thus creating a suitable environment for further pathogen development. In *S. chmielewskii*, the highest amounts of NO produced locally at 72 hpi could be explained by the involvement of NO in extensive HR, analogously to barley–powdery mildew interaction (Prats et al., 2005). Biphasic accumulation of NO was recorded in both resistant tomato genotypes on the local level in infected leaves as well as in the systemic response, namely in leaves upwards of the infected leaves. *Solanum chmielewskii*, a potential germplasm for breeding of tomatoes possessing moderate resistance in experimental conditions but field resistance to powdery mildew, exhibits pronounced changes in RNS and ROS during *O. neolycopersici* pathogenesis (Lebeda et al., 2014).

We investigated the effect of NO in *O. neolycopersici* development using translaminar application of NO modulators on leaf discs (Piterková et al., 2011). Increased NO levels activated defence pathways in the highly resistant genotype *S. habrochaites*, characterized by less extensive HR, but an accelerated pathogen development in the moderately resistant *S. chmielewskii*. NO scavenger stimulated pathogen growth in resistant genotypes, namely in *S. chmielewskii*, probably as a consequence of downregulation of NO-dependent pathways linked to HR. The animal NOS inhibitor L-NAME strongly inhibited powdery mildew germination and development, suggesting that an NOS-like enzyme could be the major source of NO in *O. neolycopersici* (Piterková et al., 2011, 2009). Since experiments using application of compounds presumably affecting NO level raised many questions on unspecific or dose-dependent effects, previously reported results require validation by intracellular analyses using additional analytical approaches (Mur et al., 2012). Histochemical staining combined with confocal microscopy was used to visualize ROS and RNS within the cytoplasm of powdery mildews in vivo. We suggested their presence in all phases of tomato powdery mildew development from conidia germination and appressorium formation, penetration to extensive mycelial growth (Piterková et al., 2011, 2009; Figure 2). Similarly, NO was observed in infection structures of barley powdery mildew, *Blumeria graminis* f. sp. *hordei*, in particular during the differentiation of appressoria (Prats et al., 2008) and development of haustoria and secondary hyphae (Prats et al., 2005). In contrast, a signal for NO detection was not detected in *O. neolycopersici* spores germinated in vitro on agar media, suggesting that a specific stimulation of NO production occurs following recognition of host surface (Figure 2). Recently, the role of NO in both compatible and incompatible interactions with powdery mildews was studied using *A. thaliana* plants infected either with the host-adapted *Golovinomyces orontii* and the nonadapted powdery mildew *Erysiphe pisi*, respectively (Schlicht & Kombrink, 2013). It was shown that *G. orontii* can apparently cope with NO produced by plant cells and even has an active role in modulating NO amounts in the infection sites, by its synthesis or degradation, leading to successful infection of host tissues. These results are in agreement with proposed multiple roles of NO in powdery mildew development, its recognition by plants and expression of host defence mechanisms. Nevertheless, the balance among NO, ROS and other molecules, such as reactive carbonyl species, remain to be better understood.

6. CONCLUSIONS

NO, in a tight connection with RNS and ROS, fulfils many divergent roles in plant responses to biotic stress stimuli. In parallel, NO and ROS are produced by plant pathogens, namely during the early phase of the pathogenesis including spore germination, oriented hypha growth and penetration of host plant cells. Within mutual interaction of plants and their biotrophic parasites, it is challenging to track the NO metabolism inside pathogen structures apart from the host tissues. However, current results support the view of NO belonging to key regulatory molecules in the life and infection cycles of both downy and powdery mildews, with multivalent and dose-dependent effects. Both plant and pathogen produce NO during their interaction with diverse spatiotemporal variability, influenced by plant genotype, pathogen race and feeding strategy and also by environmental variables. Fine balancing of NO, ROS and hormone levels in downy and powdery mildew development seems to be crucial for successful invasion to host cells. Future research can potentially address the relation of NO signaling to other pathways of pathogen metabolism during colonization of host tissues and fructification, as well as attempt to obtain a precise characterization of apparently divergent sources of NO production in oomycetes and fungi.

CONFLICT OF INTEREST

The authors declare that they have no conflict of interest.

ACKNOWLEDGEMENTS

This research was supported by Grant Agency of the Czech Republic (P501/12/0590), Ministry of Education, Youth and Sports of the Czech Republic (MSM 6198959215) and Internal Grant Agency of the Palacký University in Olomouc (PrF-2016-001). Downy and powdery mildews used in this study come from UPOC collection funded by Ministry of Agriculture of the Czech Republic as a part of the National Programme of Genepool Conservation of Microorganisms and Small Animals of Economic Importance.

REFERENCES

Arasimowicz-Jelonek, M., & Floryszak-Wieczorek, J. (2014). Nitric oxide: an effective weapon of the plant or the pathogen? *Molecular Plant Pathology, 15*, 406–416.

Arasimowicz, M., & Floryszak-Wieczorek, J. (2007). Nitric oxide as a bioactive signalling molecule in plant stress responses. *Plant Science, 172*, 876–887.

Asai, S., Mase, K., & Yoshioka, H. (2010). Role of nitric oxide and reactive oxide species in disease resistance to necrotrophic pathogens. *Plant Signaling and Behaviour, 5*, 872–874.

Astier, J., & Lindermayr, C. (2012). Nitric oxide-dependent posttranslational modification in plants: an update. *International Journal of Molecular Science, 13*, 15193–15208.

Bolwell, G. P., & Daudi, A. (2009). Reactive oxygen species in plant–pathogen interactions. In L. A. del Río, & A. Puppo (Eds.), *Reactive oxygen species in plant signalling* (pp. 113–133). Berlin, Heidelberg: Springer-Verlag.

Brunner, P. C., Torriani, S. F., Croll, D., Stukenbrock, E. H., & McDonald, B. A. (2013). Coevolution and life cycle specialization of plant cell wall degrading enzymes in a hemibiotrophic pathogen. *Molecular Biology and Evolution, 30*, 1337–1347.

Catalá, M., Gasulla, F., Pradas del Real, A., García-Breijo, F., Reig-Armiñana, J., & Barreno, E. (2010). Fungal-associated NO is involved in the regulation of oxidative stress during rehydration in lichen symbiosis. *BMC Microbiology, 10*, 297–300.

Chaki, M., Fernández-Ocaña, A. M., Valderrama, R., Carreras, A., Esteban, F. J., Luque, F., et al. (2009). Involvement of reactive nitrogen and oxygen species (RNS and ROS) in sunflower–mildew interaction. *Plant Cell and Physiology, 50*, 665–679.

Conrath, U., Amoroso, G., Kohle, H., & Sultemeyer, D. F. (2004). Noninvasive online detection of nitric oxide from plants and some other organisms by mass spectrometry. *Plant Journal, 38*, 1015–1022.

Corpas, F. J., Hayashi, M., Mano, S., Nishimura, M., & Barroso, J. B. (2009). Evidence supporting the existence of L-arginine-dependent nitric oxide synthase activity in plants. *New Phytologist, 184*, 9–14.

Delledonne, M., Polverari, A., & Murgia, I. (2003). The functions of nitric oxide-mediated signalling and changes in gene expression during the hypersensitive response. *Antioxidants & Redox Signaling, 5*, 33–41.

Delledonne, M., Xia, Y., Dixon, R. A., & Lamb, C. (1998). Nitric oxide functions as a signal in plant disease resistance. *Nature, 394*, 585–588.

Delledonne, M., Zeier, J., Marocco, A., & Lamb, C. (2001). Signal interactions between nitric oxide and reactive oxygen intermediates in the plant hypersensitive disease resistance response. *Proceedings of the National Academy of Sciences of the United States of America, 98*, 13454–13459.

El-Shetehy, M., Wang, C., Shine, M. B., Yu, K., Kachroo, A., & Kachroo, P. (2015). Nitric oxide and reactive oxygen species are required for systemic acquired resistance in plants. *Plant Signaling & Behavior, 10*, e998544.

Foresi, N., Correa-Aragunde, N., Parisi, G., Calo, G., Salerno, G., & Lamattina, L. (2010). Characterization of a nitric oxide synthase from the plant kingdom: NO generation from

the green alga *Ostreococcus tauri* is light irradiance and growth phase dependent. *Plant Cell, 22*, 3816–3830.
Gessler, N. N., Averyanov, A. A., & Belozerskaya, T. A. (2007). Reactive oxygen species in regulation of fungal development. *Biochemistry, 72*, 1091–1109.
Glazebrook, J. (2005). Contrasting mechanisms of defense against biotrophic and necrotrophic pathogens. *Annual Review of Phytopathology, 43*, 205–227.
Heller, J., & Tudzynski, P. (2011). Reactive oxygen species in phytopathogenic fungi: signaling, development, and disease. *Annual Review of Phytopathology, 49*, 369–390.
Hichri, I., Boscari, A., Castella, C., Rovere, M., Puppo, A., & Brouquisse, R. (2015). Nitric oxide: a multifaceted regulator of the nitrogen-fixing symbiosis. *Journal of Experimental Botany, 66*, 2877–2887.
Holzmeister, C., Gaupels, F., Geerlof, A., Sarioglu, H., Sattler, M., Durner, J., et al. (2015). Differential inhibition of Arabidopsis superoxide dismutases by peroxynitrite-mediated tyrosine nitration. *Journal of Experimental Botany, 66*, 989–999.
Huang, C.-C., Groot, T., Meijer-Dekens, F., Niks, R. E., & Lindhout, P. (1998). The resistance to powdery mildew (*Oidium lycopersicum*) in *Lycopersicon* species is mainly associated with hypersensitive response. *European Journal of Plant Pathology, 104*, 399–407.
Kiss, L., Cook, R. T. A., Saenz, G. S., Cunnington, J. H., Takamatsu, S., Pascoe, I., et al. (2001). Identification of two powdery mildew fungi, *Oidium neolycopersici* sp. nov. and *O. lycopersici*, infecting tomato in different parts of the world. *Mycological Research, 105*, 684–697.
Kunert, J. (1995). Effect of nitric oxide donors on survival of conidia, germination and growth of *Aspergillus fumigatus* in vitro. *Folia Microbiologica, 40*, 238–244.
Lazar, E. E., Wills, R. B. H., Ho, B. T., Harris, A. M., & Spohr, L. J. (2008). Antifungal effect of gaseous nitric oxide on mycelium growth, sporulation and spore germination of the postharvest horticulture pathogens, *Aspergillus niger, Monilinia fructicola* and *Penicillium italicum. Letters of Applied Microbiology, 46*, 688–692.
Lebeda, A., & Mieslerová, B. (2002). Variability in pathogenicity of *Oidium neolycopersici* on *Lycopersicon* species. *Journal of Plant Disease and Protection, 109*, 129–141.
Lebeda, A., Mieslerová, B., Petřivalský, M., Luhová, L., Špundová, M., Sedlářová, M., et al. (2014). Resistance mechanisms of wild tomato germplasm to infection of *Oidium neolycopersici. European Journal of Plant Pathology, 138*, 569–596.
Lebeda, A., & Pink, D. A. C. (1998). Histological aspects of the response of wild *Lactuca* spp. and their hybrids, with *L. sativa* to lettuce downy mildew (*Bremia lactucae*). *Plant Pathology, 47*, 723–736.
Lebeda, A., Pink, D. A. C., & Astley, D. (2002). Aspects of the interactions between wild *Lactuca* spp. and related genera and lettuce downy mildew (*Bremia lactucae*). In P. T. N. Spencer-Phillips, U. Gisi, & A. Lebeda (Eds.), *Advances in downy mildew research* (pp. 85–117). Dordrecht: Kluwer.
Lebeda, A., Pink, D. A. C., & Mieslerová, B. (2001). Host-parasite specificity and defense variability in the *Lactuca* spp. – *Bremia lactucae* pathosystem. *Journal of Plant Pathology, 83*, 25–35.
Lebeda, A., Sedlářová, M., Lynn, J., & Pink, D. A. C. (2006). Phenotypic and histological expression of different genetic backgrounds in interactions between lettuce, wild *Lactuca* spp., *L. sativa* × *L. serriola* hybrids and *Bremia lactucae*. *European Journal of Plant Pathology, 115*, 431–441.
Lebeda, A., Sedlářová, M., Petřivalský, M., & Prokopová, J. (2008). Diversity of defence mechanisms in plant-oomycete interactions: a case study of *Lactuca* spp.-*Bremia lactucae*. *European Journal of Plant Pathology, 122*, 71–89.
Lehmann, S., Serrano, M., L'Haridon, F., Tjamos, S. E., & Metraux, J. P. (2015). Reactive oxygen species and plant resistance to fungal pathogens. *Phytochemistry, 112*, 54–62.
Leitner, M., Vandelle, E., Gaupels, F., Bellin, D., & Delledonne, M. (2009). NO signals in the haze: nitric oxide signalling in plant defence. *Current Opinion in Plant Biology, 12*, 451–458.

Lin, A., Wang, Y., Tang, J., Xue, P., Li, C., Liu, L., et al. (2012). Nitric oxide and protein S-nitrosylation are integral to hydrogen peroxide-induced leaf cell death in rice. *Plant Physiology, 158*, 451–464.

Lipka, E., & Müller, S. (2014). Nitrosative stress triggers microtubule reorganization in *Arabidopsis thaliana*. *Journal of Experimental Botany, 65*, 4177–4189.

Maier, J., Hecker, R., Rockel, P., & Ninnemann, H. (2001). Role of nitric oxide synthase in the light-induced development of sporangiophores in *Phycomyces blakesleeanus*. *Plant Physiology, 126*, 1321–1330.

Martínez-Ruiz, A., Cadenas, S., & Lamas, S. (2011). Nitric oxide signaling: classical, less classical, and nonclassical mechanisms. *Free Radicals and Biology Medicine, 51*, 17–29.

Mengel, A., Chaki, M., Shekariesfahlan, A., & Lindermayr, C. (2013). Effect of nitric oxide on gene transcription – S-nitrosylation of nuclear proteins. *Frontiers in Plant Science, 4*, 293.

Micali, C., Göllner, K., Humphry, M., Consonni, C., & Panstruga, R. (2008). The powdery mildew disease of arabidopsis: a paradigm for the interaction between plants and biotrophic fungi. *The Arabidopsis Book, 6*, e0115.

Mieslerová, B., Lebeda, A., & Chetelat, R. T. (2000). Variation in response of wild *Lycopersicon* and *Solanum* spp. against tomato powdery mildew (*Oidium lycopersici*). *Journal of Phytopathology, 148*, 303–311.

Mieslerová, B., Lebeda, A., & Kennedy, R. (2004). Variation in *Oidium neolycopersici* development on host and nonhost plant species and their tissue defence response. *Annals of Applied Biology, 144*, 237–248.

Mlíčková, K., Luhová, L., Lebeda, A., Mieslerová, B., & Peč, P. (2004). Reactive oxygen species generation and peroxidase activity during *Oidium neolycopersici* infection on *Lycopersicon* species. *Plant Physiology and Biochemistry, 42*, 753–761.

Mur, L. A., Carver, T. L., & Prats, E. (2006). NO way to live; the various roles of nitric oxide in plant–pathogen interactions. *Journal of Experimental Botany, 57*, 489–505.

Mur, L. A., Mandon, J., Cristescu, S. M., Harren, F. J., & Prats, E. (2011). Methods of nitric oxide detection in plants: a commentary. *Plant Science, 181*, 509–519.

Mur, L. A. J., Kenton, P., Lloyd, A. J., Ougham, H., & Prats, E. (2008). The hypersensitive response; the centenary is upon us but how much do we know? *Journal of Experimental Botany, 59*, 501–520.

Mur, L. A. J., Mandon, J., Persijn, S., Cristescu, S. M., Moshkov, I. E., Novikova, G. V., et al. (2012). Nitric oxide in plants: an assessment of the current state of knowledge. *Annals of Botany Plants, 5*. pls052.

Mur, L. A. J., Prats, E., Pierre, S., Hall, M. A., & Hebelstrup, K. H. (2013). Integrating nitric oxide into salicylic acid and jasmonic acid/ethylene plant defense pathways. *Frontiers in Plant Science, 4*, 215.

Panstruga, R. (2003). Establishing compatibility between plants and obligate biotrophic pathogens. *Current Opinion in Plant Biology, 6*, 320–326.

Petřivalský, M., Kočířová, J., Sedlářová, M., Piterková, J., Luhová, L., & Lebeda, A. (2007). On the role of nitric oxide in *Bremia lactucae* pathogenesis on *Lactuca sativa*. In A. Lebeda, & P. T. N. Spencer-Phillips (Eds.), *Advances in downy mildew research* (Vol. 3, pp. 175–184). Olomouc and Kostelec na Hané, Czech Republic: Palacký University in Olomouc and JOLA Publishers.

Pieterse, C. M., van't Klooster, J., van den Berg-Velthuis, G. C., & Govers, F. (1995). NiaA, the structural nitrate reductase gene of *Phytophthora infestans*: isolation, characterization and expression analysis in *Aspergillus nidulans*. *Current Genetics, 27*, 359–366.

Piterková, J., Hofman, J., Mieslerová, B., Sedlářová, M., Luhová, L., Lebeda, A., et al. (2011). Dual role of nitric oxide in *Solanum* spp.–*Oidium neolycopersici* interactions. *Environmental and Experimental Botany, 74*, 37–44.

Piterková, J., Petřivalský, M., Luhová, L., Mieslerová, B., Sedlářová, M., & Lebeda, A. (2009). Local and systemic production of nitric oxide in tomato responses to powdery mildew infection. *Molecular Plant Pathology, 10*, 201–213.

Planchet, E., Sonoda, M., Zeier, J., & Kaiser, W. M. (2006). Nitric oxide (NO) as an intermediate in the cryptogein induced hypersensitive response – a critical re-evaluation. *Plant Cell & Environment, 29*, 59–69.

Prats, E., Carver, T. L. W., & Mur, L. A. J. (2008). Pathogen-derived nitric oxide influences formation of the appressorium infection structure in the phytopathogenic fungus *Blumeria graminis*. *Research in Microbiology, 159*, 476–480.

Prats, E., Mur, L. A. J., Sanderson, R., & Carver, T. L. W. (2005). Nitric oxide contributes both to papilla-based resistance and the hypersensitive response in barley attacked by *Blumeria graminis* f. sp. *hordei*. *Molecular Plant Pathology, 6*, 65–78.

del Río, L. A., Corpas, F. J., & Barroso, J. B. (2004). Nitric oxide and nitric oxide synthase activity in plants. *Phytochemistry, 65*, 783–792.

Rümer, S., Krischke, M., Fekete, A., Mueller, M. J., & Kaiser, W. M. (2012). DAF-fluorescence without NO: elicitor treated tobacco cells produce fluorescing DAF-derivatives not related to DAF-2 triazol. *Nitric Oxide, 27*, 123–135.

Saito, S., Yamamoto-Katou, A., Yoshioka, H., Doke, N., & Kawakita, K. (2006). Peroxynitrite generation and tyrosine nitration in defense responses in tobacco BY-2 cells. *Plant Cell Physiology, 47*, 689–697.

Samalova, M., Johnson, J., Illes, M., Kelly, S., Fricker, M., & Gurr, S. (2013). Nitric oxide generated by the rice blast fungus *Magnaporthe oryzae* drives plant infection. *New Phytologist, 197*, 207–222.

Scheler, C., Durner, J., & Astier, J. (2013). Nitric oxide and reactive oxygen species in plant biotic interactions. *Current Opinion in Plant Biology, 16*, 534–539.

Schlicht, M., & Kombrink, E. (2013). The role of nitric oxide in the interaction of *Arabidopsis thaliana* with the biotrophic fungi, *Golovinomyces orontii* and *Erysiphe pisi*. *Frontiers in Plant Science, 4*, 351.

Sedlářová, M., Binarová, P., & Lebeda, A. (2001). Changes in microtubular alignment in *Lactuca* spp. (Asteraceae) epidermal cells during early stages of infection by *Bremia lactucae* (Peronosporaceae). *Phyton, 41*, 21–34.

Sedlářová, M., Lebeda, A., & Pink, D. A. C. (2001). The early stages of interaction between effective and non-effective race-specific genes in *Lactuca sativa*, wild *Lactuca* spp. and *Bremia lactucae* (race NL16). *Journal of Plant Disease and Protection, 108*, 477–489.

Sedlářová, M., Luhová, L., Petřivalský, M., & Lebeda, A. (2007). Localization and metabolism of reactive oxygen species during *Bremia lactucae* pathogenesis in *Lactuca sativa* and wild *Lactuca* spp. *Plant Physiology and Biochemistry, 45*, 607–616.

Sedlářová, M., Petřivalský, M., Piterková, J., Kočířová, J., Luhová, L., & Lebeda, A. (2011). Influence of nitric oxide and reactive oxygen species on development of lettuce downy mildew in *Lactuca* spp. *European Journal of Plant Pathology, 129*, 267–280.

Seifi, A., Gao, D., Zheng, Z., Pavan, S., Visser, R. G. F., Wolters, A.-M. A., & Bai, Y. (2014). Genetics and molecular mechanisms of resistance to powdery mildews in tomato (*Solanum lycopersicum*) and its wild relatives. *European Journal of Plant Pathology, 138*, 641–665.

Servent, D., Ducrocq, C., Henry, Y., Guissani, A., & Lenfant, M. (1991). Nitroglycerin metabolism by *Phanerochaete chrysosporium*: evidence for nitric oxide and nitrite formation. *Biochimica et Biophysica Acta, 1074*, 320–325.

Song, N. K., Jeong, C. S., & Choi, H. S. (2000). Identification of nitric oxide synthase in *Flammulina velutipes*. *Mycologia, 92*, 1027–1032.

Stöhr, C., & Ullrich, W. R. (2002). Generation and possible roles of NO in plant roots and their apoplastic space. *Journal of Experimental Botany, 53*, 2293–2303.

Tada, Y., Mori, T., Shinogi, T., Yao, N., Takahashi, S., & Betsuyaku, S. (2004). Nitric oxide and reactive oxygen species do not elicit hypersensitive cell death but induce apoptosis in the adjacent cells during the defence response of oat. *Molecular Plant—Microbe Interactions, 17*, 245–253.

Takemoto, D., & Hardham, A. R. (2004). The cytoskeleton as a regulator and target of biotic interactions in plants. *Plant Physiology, 136*, 3864–3876.

Terrile, M. C., Mansilla, A. Y., Albertengo, L., Rodríguez, M. S., & Casalongué, C. A. (2015). Nitric-oxide-mediated cell death is triggered by chitosan in *Fusarium eumartii* spores. *Pest Management Science, 71*, 668–674.

Tománková, K., Luhová, L., Petřivalský, M., Peč, P., & Lebeda, A. (2006). Biochemical aspects of reactive oxygen species formation in the interaction between *Lycopersicon* spp. and *Oidium neolycopersici*. *Physiological and Molecular Plant Pathology, 68*, 22–32.

Trapet, P., Kulik, A., Lamotte, O., Jeandroz, S., Bourque, S., Nicolas-Francès, V., et al. (2015). NO signaling in plant immunity: a tale of messengers. *Phytochemistry, 112*, 72–79.

Turrion-Gomez, J. L., & Benito, E. P. (2011). Flux of nitric oxide between the necrotrophic pathogen *Botrytis cinerea* and the host plant. *Molecular Plant Pathology, 12*, 606–616.

Turrion-Gomez, J. L., Eslava, A. P., & Benito, E. P. (2010). The flavohemoglobin BCFHG1 is the main NO detoxification system and confers protection against nitrosative conditions but is not a virulence factor in the fungal necrotroph *Botrytis cinerea*. *Fungal Genetics and Biology, 47*, 484–496.

Vandelle, E., & Delledonne, M. (2011). Peroxynitrite formation and function in plants. *Plant Science, 181*, 534–539.

Wang, J., & Higgins, V. J. (2005). Nitric oxide has a regulatory effect in the germination of conidia of *Colletotrichum coccodes*. *Fungal Genetics and Biology, 42*, 284–292.

Wang, Y., Loake, G., & Chu, C. (2013). Cross-talk of nitric oxide and reactive oxygen species in plant programed cell death. *Frontiers in Plant Science, 4*, 314.

Wendehenne, D., Durner, J., & Klessig, D. F. (2004). Nitric oxide: a new player in plant signalling and defence responses. *Current Opinion in Plant Biology, 7*, 449–455.

Wilson, I. D., Neill, S. J., & Hancock, J. T. (2008). Nitric oxide synthesis and signalling in plants. *Plant Cell & Environment, 31*, 622–631.

Xu, S., Guerra, D., Lee, U., & Vierling, E. (2013). S-nitrosoglutathione reductases are low-copy number, cysteine-rich proteins in plants that control multiple developmental and defense responses in Arabidopsis. *Frontiers in Plant Science, 4*, 430.

Yamasaki, H. (2000). Nitrite-dependent nitric oxide production pathway: implications for involvement of active nitrogen species in photoinhibition in vivo. *Philosophical Transaction of Royal Society London B, 355*, 1477–1488.

Yamasaki, H., & Cohen, M. F. (2006). NO signal at the crossroads: polyamine-induced nitric oxide synthesis in plants? *Trends in Plant Science, 11*, 522–524.

Yemets, A. I., Krasylenko, Y. A., Lytvyn, D. I., Sheremet, Y. A., & Blume, Y. B. (2011). Nitric oxide signalling via cytoskeleton in plants. *Plant Science, 181*, 545–554.

Yoshioka, H., Asai, S., Yoshioka, M., & Kobayashi, M. (2009). Molecular mechanisms of generation for nitric oxide and reactive oxygen species, and role of the radical burst in plant immunity. *Molecules and Cells, 28*, 321–329.

Yu, M., Lamattina, L., Spoel, S. H., & Loake, G. J. (2014). Nitric oxide function in plant biology: a redox cue in deconvolution. *New Phytologist, 202*, 1142–1156.

Zaninotto, F., La Camera, S., Polverari, A., & Delledonne, M. (2006). Cross talk between reactive nitrogen and oxygen species during the hypersensitive disease resistance response. *Plant Physiology, 141*, 379–383.

Zeier, J., Delledonne, M., Mishina, T., Severi, E., Sonoda, M., & Lamb, C. (2004). Genetic elucidation of nitric oxide signalling in incompatible plant-pathogen interactions. *Plant Physiology, 136*, 2875–2888.

Zemojtel, T., Frohlich, A., Palmieri, M. C., Kolanczyk, M., Mikula, I., Wyrwicz, L. S., et al. (2006). Plant nitric oxide synthase: a never-ending story? *Trends in Plant Science, 11*, 524–525.

Zhang, Z., Wang, J., Chai, R., Qiu, H., Jiang, H., Mao, X., et al. (2015). An S-(hydroxymethyl) glutathione dehydrogenase is involved in conidiation and full virulence in the rice blast fungus *Magnaporthe oryzae*. *PLoS One, 10*, e0120627.

Zhou, S., Fushinobu, S., Kim, S., Nakanishi, Y., Wakagi, T., & Shoun, H. (2010). *Aspergillus oryzae* flavohemoglobins promote oxidative damage by hydrogen peroxide. *Biochemical and Biophysical Research Communications, 394*, 558–561.

CHAPTER FOURTEEN

NO and Ca^{2+}: Critical Components of Cytosolic Signaling Systems Involved in Stomatal Immune Responses

Yi Ma[1], Gerald A. Berkowitz[1]
Department of Plant Science and Landscape Architecture, University of Connecticut, Storrs, CT, USA
[1]Corresponding authors: E-mail: yi.ma@uconn.edu; gerald.berkowitz@uconn.edu

Contents

Abstract

Nitric oxide (NO) is a secondary messenger involved in a wide range of signal transduction pathways, including plant immune responses. NO production can be triggered by various microbial pathogens and endogenous defense signaling molecules. Some phyllosphere pathogens invade plant leaves through the leaf surface stomatal pore formed by a pair of guard cells. To prevent the first-line attack, a series of innate immune signaling cascades occur in the guard cells that trigger stomatal closure. NO is one of the critical components involved in stomatal closure, evidenced by using pharmacological reagents and genetic resources. NO biosynthesis in plant cells is thought to occur by two pathways. One is the L-arginine-dependent pathway and the other is the nitrate reduction pathway. However, enzymes responsible for NO synthesis in the L-arginine-dependent pathway are still unidentified. Therefore the currently available NO synthesis related mutants are *nia1*, *nia2* (the two genes encoding nitrate reductase) and *noa1*, an NO-associated gene mistakenly characterized as the NO

Advances in Botanical Research, Volume 77
ISSN 0065-2296
http://dx.doi.org/10.1016/bs.abr.2015.11.001

synthase previously. Studies on stomatal defense signaling have demonstrated an interdependency between NO and several other secondary messengers, such as Ca^{2+}, reactive oxygen species and cGMP and the gasotransmitter H_2S. These findings indicate that NO is downstream from those signaling molecules. However, recent studies have shown a distinct function of NO that NO-derived moieties bind to certain molecules, resulting in S-nitrosylation of cysteine and tyrosine nitration. NO-associated modification has been identified in several molecules involved in guard cell signaling, indicating a feed-forward role that NO could play in the guard cell. Perspectives and insights on how NO contributes to stomatal immunity through its interplay with other signaling molecules using genetic approaches are proposed and discussed.

1. INTRODUCTION

Nitric oxide (NO) is a free radical gas identified as a ubiquitous secondary messenger involved in a plethora of physiological processes in plants, such as seed germination, flowering, pollen tube growth, root growth, stomatal closure, fruit ripening and senescence (Besson-Bard, Pugin, & Wendehenne, 2008; Simontacchi, García-Mata, Bartoli, Santa-María, & Lamattina, 2013). In the past two decades, an extensive body of evidence has supported the essential role of NO in plant defense responses to invading pathogens by triggering hypersensitive response (HR) or upregulating defense gene expression (Dangl, 1998; Durner, Wendehenne, & Klessig, 1998; Jeandroz et al., 2013; Ma, 2011; Ma, Zhao, Walker, & Berkowitz, 2013; Polverari et al., 2003; Romero-Puertas, Perazzolli, Zago, & Delledonne, 2004; Trapet et al., 2014; Wendehenne, Pugin, Klessig, & Durner, 2001; Zottini & Costa, 2009). Although NO generation contributing to HR development is due to L-Arginine (L-Arg)-dependent nitric oxide synthase (NOS) activity (Delledonne, Xia, Dixon, & Lamb, 1998; Zhang, Czymmek, & Shapiro, 2003), the exact NO enzymatic biosynthesis pathway is still a controversial topic (Moreau, Lindermayr, Durner, & Klessig, 2010). Nevertheless, the application of a mammalian NOS inhibitors diminishes HR in Arabidopsis and *Nicotiana* spp. (Delledonne et al., 1998; Huang & Knopp, 1998). Inoculation with an avirulent pathogen elicits an NO burst in Arabidopsis leaf tissue (Zeier et al., 2004; Zhang et al., 2003) and this NO generation is blocked by an NOS inhibitor (Zhang et al., 2003).

Another critical and early component in plant immune signaling cascades is Ca^{2+}, whose cytosolic concentration is transiently elevated causing a Ca^{2+} burst. The perception of evolutionarily conserved components of microbial invaders known as microbe-associated molecular patterns (MAMPs), such as flagellin, elongation factor-Tu (EF-Tu), lipopolysaccharide (LPS), chitin or damage-associated molecular patterns (DAMP) by plant cells triggers a rapid cytosolic Ca^{2+} ($[Ca^{2+}]_{cyt}$) elevation, which is required to activate downstream molecular and physiological processes (Ali et al., 2007; Aslam et al., 2009; Jeworutzki et al., 2010; Kwaaitaal, Huisman, Maintz, Reinstädler, & Panstruga, 2011; Ma, Walker, Zhao, & Berkowitz, 2012; Ma et al., 2013). It has been well established that there is a connection between $[Ca^{2+}]_{cyt}$ elevation and NO synthesis during plant innate immune response (Ali et al., 2007; Lamotte et al., 2004; Lecourieux, Ranjeva, & Pugin, 2006). NO also forms an interaction network with other defense signaling molecules, such as reactive oxygen species (ROS), cyclic nucleotides, mitogen-activated protein kinases (MAPKs; MPK is used when describing specific genes) and phosphatidic acid, to regulate the defense responses in plant cells (Gaupels, Kuruthukulangarakoola, & Durner, 2011). Thus, NO may play a role in immune signaling upstream from the Ca^{2+} burst in post-translational modification (PTM) of signaling proteins, and also may be involved in the immune cascade downstream from the Ca^{2+} burst which leads to NO generation.

Stomata are minute pores formed by a pair of guard cells located at the leaf epidermis. They function like 'mouths' in plants for regulating gas exchange and transpiratory water loss between the exterior environment and leaf surface. Additionally, stomata act as gateways for the entry of pathogens. The entry of some pathogenic microbes into the plant interior is controlled by stomata movement. Stomatal opening and closing occurs through turgor pressure changes in the pairs of guard cells forming the stomatal pore. Changes in guard cell turgor pressure can be regulated by the plant hormone abscisic acid (ABA) and cytosolic secondary messengers such as NO (García-Mata & Lamattina, 2013; Gayatri, Agurla, & Raghavendra, 2013; Hancock, Neill, & Wilson, 2011; Yoshioka, Mase, Yoshioka, Kobayashi, & Asai, 2011).

This chapter discusses and updates knowledge on recent findings of the interdependent relationship of NO and Ca^{2+} generation and how these signaling molecules contribute to stomatal innate immune responses to pathogens and danger signals.

2. NO AND Ca^{2+} INVOLVE IN PLANT INNATE IMMUNITY

Plant defense to biotic stress is considered to be initiated by the recognition of exogenous MAMPs or endogenous DAMPs by plasma membrane localized receptors (Gómez-Gómez & Boller, 2000; Gómez-Gómez,

Felix, & Boller, 1999; Miya et al., 2007; Yamaguchi, Huffaker, Bryan, Tax, & Ryan, 2010; Yamaguchi, Pearce, & Ryan, 2006; Zipfel et al., 2006). LPS is an MAMP derived from the outer membrane of Gram-negative bacteria that activates plant cell defenses; however its plant receptor is still unrevealed. It has been generally accepted that $[Ca^{2+}]_{cyt}$ elevation is a critical step during plant innate immune response. Upon ligand and receptor binding, a cytosolic Ca^{2+} burst is triggered within minutes, which is thought to be upstream of other steps of immune signaling, such as MAPK phosphorylation cascades, ROS and NO generation, gene regulation as well as HR (Gao, Cox, & He, 2014; Lecourieux et al., 2006). Very recently, a bulb-type (B-type) lectin S-domain-1 receptor like kinase was identified through screening an ethyl methanesulfonate mutagenized Arabidopsis population for mutants showing low or no $[Ca^{2+}]_{cyt}$ induction upon LPS treatment. One mutant was identified and named as lipooligosaccharide-specific reduced elicitation (LORE). The *lore* mutant showed no $[Ca^{2+}]_{cyt}$ elevation and ROS generation after LPS treatment. It was also more susceptible to *Pseudomonas syringae* pv. *tomato* DC3000 (*Pst* DC3000) compared to the wild type (wt) and lost the ability to gain immunization triggered by LPS pretreatment (Ranf et al., 2015). These results suggest the possibility that LORE is a candidate for an LPS receptor.

Families of Ca^{2+} conducting channels reported in plants include cyclic nucleotide gated channels (Ali et al., 2007; CNGCs, Leng, Mercier, Yao, & Berkowitz, 1999), glutamate receptors (GLRs, Qi, Stephens, & Spalding, 2006), inositol-1,4,5-phosphate (InsP3) activated channels and recently annexins (Davies, 2014), whose function in plant immune responses is still ambiguous. Despite this knowledge of the molecules that facilitate Ca^{2+} conductance across plant cell membranes, there is still no conclusive understanding of the route of cellular Ca^{2+} movement after cells perceive stress signals. A reasonable amount of genetic and pharmacological evidence indicates that some CNGCs and GLRs are involved in $[Ca^{2+}]_{cyt}$ elevation during plant defense responses to MAMPs or DAMPs. CNGC2 and CNGC4 are the best characterized channel proteins involved in plant defense responses to bacteria pathogens. *dnd1*, a CNGC2 truncation mutant, attenuates LPS- and plant elicitor peptide (Pep, a DAMP)-induced $[Ca^{2+}]_{cyt}$ increase (Ma et al., 2009; Ma et al., 2012; Qi et al., 2010). *CNGC2* mutation, on the other hand, does not affect flg22- (a peptide corresponding to the active epitope of the bacterial flagellin protein) (Ma et al., 2012) and elf18 (another peptide from the bacterial MAMP EF-Tu; Ma et al., unpublished)-induced $[Ca^{2+}]_{cyt}$ increase, indicating that the $[Ca^{2+}]_{cyt}$

elevation induced by different ligands has discrete sources. The in-frame chimeric CNGC11/12 (a functional chimeric CNGC comprised of the front half of *AtCNGC11* and the second half of *AtCNGC12*) is found to activate multiple pathogen defense responses, however the function of these two channel proteins in Ca^{2+}-mediated resistance signaling is unclear although they are functional Ca^{2+} channels as confirmed by complementation of mutant yeast growth on ion-selective media (Abdel-Hamid et al., 2013; Chin et al., 2010).

The GLR family contains 20 members as the CNGC family in Arabidopsis. Much less study has been conducted on GLRs than CNGCs in plant defense signaling. To date, no Ca^{2+} current or $[Ca^{2+}]_{cyt}$ burst observed in plant cells in response to MAMPs or DAMPs has been definitely linked to GLR function. In the past couple of years, genetic evidence of GLRs involvement in plant immune response (Li et al., 2013; Manzoor et al., 2013) complement previous pharmacological studies (Kwaaitaal et al., 2011; Vatsa et al., 2011). *Atglr3.3* null mutant demonstrated increased susceptibility to the *P. syringae*, correlating with defective activation of defense gene expression in response to infection. Glutathione enhanced the immunity to *P. syringae* and suppressed bacterial growth in an *AtGLR3.3* dependent manner (Li et al., 2013). Manzoor et al. (2013) showed that a DAMP oligogalacturonide (OG)-induced $[Ca^{2+}]_{cyt}$ elevation was highly impaired by glutamate antagonists whereas *atglr3.3* mutant did not affect OG-induced $[Ca^{2+}]_{cyt}$ elevation. In addition, *AtGLR3.3* was also found to be required for basal resistance to oomycete *Hyaloperonospora arabidopsidis* and failed to activate defense gene expression in response to infection (Manzoor et al., 2013). This indicates that there may be functional redundancy among clade 3 GLRs in conducting OG-induced Ca^{2+} or an internal Ca^{2+} source contributes to the process.

InsP3 signaling in animals has been studied extensively. In guard cells, InsP3 and inositol hexaphosphate (InsP6) were characterized to activate plasma membrane Ca^{2+} channels and inactivate inward K^+ channels, leading to stomatal closure (Blatt, Thiel, & Trentham, 1990; Gilroy, Read, & Trewavas, 1990; Lemtiri-Chlieh et al., 2003). However, since no InsP3 receptor has been identified in plants, InsP3 as a signal molecule involved in internal Ca^{2+} mobilization and plant defense responses is still controversial because of the abundant intermediates in the inositol phosphate biosynthesis pathway (Laha et al., 2015). InsP3 was first characterized as a signal molecule playing a role in gravitropism and negatively regulating drought stress responses (Perera, Hung, Brady, Muday, & Boss, 2006; Perera, Hung, Moore,

Stevenson-Paulik, & Boss, 2008). Later, Ma et al. (2012) found that InsP3 is responsible for flg22 induced $[Ca^{2+}]_{cyt}$ elevation which was not observed after applying Pep or elf18 (Ma et al., unpublished). Recent findings with transgenic Arabidopsis plants constitutively expressing the human type I inositol polyphosphate 5-phosphatase (InsP5-ptase) (the enzyme hydrolyzes InsP3 and thus prevents InsP3 generation during a signaling cascade), showed impaired $[Ca^{2+}]_{cyt}$ induction and defense gene expression but not ROS production, indicating the importance of InsP3 in flg22 signaling. This work indicates that InsP3 may act downstream of ROS production (Hung, Aspesi, Hunter, Lomax, & Perera, 2014). On the other hand, InsP3 delays $[Ca^{2+}]_{cyt}$ increase induced by avirulent *P. syringae avrRpm1* and responses involved in systemic acquired resistance. This indicates that InsP3 may have a more specific role in flg22-induced defense signaling.

As mentioned above, $[Ca^{2+}]_{cyt}$ elevation is an upstream and rapid cellular response to pathogen invasion or danger signals derived from adjacent cells. NO as well is a signaling molecule during immune signaling cascades and in most cases NO production is Ca^{2+}-dependent. Previous work on MAMP and DAMP signaling has gathered substantial evidence that NO is a component in immune signaling cascades (Gerber, Zeidler, Durner, & Dubery, 2004; Ma et al., 2009; Ma et al., 2013; Rasul et al., 2012; Sun & Li, 2013; Sun, Nie, & Xing, 2012; Zuppini et al., 2004) despite the absence of knowledge on key enzymes of the L-Arg-dependent NO biosynthetic pathway in plants. Plant defense response to perception of pathogen presence is conceived of as occurring through two immune cascades, MTI (MAMP triggered immunity), which activates defense gene expression, and ETI (effector triggered immunity), which leads to programmed cell death and HR. Studies have showed that NO plays a determinant role in the onset of HR (Chen, Vandelle, Bellin, & Delledonne, 2014; Delledonne et al., 1998; Durner et al., 1998; Mur, Kenton, Lloyd, Ougham, & Prats, 2008). NO accumulation was increased in the NO-oxidizing haemoglobins mutant *glb1* while reduced in the 35S-GLB1 overexpression line upon infection by both *P. syringae AvrRpm1* and *Botrytis cinerea*. These observations are consistent with the same NO accumulation dynamic patterns induced by salicylate (SA) and jasmonate/ethylene (JA/Et) respectively. This indicates the function of NO as a partner in the defense response promoting its potentiation, by controlling the biphasic ethylene formation during the HR in plants subjected to pathogens (Mur et al., 2012).

NO as a regulator in plant defense response has been demonstrated using genetic tools and pharmacological reagents. NO donor sodium nitroprusside

(SNP) treatment induced visible hypersensitive cell death after 24 h (Polverari et al., 2003). In a report demonstrating an approach for *in planta* NO detection, L-NAME, a mammalian NOS inhibitor, and an NO scavenger cPTIO, were used to examine how NO depletion affects tobacco cell death caused by avirulent and virulent bacterial pathogens. Co-infiltration of the two chemicals with the avirulent strain *Ps* pv. *phaseolicola* delayed the HR by at least 3 h, implying that NO controls cell death kinetics. Furthermore, although bacteria growth was not changed by the two chemicals after co-infiltration with the virulent *Ps* pv. *tabaci*, necrosis in the leaves was significantly increased and bacteria escape from the lesion occurred to a greater extent (Mur, 2005). These results indicate that NO plays a role in ETI.

Although NO enzymatic pathway is still obscure, NO-related mutants and transgenic plant materials have been developed to explore how NO regulates plant defense activities. NO-associated1 (NOA1) was originally identified as NOS while it was later discovered that it does not have this activity (Crawford et al., 2006; Guo, Okamoto, & Crawford, 2003; Moreau, Lee, Wang, Crane, & Klessig, 2008; Zemojtel et al., 2006). However, as NO elevation in response to some signals is impaired in the *noa1* mutant (Guo et al., 2003), it is still a useful tool for studying NO-mediated signaling transduction. *Atnoa1* shows impaired defense gene (*MPK3* and *WRKY33*) upregulation induced by Pep or flg22, and Pep-triggered immunity against *Pst* DC3000 is compromised in the mutant, indicating that disrupted NO function makes plants more susceptible to pathogen attack; treatment with cPTIO also showed similar impairment in gene expression (Ma et al., 2013). An interesting piece of work showed that a genetic mutation in the oleic acid-synthesizing gene, suppressor of SA insensitivity of npr1-5 (SSI2), constitutively activated defense signaling. The decreased oleic acid levels in the *ssi2* mutant hyper-accumulated chloroplastic NO, and the loss function of *NOA1* and either of the two *NIA* genes (encoding nitrate reductase in NO synthesis pathway) partially restored the defense phenotype (Mandal et al., 2012). However, overexpression of *NOA1* did not enhance NO-mediated defense responses in wt plants unless oleic acid levels were lowered. They further demonstrated that oleic acid physically interacted with NOA1, leading to the degradation of NOA1 in a protease-dependent manner. These results suggest that oleic acid levels modulate NO accumulation, and thus NO-associated signaling, by regulating NOA1 levels. Study on oil palm (*Elaeis guineensis*) root tissues treated with *Ganoderma boninense*, the causal agent of basal stem rot disease, showed that NO burst occurred prior to increased *EgNOA1* expression, indicating that NO production is independent of *EgNOA1* but may contribute to increased *EgNOA1* transcription, which could be important for the resistance to this fungal pathogen (Kwan, Meon, Ho, & Wong, 2015). Expression of a bacterial nitric oxide dioxygenase in Arabidopsis decreased NO levels leading to H_2O_2 reduction, plant antioxidative machinery inhibition, decreased SA production, impaired development of HR-related cell death and a transcriptional delay in pathogenesis-related protein 1 (PR1) (Zeier et al., 2004). These findings showed that NO is involved in MTI and ETI induced by a broad range of pathogens.

NO is not only thought to trigger the development of hypersensitive cell death and defense gene upregulation, but also to modulate transcriptional reprogramming. Several medium- and large-scale transcriptome profiling analyses have provided valuable information on putative NO-responsive defense-related genes (Palmieri et al., 2008; Parani et al., 2004; Polverari et al., 2003). Whole genome microarray applied to Arabidopsis cell suspension cultures or plants treated with an NO donor NOR-3 or gaseous NO identified some common transcription factor-binding sites that bind to octopine synthase element-like sequences (these sequences are important for the expression of specific and pathogenesis-related genes) and WRKY family genes that regulate SA or JA signaling pathway (Palmieri et al., 2008). ATH1 microarray analysis of plants treated with SNP and NO identified NO specific responsive genes upregulated by NO but downregulated by cPTIO, including genes encoding disease-resistance proteins such as a mitogen-activated kinase MKK5 (Parani et al., 2004). More analyses of NO biosynthesis or turnover related mutants, such as *nia1*, *nia2*, *noa1* or *glb1* may be useful in future studies to explore NO associated transcriptional profiling.

3. NO AND Ca^{2+} SIGNALING IN STOMATAL INNATE IMMUNITY

Stomata are generally the entry port for bacteria invasion. Other phyllosphere microbes, such as fungi and oomycetes can also disrupt stomatal movement during infection. In order to fight the invaders at the first line of defense, guard cells actively regulate their own movement to shut the gate for the strangers (Melotto, Underwood, Koczan, Nomura, & He, 2006). Guard cells restrict pathogen entry by either closing the stomata pore or inhibiting the opening caused by virulence factors, such as coronatine (COR), a JA-mimicking phytotoxin (Melotto et al., 2006). Stomatal aperture changes due to the dynamic ion fluxes which causes turgor changes in guard cells. The turgor pressure of guard cells is reduced by the efflux of potassium ions and anions upon the perception of MAMPs and ABA, driving the closure of the stomata.

3.1 Stomatal Movement upon Perception of Pathogen Invasion

ABA is the key phytohormone that modulates stomata aperture. The core ABA signaling module contains pyrabactin resistance1/PYR1-LIKE/regulatory components of ABA receptor (PYR1/PYL/RCAR), a cluster of protein phosphatases 2C (PP2C) and three SNF1-related protein kinases. Secondary messengers such as Ca^{2+}, NO and ROS are also important regulators of stomatal closure in response to microbial stresses that may or may not involve ABA. Almost every MAMP identified in foliar microbes is able to promote stomatal closure, including flg22, elf18 or elf26 and LPS from bacteria, chitin and chitosan (a deacetylated derivative of chitin) from fungi and yeast (*Saccharomyces cerevisiae*) elicitors (YEL). Flg22, LPS and *Pst avrRpt2*$^{+}$ triggers NO generation in guard cells (Ali et al., 2007; Ma et al., 2009; Ma, Smigel, Tsai, Braam, & Berkowitz, 2008; Ma et al., 2013). LPS-induced NO generation requires the calcium channel CNGC2 and the calmodulin (CaM) gene *CML24*, indicating that Ca^{2+} is upstream of NO generation in response to LPS and *Pst avrRpt2*$^{+}$. In addition, a type of DAMP, the OGs were found to trigger stomatal closure and inhibit stomatal opening (Arnaud & Hwang, 2015). Another DAMP, the Peps, were found to induce ROS and NO generation in guard cells after treatment of epidermal peels (Ma et al., 2013), but there is no direct evidence suggesting a role of Pep/PEPR (Pep receptor) in stomatal closure. The elicitor-induced stomatal closure signals through the cognate receptors of the elicitors. The series of pattern recognition receptors (PRR), flagellin sensing2 (FLS2), the cognate receptor for flg22, EF-Tu receptor (EFR), the cognate receptor for EF-Tu, and chitin elicitor receptor kinase1 (CERK1), the cognate receptor for chitin, are all expressed in guard cells (Liu et al., 2009; Melotto et al., 2006; Mersmann, Bourdais, Rietz, & Robatzek, 2010). Use of the *fls2* mutant demonstrated that the flg22 receptor FLS2 was critical for stomatal closure in response to wt or COR-deficient *Pst* bacteria, while the *fls2* mutant still responded to LPS and non-pathogenic *Escherichia coli* (which also contain flagellin protein) to induce stomatal closure, suggesting that *fls2* is more specific to flagellin derived from *Pst*. Till now, there is still no direct genetic evidence of stomata movement induced by the binding of other known MAMP ligands to their cognate receptors. Overexpression of RCAR3 enhanced Arabidopsis resistance to surface applied *Pst DC3000* through the inhibition of COR-induced stomatal reopening (Lim, Luan, & Lee, 2014). However, stomatal movement does not correlate with SA-induced defense gene upregulation, indicating that ABA-induced stomatal closure and SA-dependent defense pathway are independent of each other. This makes sense because stomatal movement in stomatal defense response is mainly an electrophysiological signaling process, and defense gene regulation is a molecular and biochemical event. Also, ABA only plays a positive role in plant defense in guard cells by controlling stomatal closure and the entry of pathogens, but it negatively regulate post-invasion plant defense by antagonizing SA-induced defense responses (Lim, Baek, Jung, Kim, & Lee, 2015). Nevertheless, both defense pathways require $[Ca^{2+}]_{cyt}$ elevation, NO and ROS generation.

A 'sticky' protein, brassinosteroid insensitive1 (BRI1)-associated receptor-like kinase (BAK1), was found to be the co-receptor of the MAMP or DAMP receptors upon elicitation by flg22, elf18 or Pep1, leading to the phosphorylation of the receptors and BAK1 (Chinchilla et al., 2007; Krol et al., 2010). Later work indicated that a receptor-like cytoplasmic kinase, Botrytis-induced kinase (BIK1), also interacts with all of the PRR complexes mentioned above, indicating that BIK1 may be an ubiquitous player in defense signaling pathways (Arnaud & Hwang, 2015; Lu et al., 2010). Notably, recent findings showed that BIK1 is required for flg22 and *Pst hrcC*$^-$-induced stomatal closure and resistance to surface inoculated *Pst hrcC*$^-$; BIK1 also positively regulate MAMPs by phosphorylation (Li et al., 2014; Lu et al., 2010; Zhang et al., 2010). In addition, BIK1 also phosphorylates and activates an NADPH oxidase, RbohD, for H_2O_2 production in a Ca^{2+} independent manner. These findings indicate that BIK1 is an essential component in MAMP-induced immune signaling. Liu et al. (2013) revealed that PEPR1 directly phosphorylates BIK1 after Pep1 treatment, and both *pepr1/2* and *bik1* mutants are susceptible to *B. cinerea* inoculation. Based on the evidence that BIK1 is working with FLS2 in the guard cells to control stomatal closure, BIK1 is phosphorylated by PEPR1 induced by Pep1 elicitation, and Pep is able to induce ROS and NO generation in the guard cells (Ma et al., 2013), it is reasonable to speculate that the Pep/PEPR1 associated immune signaling pathway could be present in guards cells where BIK1 is activated by PEPR1 leading to the phosphorylation of RbohD and subsequent ROS generation, a signal molecule for stomatal closure and innate immunity defense in the guard cells. The majority of components in Pep-induced innate immune signaling pathway overlaps with those induced by flg22, including $[Ca^{2+}]_{cyt}$ elevation, protein phosphorylation, ROS and NO production and defense gene upregulation, and there is interdependence between these two pathways (Ma et al., 2012). It has been proposed that Pep is working as an amplifier to boost plant defense responses to invaders. Pep in the guard cells may function as a self-defense signal for plants to inhibit invasion at the first line. The involvement of Pep in promoting stomatal closure needs further investigation.

3.2 The Role of Ca^{2+} in NO Generation and Stomatal Signaling

$[Ca^{2+}]_{cyt}$ elevation is thought to be the initial cellular response when cell perceives pathogen or danger signals. At the whole plant scale, Ca^{2+} has been shown to be pivotal for plant innate immunity (Ma et al., 2008; Ma et al., 2013; Ranf et al., 2014; Ranf, Eschen-Lippold, Pecher, Lee, & Scheel, 2011). As the receptors for MAMPs and DAMPs are localized in the guard cells, similar Ca^{2+} induction would be observed and the $[Ca^{2+}]_{cyt}$ elevation can be diminished by pharmacological chemical treatments that affects stomatal aperture. Light induced opening was inhibited and midday stomatal closure was enhanced by OGs. However, the effect of OGs was compromised by an extracellular Ca^{2+} chelator EGTA, indicating that Ca^{2+} is required for OGs induced stomatal movement (Lee et al., 1999).

3.2.1 Involvement of Ca^{2+} Channels in NO Production in the Guard Cell

Recently, ratiometric fluorescence imaging using a calcium reporter Yellow Cameleon 3.6 showed flg22-induced guard cell $[Ca^{2+}]_{cyt}$ elevation. The Ca^{2+} signal was abolished by the treatment with Phospholipase C (PLC) inhibitors neomycin and U73122 (Thor & Peiter, 2014). PLC catalyzes the synthesis of InsP3, thus this result indicates that InsP3 is involved in flg22-induced $[Ca^{2+}]_{cyt}$ in guard cells, consistent with the findings that InsP3 is involved in flg22-mediated innate immunity (Hung et al., 2014; Zhao et al., unpublished data). Transgenic plants expressing human InsP5-ptase demonstrate altered stomatal movement in response to ABA, suggesting a role for InsP3 in guard cell signaling (Perera et al., 2008). Ma et al. (2013) have proposed that flg22-induced $[Ca^{2+}]_{cyt}$ changes could use both internal and external Ca^{2+} pools. The use of an internal Ca^{2+} release inhibitor TMB-8 further confirmed that flg22-induced $[Ca^{2+}]_{cyt}$ elevation in the guard cells requires internal Ca^{2+} sources. In addition, EGTA and lanthanum completely abolished guard cell Ca^{2+} oscillation implicating a requirement for an apoplastic Ca^{2+} source as well (Thor & Peiter, 2014). Interestingly, GLR inhibitors and mammalian adenylyl cyclase inhibitors did not abolish flg22 induced $[Ca^{2+}]_{cyt}$, suggesting that GLRs and CNGCs may not be responsible for flg22 induced guard cell $[Ca^{2+}]_{cyt}$ oscillation.

Members of the CNGC family could function in guard cells to mediate elicitor- or pathogen-induced immune signaling. To understand whether CNGCs are involved in $[Ca^{2+}]_{cyt}$ elevation in guard cells, both cytosolic loaded cGMP and extracellularly applied membrane permeable 8-bromo-cGMP activated hyperpolarization-induced inward-conducting currents in wt guard cells (Wang et al., 2013). The cGMP-activated currents were highly compromised by Ca^{2+} channel blockers such as lanthanum and gadolinium. In other studies, cAMP was found to activate hyperpolarization-induced Ca^{2+} currents in guard cells (Lemtiri-Chlieh & Berkowitz, 2004). The cAMP-dependent Ca^{2+} current was reduced in the *dnd1* mutant (Ali et al., 2007), implicating CNGC2 as a component of guard cell CNGC channels. However, Wang et al. (2013) did not observe any differences between wt and *cngc2* when the activating ligand was cGMP. CNGC2 has been found to also be required for cAMP-, and LPS-dependent NO generation in guard cells, suggesting CNGC2 involvement in guard cell Ca^{2+} signals initiating immune responses. Studies linking CNGC5 and CNGC6 with guard cell immune responses have not been undertaken. The differences in CNGC5, CNGC6 and CNGC2 involvement with cGMP versus cAMP dependent currents might be due to differences in CNGC affinity for the different cyclic nucleotides. CNGCs native to animal cell membranes have different responses to cAMP and cGMP (Craven & Zagotta, 2006).

It has been reported that the *glr3.3* mutant has enhanced susceptibility to *Pst* DC3000 correlating with significantly reduced *Pst* DC3000-induced defense genes expression, implying that GLR3.3 is involved in plant defense to pathogens. On the other hand, at present there is neither any report showing functional GLR channels in guard cells nor direct evidence for their involvement in MAMP- or DAM-induced $[Ca^{2+}]_{cyt}$ burst and defense signaling cascades. Two possibilities for GLRs' function in the guard cells are as follows: (1) GLRs may not be involved in MAMP or DAMP induced $[Ca^{2+}]_{cyt}$ burst and stomatal closure, (2) GLRs are not functioning in the guard cells to regulate stomatal closure. Unlike innate immune responses in mesophyll cells which cause defense gene expression and HR, stomatal innate immune responses primarily restrict the entry of foliar pathogens from the epidermis into inner layers of leaf cells. This restriction is achieved by regulating stomatal movement, inducing either stomatal closure or inhibition of stomatal opening. GLRs may be associated with protein complexes different from the ones involved in pathogen defense in the guard cells. Functional characterization of GLR genes would help to understand their roles in guard cells if certain GLRs are bona fide guard cell proteins.

3.2.2 Protein Kinases Regulate Stomatal Movement

SNF1-related protein kinase 2.6 (SnRK2.6), usually known as open stomata1 (OST1), is a key regulator of ABA signaling in guard cells. ABA accumulated under stress binds to PYR1/PYL/RCAR, which inhibits PP2C. This leads to the release of OST1 from the inhibition by phosphatase ABA insensitive1 (ABI1) and the activation of OST1. OST1 phosphorylates and activates guard cell slow anion channel-associated1 (SLAC1) independently of Ca^{2+}, allowing outward Cl^{-} current through the SLAC1 channel. Cl^{-} efflux from the guard cell cytosol through SLAC1 results in membrane depolarization which activates outward conductance of K^{+}. These solute efflux-based osmotic changes lead to water loss, which reduces the turgor pressure within the guard cells causing stomatal closure (Geiger et al., 2009; Vahijsalu et al., 2008). This process does not involve cytosolic Ca^{2+}. Recent studies showed that SLAC1 and SLAC1 homolog3 (SLAH3) responded to flg22 treatment, and *slac1/slah3* double mutant lost flg22-induced anion channel activity. In addition, *slac1/slah3* stomata did not close in response to flg22 and stomatal closure was highly impaired in *ost1-2*. These results suggest that SLAC1 and SLAH3 function redundantly in flg22-induced stomatal closure, and OST1 is partially involved. However, *abi1-1* responded normally to flg22, indicating that flg22-induced stomatal closure does not require ABI1. How OST1 is activated in flg22-induced stomatal signaling still needs investigation (Guzel Deger et al., 2015).

In Arabidopsis, Ca^{2+}-dependent protein kinases (CPK) function in ABA- and elicitor-induced stomatal closure (Sawinski, Mersmann, Robatzek, & Böhmer, 2013). Recent studies identified CPK8 as a positive regulator in drought stress tolerance (Zou et al., 2015). CPK8 phosphorylates and activates CATALASE3 (CAT3) in ABA- and H_2O_2-mediated signaling transduction in the guard cells. CAT can function as an ROS scavenger to protect cells from oxidative damage. ABA-, Ca^{2+}- and H_2O_2-induced stomatal closure was impaired in *cpk8* mutants, indicating that CPK8 and CAT3 are downstream of these molecules in this process. It is well known that NO works downstream of ABA and ROS to regulate stomatal closure in response to drought tolerance (Leon, Castillo, Coego, Lozano-Juste, & Mir, 2014).

It would be interesting to learn whether NO-induced stomatal closure is affected by CPK8 and CAT3. A new study showed that activation of ABA and Ca^{2+} insensitive S-type anion channel current occurs in the *cpk5/6/11/23* quadruple mutant, and this lack of CPK activation of Cl^- efflux in the mutant impairs stomatal closure. This indicates that these CPKs are required for ABA-induced Ca^{2+}-dependent stomatal signaling (Brandt et al., 2015). Furthermore, CPK activities are not directly regulated by ABA and PP2Cs in that PP2Cs bypass CPKs to directly interact and dephosphorylate SLAC1, leading to inactivation of SLAC1. This is reasonable because ABA-triggered inhibition of PP2Cs activates plasma membrane Ca^{2+} channels, causing $[Ca^{2+}]_{cyt}$ burst and activation of CPKs (Murata, Pei, Mori, & Schroeder, 2001). Several CPKs, CPK5, CPK6, CPK11 were shown to be redundant positive regulators involved in flg22-triggered defense responses including the oxidative burst (Boudsocq et al., 2010). It is still unknown whether CPKs behave similarly in guard cells. Recently, a positive role of CPK6 in YEL-induced stomata movement was revealed using *cpk6* null mutants (Ye et al., 2013). These studies showed that CPK6 is essential for YEL-mediated processes, including activation of nonselective inward Ca^{2+} current and S-type anion channels (SLAC1 and SLAH3) as well as inhibition of inward-rectifying K^+ channels. CPK6 also functions in the oxidative burst triggered by YEL. CPK6 works with CPK5 and CPK11 at the whole plant level to regulate flg22-triggered plant innate immunity, thus it is plausible that CPK6 is a positive regulator as well in the guard cell to regulate ion channels in response to MAMPs or DAMPs.

CPKs were found to directly regulate ion channel proteins. Elevation of cytosolic free Ca^{2+} concentration results in a rapid activation of S-type anion channels in guard cells (Chen, Hills, Lim, & Blatt, 2010; Stange, Hedrich, & Roelfsema, 2010). Split YFP-based protein—protein interaction assay identified CPK21 and 23 as interacting partners of SLAC1 (Geiger et al., 2010). Interestingly, CPK21 activates SLAC1 in a Ca^{2+}-dependent manner but CPK23 activates SLAC1 independently of Ca^{2+}. This indicates that stomatal closure induced by ligands converge at specific $[Ca^{2+}]_{cyt}$ levels. CPK13 expressed in guard cells phosphorylates the K^+ channel KAT2 and inhibits light-induced stomatal opening (Ronzier et al., 2014). In vitro kinase assay indicates that CPK13 is a Ca^{2+}-independent protein kinase, thus it is unclear whether its function in controlling stomatal movement requires Ca^{2+} or not. It is already known that CPKs have different Ca^{2+} requirement for their kinase activities possibly due to their binding affinities to Ca^{2+} (Boudsocq & Sheen, 2013; Scherzer, Maierhofer, Al-Rasheid,

Geiger, & Hedrich, 2012). A recent study in *Xenopus oocytes* showed that CNGC18-mediated Ca^{2+} influx requires CPK32 in regulating pollen tube growth (Gao, Fei, Dong, Gu, & Wang, 2014; Zhou, Lan, Jiang, Fang, & Luan, 2014), indicating that Ca^{2+}-dependent CPKs could function in a feed-forward mechanism to regulate CNGCs to promote more Ca^{2+} influx in certain signaling pathways.

MAPKs control stomatal movement in response to ABA and H_2O_2 (Liu, Liu, Zhang, & Li, 2010). MAPK cascades are critical signaling modules in MTI and this machinery exists in guard cells (Meng & Zhang, 2013). MAMP- or DAMP-triggered and Ca^{2+}-dependent MPK3 and MPK6 activation has been extensively studied using both pharmacological and genetic approaches, and both positively regulate flg22-triggered stomatal closure (Arnaud & Hwang, 2015). In addition, MPK6 was reported to regulate H_2O_2-mediated NO biosynthesis in lateral root development (Wang, Du, Li, Ren, & Song, 2010). Recently, a study in rice uncovered a novel regulatory mechanism of MPK phosphorylation (Xie, Chen, Wang, & Yang, 2014). The work showed that CPK18 activates MPK5 by phosphorylating the two threonine residues largely conserved in MPKs of land plants, which identified a direct connection between Ca^{2+} signaling and MAPK machinery. Based on these findings, MPKs could work in guard cells to modulate NO generation in a Ca^{2+}-dependent manner in response to elicitors, eventually leading to stomatal closure.

3.3 NO as a Secondary Messenger Regulating Stomatal Innate Immunity

NO and ROS can act as antimicrobial substances to impede infection of plants by microbial pathogens (Boccara et al., 2005; Stamler, Lamas, & Fang, 2001; Torres, 2010). However, these molecules are mainly considered as components of signaling pathways when the cells are under biotic or abiotic stresses. NO, ROS and $[Ca^{2+}]_{cyt}$ are critical secondary messengers in MAMP-, DAMP- and ABA-induced stomatal closure. NO production can be detected in guard cells after application of LPS, flg22, Pep, cryptogein, chitosan, YEL as well as *Pst avrRpt2*$^{+}$ (Agurla, Gayatri, & Raghavendra, 2014; Ma et al., 2008; Ma et al., 2013), implying that NO is involved in both MTI and ETI (Table 1). When plants are under stress, both ROS and reactive nitrogen species (RNS) are generated, indicating that there is a certain extent of interdependent signaling between these two groups of secondary messengers. Genetic and pharmacological studies have shown that NO act downstream of ROS in ABA-dependent stomatal closure

Table 1 Biotic Stress Elicitors Induce NO Production in the Guard Cells

Elicitors	Source	Plant Species	References
Flg22	The N-terminal 22 amino acids peptide of the bacterial flagella protein, flagellin	*Arabidopsis thaliana*	Melotto et al. (2006) and Ma et al. (2013)
LPS	Glycolipid component of outer membrane of Gram-negative bacteria	*A. thaliana*	Melotto et al. (2006), Ali et al. (2007) and Ma et al. (2008)
Peps	Endogenous plant elicitor peptides	*A. thaliana*	Ma et al. (2013)
OG	Cell wall degradation product under biotic and abiotic stress	*Lycopersicon esculentum* *Commelina communis*	Lee et al. (1999)
		A. thaliana	Rasul et al. (2012)
Chitosan	Deacetylated derivative of chitin fragments from fungal cell wall	*L. esculentum, C. communis*	Lee et al. (1999)
		Pisum sativum	Srivastava et al. (2009)
		A. thaliana	Khokon, Uraji, et al. (2010)
YEL	Yeast elicitor from yeast extract	*A. thaliana*	Khokon, Hossain, et al. (2010)
Harpin	*Xanthomonas oryzae*	*Nicotiana benthamiana*	Zhang et al. (2009, 2010, 2012)
INF1	*Phytophthora infestans*	*N. benthamiana*	Zhang et al. (2009)
Boehmerin	*P. infestans*	*N. benthamiana*	Zhang et al. (2009, 2010, 2012)
Nep1	*Magnaporthe oryzae*	*N. benthamiana*	Zhang et al. (2010, 2012)
E. coli O157:H7	Human pathogen	*A. thaliana*	Melotto et al. (2006)

(Garcia-Mata & Lamattina, 2007; Khokon, Hossain, et al., 2010; Neill, Desikan, Clarke, & Hancock, 2002). A recent study on wheat infected with *Puccinia triticina* showed that small amount of NO accumulation was first detected by DAF-FM DA, a cell-permeable fluorescent probe for the detection of NO in the guard cells 4 h after inoculation at the infection site. NO accumulation expanded continuously with time and NO was predominantly accumulated in the cells undergoing an HR. The compatible strain only caused a small amount of NO accumulation in the guard cells where the pathogenic contact occurred 4 h after inoculation, but no NO was observed thereafter. Pretreatment with cPTIO, the NADPH oxidase inhibitor imidazole, and EGTA, delayed NO production induced by the incompatible strain and suppressed HR development. These observations also indicate that Ca^{2+} and ROS act upstream of NO to induce the HR (Qiao et al., 2015). However, much evidence indicates that the relationship between stress-induced NO, ROS and Ca^{2+} is reciprocal.

3.3.1 NO is Required for Ligand-Induced Stomatal Closure

Cytosolic NO increase is a common event during stomatal closure triggered by various cellular and environmental cues (Gayatri et al., 2013). Due to the lack of the knowledge on L-Arg-dependent NO synthesis enzyme(s) in plants, studies on NO were performed using mammalian NOS inhibitors and NO scavengers. This work showed that NO released by NO donors triggers stomatal closure in different plant species (Leon et al., 2014). Consistently, cPTIO and L-NAME treatments attenuate but did not abolish ABA-induced stomatal closure, indicating a possible NO-independent pathway for ABA-triggered stomatal closure, although the possibility of inadequate function of these chemicals cannot be ruled out. NO is involved not only in ABA-induced stomatal closure, but also in ABA-inhibited light-induced stomatal opening. L-NAME and cPTIO compromised ABA mediated inhibition of light-induced stomatal opening in *Vicia faba*, implying that this process requires NOS-like dependent NO generation (Garcia-Mata & Lamattina, 2007). Despite the inhibitory effect of ABA, the treatment of a cell permeable Ca^{2+} chelator BAPTA-AM restored light-induced stomatal opening, while BAPTA-AM failed to compromise NO-mediated stomatal closure. Externally added Ca^{2+} also increased NO generation. These observations indicate that Ca^{2+} is downstream of ABA but upstream of NO in the inhibition of light-induced stomatal opening. Channel proteins involved in ABA mediated $[Ca^{2+}]_{cyt}$ elevation is still not well studied. Recent study showed that *cngc5cngc6* double mutant still maintains ABA-activated

Ca^{2+}-permeable cation channel currents, ruling out these two CNGCs as ABA-associated Ca^{2+}-permeable channels (Wang et al., 2013).

As a key regulator controlling guard cell osmotic pressure and turgor, ABA has been well documented to be associated with abiotic stresses. In the past several years, guard cells and ABA were also demonstrated to be imperative for biotic stresses (Asselbergh, Vleesschauwer, & Höfte, 2008; Lim et al., 2015; Ton, Flors, & Mauch-Mani, 2009). In response to MAMPs flg22 or LPS, Arabidopsis guard cells close in an ABA- and NO-dependent manner (Melotto et al., 2006). The *ost1* mutant failed to respond to flg22 and LPS in inducing stomatal closure, indicating that ABA signaling components are downstream of MAMP-receptor perception. This is consistent with the finding that ABA-mediated stomatal closure is normal in *fls2* mutant (Zeng & He, 2010). Treatment with L-NAME inhibited MAMP- and microbe (including both pathogenic and non-pathogenic)-induced stomatal closure indicating that NO is required for MAMP-triggered stomatal movement. Whether Ca^{2+} also plays a role in the above mentioned process is unclear. It is well understood that MAMP-induced $[Ca^{2+}]_{cyt}$ elevation is an initial point upstream of the signaling cascades, including NO and ROS production. However, $[Ca^{2+}]_{cyt}$ was thought to be downstream of NO and ROS in guard cells (Gayatri et al., 2013). Pharmacological study in Arabidopsis positions cGMP, and also likely Ca^{2+}, as downstream of NO and ROS during ABA-mediated stomatal closure (Dubovskaya et al., 2011). This also suggests a role of CNGCs in ABA-mediated stomatal closure although CNGC5 and 6 have been excluded (Wang et al., 2013). These findings on the interplay among the secondary messengers to some extent conflict with previous work by Garcia-Mata and Lamattina (2007). However, it is reasonable to postulate that the $[Ca^{2+}]_{cyt}$ requirement in ABA-induced stomatal closure and inhibition of light-induced stomatal opening are discrete. More genetic evidence is necessary to further validate results obtained from pharmacological studies.

Methyl jasmonate (MeJA) and ABA signaling in guard cells contain several common signaling components, and the mutants in one hormone signaling pathway compromise stomatal closure induced by the other hormone. This indicates that there is interplay between these two hormone response pathways in regulating stomatal movement (Saito, Nakamura, Mori, & Murata, 2009; Sawinski et al., 2013). Cyclic adenosine 5'-diphosphoribose (cADPR) is another secondary messengers in ABA-induced stomatal closure. Recent study showed that cADPR synthesis inhibitor nicotinamide (NA), and cGMP synthesis inhibitor LY83583 (6-anilino-5,8-quinolinedione) inhibited MeJA-induced stomatal closure by suppressing ROS accumulation and NO production in guard cells. As expected, MeJA-induced $[Ca^{2+}]_{cyt}$ elevation in guard cells was also impaired by NA and LY83583, suggesting that CNGC could be involved in mediating Ca^{2+} movement. These results suggest that during MeJA-mediated stomatal closure, cADPR and cGMP act upstream of ROS accumulation and NO production in Arabidopsis guard cells (Hossain et al., 2014). The different positions of signaling components in these two pathways may be explained by a feed-forward interaction (Figure 1).

BR is essential for various plant growth and development processes. Similar to flg22 and Pep, BR binds to its cognate plasma membrane receptor BRI1 causing $[Ca^{2+}]_{cyt}$ elevation mainly through CNGC2 (Zhao, Qi, & Berkowitz, 2013). The most bioactive BR form, brassinolide (BL), mediates BR-induced stomatal movement (Misra, Acharya, Granot, Assmann, & Chen, 2015), but this process is not fully understood. A new report using a series of mutants and inhibitors showed that eBL induced NO and ROS production in the guard cells (Shi et al., 2015). NIA1 but not NIA2 is the primary enzyme for NO production. As stomatal closure is not completely compromised in *nia1-2nia2-5* double mutant, another enzyme, possibly the unknown NOS, could have also been responsible for NO production. In addition, eBL-induced ROS is produced by RbohF but not catalyzed by RbohD as seen in ABA- or MAMP-mediated stomatal signaling pathways. Interestingly, eBL-mediated stomatal closure involves the induction of ethylene synthesis, which activates Gα protein. This activation promotes downstream NO and ROS production. It was demonstrated that Gα protein also mediates flg22 inhibition of inward K^+ currents occurring during stomatal closure (Zhang, He, & Assmann, 2008). These findings indicate that BR- and flg22-induced stomatal closure may merge at the level of Gα protein. The interplay between BR and PTI has been demonstrated recently but further investigation is required to elaborate the interaction between the two pathways (Lozano-Durán & Zipfel, 2014).

NO and ROS are usually coupled in cells under stress conditions. ROS can be produced by plasma membrane Rboh proteins and apoplastic peroxidases. *Rboh* genes have been primarily studied in Arabidopsis. RbohD generated ROS is involved in plant innate immune responses (Kadota, Shirasu, & Zipfel, 2015). In *Nicotiana benthamiana*, virus-induced gene silencing of *NbRbohA* and *NbRbohB* compromised elicitor-induced NO production, resulting in impaired stomatal closure (Zhang, Fang, Zhang, Wang, & Zheng, 2009). However, elicitor-induced $[Ca^{2+}]_{cyt}$ elevation in

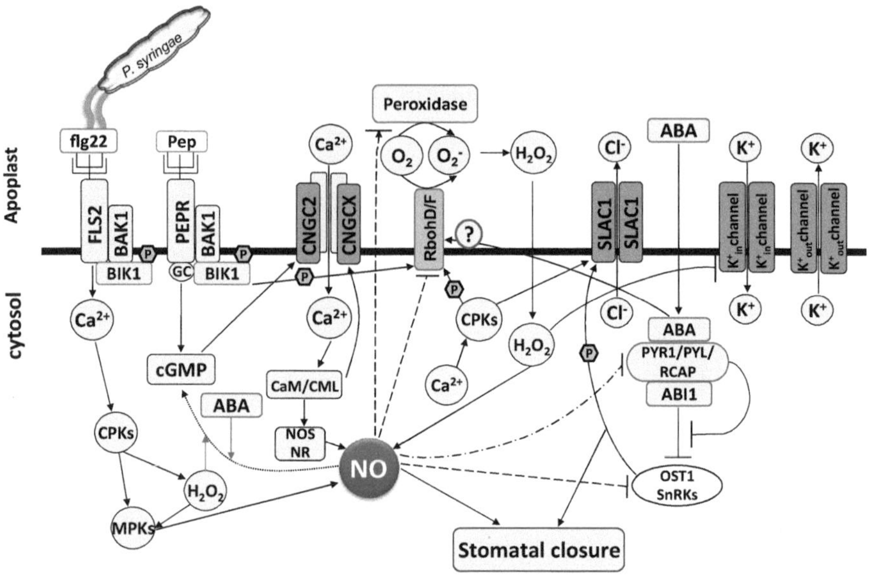

Figure 1 ***Model of MAMP- or DAMP-induced stomatal closure.*** In the guard cell, plasma membrane localized receptors FLS2 or PEPR1 perceives extracellular MAMP (flg22) or DAMP (Pep) respectively, leading to $[Ca^{2+}]_{cyt}$ elevation and activation of CPKs. Pep induced Ca^{2+} elevation is mediated by CNGCs, which form a heterotetrameric channel including CNGC2. However, Ca^{2+} channels associated with flg22/FLS2 are still unknown. BIK1 phosphorylation requires FLS2, BAK1 and PEPR1 in either pathway, and BIK1 activates RbohD through phosphorylation. RbohD is required for flg22 and Pep induced H_2O_2 generation. MPKs can be phosphorylated by CPKs. CPKs, MPKs and H_2O_2 are required for NO production, leading to elicitor induced stomatal closure. CaMs also bind Ca^{2+} and a member of the CaM-like protein family, CML24, has been shown to be required for pathogen-induced NO production probably through the regulation of NO synthesis enzymes, such as NR and/or NOS. Ca^{2+} activated CPKs also phosphorylate RbohD to promote H_2O_2 production. On the other hand, apoplastic peroxidases also make H_2O_2, but no peroxidase gene has been identified. The connection between MAMP- or DAMP-induced and ABA-induced stomatal closure is still ambiguous. However, there are certain interactions in the guard cell between components of PAMP or DAMP signaling pathways and those of the ABA signaling pathway. ABA-induced stomatal closure can be both Ca^{2+}-dependent and Ca^{2+}-independent. In the Ca^{2+}-dependent pathway, CPK21 and 23 phosphorylate and activate SLAC1 to promote stomatal closure. H_2O_2 inhibits rectifying inward K^+ channel and positively regulates stomatal closure. In addition, NO mainly plays a negative role through S-nitrosylation and nitration. S-nitrosylation of OST1 and RbohD negatively regulates stomatal closure. S-nitrosylation of APX1 was shown to positively regulate H_2O_2 production, while it was proposed to play a negative role in flg22-induced signaling pathway. Furthermore, several ABA receptors exhibit Tyr nitration and their functions in suppressing PP2C proteins are impaired. Conversely, S-nitrosylation of these proteins does not demonstrate negative regulation. In addition, ABA- and NO-induced 8-nitro-cGMP formation in the guard cell requires H_2O_2. 8-nitro-cGMP is upstream of Ca^{2+} and SLAC1 positively regulating stomatal closure. Dashed lines indicate S-nitrosylation, the dotted line indicates nitration and the dash-dotted line indicates both S-nitrosylation and nitration.

guard cells was unchanged, placing ROS upstream of NO production but downstream of Ca^{2+}. On the other hand, effects of a series of pharmacological treatments on chitosan-induced stomatal closure showed that ROS is upstream of NO production but neither is downstream of $[Ca^{2+}]_{cyt}$ elevation (Srivastava, Gonugunta, Puli, & Raghavendra, 2009). However, this result may not be definitive. With regards to the BAPTA treatment experiment in this paper, it is not sufficient to conclude that $[Ca^{2+}]_{cyt}$ elevation is downstream of ROS and NO just because BAPTA did not affect chitosan-induced ROS and NO production. More empirical work is required to examine whether disrupted ROS and NO production affects chitosan-induced $[Ca^{2+}]_{cyt}$ elevation using either mutants or chemical treatment.

NO has been shown to act downstream of ROS generated primarily by RbohD during stomatal innate immune responses to bacterial pathogens (Arnaud & Hwang, 2015; Torres & Dangl, 2005). Other signaling pathways, such as SA-, chitosan- and YEL-induced stomatal closure require mainly peroxidase generated ROS (Arnaud & Hwang, 2015). SA-induced NO production is also downstream of ROS. This has been demonstrated using salicyl-hydrooxamic acid, an inhibitor of apoplastic peroxidases (Sawinski et al., 2013). However, specific peroxidase genes involved in stomatal movement have not been identified in plants. Peroxidases form a large gene family and there could be functional redundancy of the genes, which makes it difficult to identify the corresponding mutant involved in stomatal movement triggered by different signaling molecules.

3.3.2 NO Causes Modification in Signaling Molecules

NO derived moieties have been shown to bind to other signaling molecules in animal studies. Similar NO-mediated molecular modification has been discovered only recently in plant cells. The modification alters the molecular structure and therefore, the function of the molecules in certain signaling pathways. These findings indicate the dual roles that NO is playing in signaling transduction pathways.

3.3.2.1 S-Nitrosylation of Cysteine

A unique biological activity of NO as a secondary messenger is that it binds to an amino acid causing S-nitrosylation, a redox-based reversible PTM process that forms a covalent bound between an NO group and a cysteine (Cys) residue. In the past decade, S-nitrosylation has been an emerging and rising subject in plant research and demonstrated to be a pivotal PTM event occurring during plant stress responses, including plant innate immune defense (see reviews by Corpas, Begara-Morales, Sánchez-Calvo, Chaki, & Barroso, 2015; Frederickson Matika & Loake, 2014; Gaupels et al., 2011; Yu, Yun, Spoel, & Loake, 2012).

Significant progress on the study of S-nitrosylation in plants has been made lately. S-nitrosoglutathione (GSNO) is a major bioactive form of RNS functioning as a primary NO donor. GSNO can be irreversibly degraded by GSNO reductase (GSNOR). *gsnor1-3* is a loss-of-function mutant with a significant increase of the S-nitrosothiol level, leading to NO elevation in cells (Feechan et al., 2005). Recently, an S-nitrosoproteome in Arabidopsis was generated using wt and the *gsnor1-3* mutant, providing valuable references for studying NO-associated defense signaling pathways (Hu et al., 2015). Several S-nitrosylated ROS-scavenging proteins were identified in this work. Although not detected in this work, RbohD can be S-nitrosylated in vivo to regulate HR (Yun et al., 2011). Increased NO production triggered by pathogen infection mediates further S-nitrosylation of Cys890 in RbohD, whose activity is compromised leading to ROS reduction. Most recently, a member of another ROS synthesizing gene family, ascorbate peroxidase1 (APX1), was recently identified as a new S-nitrosylated protein. Contrary to being a negative regulator in previous reports, S-nitrosylation in this study enhanced APX1 activity. This was associated with ROS generation and resistance to oxidative stress, indicating a positive role of S-nitrosylation (Yang et al., 2015). The authors suggested that S-nitrosylation of APX1 negatively regulates flg22-induced immune responses. Flg22-induced ROS production and defense gene expression were upregulated in the *apx1* mutant as compared to the wt, while there was no difference observed between wt and the *apx1* mutant complemented with APX1. These data indicate that APX1 is not required for flg22-induced ROS production. As mentioned earlier, one of the NADPH oxidases RbohD is critical for MAMP-induced ROS generation in Arabidopsis. Thus, the enhanced flg22-mediated immune responses in the *apx1* mutant may due to the activation of other peroxidases or even Rboh genes. Also, the upregulation of flg22-triggered signaling components in the *apx1* null mutant is not necessarily due to the inhibition of S-nitrosylation on APX1 function but rather the activation of other oxidative pathways. Making double or triple mutants may allow evaluation of this hypothesis. Collectively, these new findings indicate that NO production, indirectly modulated by GSNOR, can also regulate ROS homeostasis by S-nitrosylating RbohD and APX1, further supporting the interconnection between NO and ROS signaling. Unlike most of the previous findings, this cellular process positions

ROS downstream of NO, indicating a feedback loop between ROS and NO homeostasis in plant cells responding to pathogens.

In ABA-regulated signaling cascades, NO has been known as a downstream mediator of some molecular steps and to positively regulate the signaling pathway. However, the *nia1nia2noa1* triple mutant displayed hypersensitivity to ABA (Lozano-Juste & León, 2010), indicating that NO can act as an antagonist of ABA through PTM. S-nitrosylation was recently identified in ABA responsive guard cell proteins (Lang & Zuo, 2015; Wang, Du, et al., 2015; Wang, Zhu, & Lang, 2015). It was shown that excessively increased NO in the *gsnor1-3* mutant negatively regulates ABA-induced stomatal closure, which is due to the S-nitrosylation of Cys137 in OST1/SnRK2.6 following ABA exposure (Wang, Du, et al., 2015). In addition, NO also suppresses ABA-inhibited seed germination through the S-nitrosylation of SnRK2.2 and SnRK2.3, indicating that S-nitrosylation is a common step in the ABA signaling pathway. Interestingly but not surprisingly, sequence alignments revealed that the Cys890 in RbohD and the Cys137 in OST1 are highly conserved not only in other members of the gene families but also in the homologous genes in animals. This suggests that S-nitrosylation is an evolutionarily conserved mechanism that maintains the equilibrium of signaling molecules in cells challenged with stresses.

3.3.2.2 Tyrosine Nitration

The reciprocal influence of NO and ROS can also be reflected by radical chemistry. NO and superoxide anion ($O_2^{\bullet -}$) can react rapidly to form peroxynitrite ($ONOO^-$). In mammalian cells, $ONOO^-$ is a powerful oxidant involved in immunity and a key redox mediator of disease response (Radi, 2013). Recent studies in plants provided solid evidence that $ONOO^-$ is also produced in plant cells under pathogen invasion or upon perception of elicitors (Gaupels et al., 2011; Saito et al., 2009; Trapet et al., 2014). $ONOO^-$ mediates irreversible tyrosine (Tyr) nitration, another PTM involving NO intermediates. Tyr nitration was previously identified in plant cells to be involved in various physiological processes, including a role in plant immune responses (Begara-Morales et al., 2014; Corpas et al., 2015; Frederickson Matika & Loake, 2014; Trapet et al., 2014).

Besides S-nitrosylation of OST1 in the guard cell, both in vitro and in vivo Tyr nitration was recently identified in several ABA receptors, PYR1 and PYL1, 4, 5, 6, and 8, when exposed to the peroxynitrite donor 3-morpholinosydnonimine (SIN-1) (Lozano-juste et al., 2015). Nitrated

Tyr120 and 143 in PYR1 compromised ABA-induced inhibition of the phosphatase HAB1, indicating that these two Tyr residues are critical for nitration-induced desensitization of PYR1 to ABA. Tyr nitration in ABA receptors may act as a feedback mechanism in ABA signaling. Along with the primary investigation of nitration, the authors also showed S-nitrosylation of all the six receptors, although no impairment of the suppression of the HAB1 phosphatase activity was observed. This indicates that Tyr nitration and S-nitrosylation functions independently in ABA-induced signaling events. Although S-nitrosylation is thought to be a negative mechanism, it could still play a positive role in preventing phosphatase binding so that HAB1 activity is always inhibited. More precise functional analyses of Tyr nitration and S-nitrosylation, such as stomatal aperture measurement, are required to further understand the role of PTM in ABA receptors. In addition, the PTMs of ABA receptors in guard cells upon pathogen attack or elicitor perception can also be examined to get more innovative insights on stomatal innate immunity. In general, PTMs of ABA signaling components in the guard cell may be an adaptive mechanism for controlling stomatal movement in response to environmental changes.

Not only can proteins be nitrated, signaling molecules can also be modified by NO intermediates. The interaction between the important secondary messengers, cGMP and NO, in plant cells has barely been examined. Definitive identification of the enzymes responsible for cGMP synthesis in plant cells that occurs during various signaling events is still unclear, although several putative guanylyl cyclases (GC) have been identified; the physiological relevance of the reported GC activity of several receptor proteins has been challenged (Ashton, 2011; Bojar et al., 2014; Irving, Kwezi, Wheeler, & Gehring, 2012; Kwezi et al., 2007; Ludidi & Gehring, 2003; Meier et al., 2010; Qi et al., 2010). Despite the debate on plant GC activity, the function of cGMP in plant cellular signaling is clear partially due to the identification of plant CNGC channels. Recently, the possible function of the nitrated cGMP derivative 8-nitro-cGMP in guard cell signaling was assessed (Joudoi et al., 2013). The authors showed that the synthesis of 8-nitro-cGMP in guard cells is induced by ABA and NO and is ROS-dependent, which is consistent with how nitration works in PTM. Interestingly, 8-nitro-cGMP is able to, and is required to trigger stomatal closure as demonstrated using the *nogc1* mutant. NOGC was previously reported to be a novel NO-binding GC (Ludidi & Gehring, 2003). On the contrary, a membrane-permeable analog of cGMP, 8-bromo-cGMP failed to trigger stomatal closure, although an article published several months later showed that 8-bromo-cGMP is able to activate inward Ca^{2+} current in guard cells (Wang et al., 2013). However, 8-bromo-cGMP could induce stomatal opening in the dark but 8-nitro-cGMP could not. This indicates that nitrated cGMP is the active form in ABA-triggered stomatal closure and is downstream of NO and ROS production, while non-nitrated cGMP may be required for induced stomatal opening. The authors of this study further showed that Ca^{2+}, cADPR, and SLAC1 are downstream of 8-nitro-cGMP. This result is consistent with the Ca^{2+}-dependent ABA-induced stomatal closure signaling pathway. This study identified some new molecular steps of this pathway and emphasizes the significance of NO. Similarly, the function of 8-bromo-cGMP in guard cell innate immunity can also be further investigated.

3.3.2.3 Other NO Interacting Signaling Molecules

As mentioned earlier, a GC was identified to be an NO-binding signaling molecules, termed 'NOGC' (Ludidi & Gehring, 2003). NOGC is required for ABA-induced stomatal closure and 8-nitro-cGMP can complement the loss-of-function of NOGC1 to trigger stomatal closure (Joudoi et al., 2013). However, how NO and GC interact is still unknown. NO could possibly modify GC by either Tyr nitration or S-nitrosylation, or even both. Biochemical approaches such as mass spectrometry and antibody detection would provide insights into this unresolved question.

Other compounds, either from the plant cell itself or from the environment may have an influence on NO-mediated signaling events. The gas hydrogen sulfide (H_2S, See chapter "Alone NO Longer: Interactions of Nitric Oxide with Reactive Oxygen Species and Hydrogen Sulfide") has been shown to deplete ABA-induced NO production in guard cells and H_2S alone does not induce NO (Lisjak et al., 2010). However, a later study showed that increased NO production is induced by H_2S using the same approach (Honda et al., 2015). The differences between these two totally separate experiments may be due to the different incubation times used by the groups. Stomatal movement is not permanent or 'fixed'. Stomatal closure could be transitory and the guard cells could return to a turgid state after a signaling event. Therefore, when comparing results of different studies, the time at which effects of a treatment are evaluated should be taken into consideration. Another conflict is that Lisjak et al. (2010) showed H_2S-induced stomatal opening rather than stomatal closure later published by two other different labs (Honda et al., 2015; Scuffi et al., 2014). Nevertheless, all this work does suggest that NO acts downstream of H_2S in ABA signaling pathway in guard cells. This observation is further supported by using genetic resources, specifically, the *nia1nia2* double mutant and the *des1* mutant (Scuffi et al., 2014). L-Cys desulfhydrase (DES1) releases H_2S during Cys degradation. The H_2S donor sodium hydrosulfide (NaHS) could not induce stomatal closure in the *nia1nia2* double mutant, indicating that NO is required for H_2S-triggered stomatal closure. On the other hand, ABA-induced NO generation is abolished in *des1* mutant or by the H_2S scavenger hypotaurine, and is restored by NaHS. Together, these results elucidate that (1) NO and H_2S are essential components in ABA-induced stomatal closure; (2) NO and H_2S are interdependent with NO acting in the signaling cascade downstream of H_2S.

4. CONCLUDING PERSPECTIVES

As a secondary messenger, NO is not studied as extensively as ROS or Ca^{2+}. This is mainly due to the lack of information on the enzymatic synthesis pathway of NO in plant cells that are undergoing signaling cascades involving NO elevation. This is the case with NO production demonstrated in guard cells challenged with elicitors and pathogens. Although most of the studies were conducted using pharmacological compounds, mutants with NO depletion phenotype were generated (Hao et al., 2010; Lozano-Juste & León, 2010). The pair of guard cells respond to the perception of extracellular danger molecules by activating signaling pathways that result in turgor loss and, hence, stomatal closure. Molecular components required for stomatal closure during plant innate immune response mostly overlap with those required for ABA-induced stomatal closure. NO is a pivotal mediator during stomatal closure involving a feedback and feed-forward interaction with other signaling molecules, such as ROS, Ca^{2+}, cGMP and H_2S (Figure 1). NO also has a function distinct from other signaling molecules in PTM. With more and more S-nitrosylated proteins during stomatal closure being identified, the full picture of NO function in stomatal innate immune signaling will be elucidated. In addition, because some results obtained from pharmacological usage are still ambiguous and sometimes inconsistent, the employment of *nia1nia2* or *nia1nia2noa1* mutants will provide genetic approaches to facilitate our understanding of the many roles NO plays in stomatal innate immunity.

Stomatal closure, ETI and MTI share some common signaling components including NO, ROS, Ca^{2+}, CNGCs, CPKs, MPKs and defense genes. However, unlike ETI and MTI in mesophyll cells, guard cells seem to only go through stomatal closure but not hypersensitive cell death. The primary purpose of stomatal closure is to prevent further pathogen infection into cells in the inner layers, and this is a relatively rapid response compared to the onset of HR. Stomatal closure is also temporary because stomata starts to open after roughly 2 h. Also COR released from *Pst* triggers stomata opening to facilitate pathogen invasion. Thus HR could still be induced in the stomata after the first line of protection, stomatal closure. On the other hand, after bacteria pass into the leaf interior through the stomatal pore, defense signaling transduction starts in the inner mesophyll cells and may leave the leaf epidermis uninfected. Thus HR will only occur in the inner cells but not in the guard cells, and HR is not necessary in guard cells. It is interesting that fungal elicitors can cause stomatal closure while fungi usually have a different infection mechanism than bacteria. Some fungal species can use stomatal pores to disperse spores into leaf cells for infection.

Another interesting question is how the signals are transmitted and synchronized in the guard cells because stomatal closure requires the signaling pathway to occur simultaneously. It is shown that ROS can be transmitted cell-to-cell and systemically (Song, Miao, & Song, 2014). At present mechanisms facilitating the intercellular movement of NO are not delineated. Differently, NO is a gasotransmitter so its mobility is more difficult to detect, and it is not clear how NO is perceived by cells if NO could be transmitted between cells. In addition, Ca^{2+} was also suggested to move between cells and probably systemically. Therefore, it is possible that these secondary messengers act as signal transmitters between guard cells. ROS and Ca^{2+} may be used to examine intercellular signal movement. ROS has cellular and apoplastic enzymatic synthesis pathways, NO could also derive from separate pathways, and ligand-induced $[Ca^{2+}]_{cyt}$ elevation have internal and external sources. It is possible that during defense response, there is interaction between these different pathways for different response purposes. Flg22-induced $[Ca^{2+}]_{cyt}$ elevation was reported to involve both internal and external sources and ROS generation also uses both Rboh and peroxidases. Mechanisms leading to NO generation during immune responses are still not clearly unravelled, but working with currently available genetic tools suggests some hints for multiple pathways.

NO-associated PTM is critical for stomatal signaling and usually negatively regulates stomatal closure induced by extracellular cues. The threshold of NO level in the cells that can turn positive NO effects on signaling to negative regulation is not clear. In addition, how pathogens deactivate

GSNOR to increase NO in the guard cell to suppress stomatal closure is still unknown. More empirical evidence using *gsnor1* and *nia1nia2noa1* mutants in MAMP- or DAMP-induced defense signaling would provide more informative insights on the impact of S-nitrosylation and nitration on immune signaling pathways.

REFERENCES

Abdel-Hamid, H., Chin, K., Moeder, W., Shahinas, D., Gupta, D., & Yoshioka, K. (2013). A suppressor screen of the chimeric AtCNGC11/12 reveals residues important for intersubunit interactions of cyclic nucleotide-gated ion channels. *Plant Physiology, 162*(3), 1681–1693. http://dx.doi.org/10.1104/pp.113.217539.

Agurla, S., Gayatri, G., & Raghavendra, A. S. (2014). Nitric oxide as a secondary messenger during stomatal closure as a part of plant immunity response against pathogens. *Nitric Oxide: Biology and Chemistry/Official Journal of the Nitric Oxide Society, 43*, 89–96. http://dx.doi.org/10.1016/j.niox.2014.07.004.

Ali, R., Ma, W., Lemtiri-Chlieh, F., Tsaltas, D., Leng, Q., von Bodman, S., et al. (2007). Death don't have no mercy and neither does calcium: Arabidopsis CYCLIC NUCLEOTIDE GATED CHANNEL2 and innate immunity. *The Plant Cell, 19*(3), 1081–1095. http://dx.doi.org/10.1105/tpc.106.045096.

Arnaud, D., & Hwang, I. (2015). A sophisticated network of signaling pathways regulates stomatal defenses to bacterial pathogens. *Molecular Plant, 8*(4), 566–581. http://dx.doi.org/10.1016/j.molp.2014.10.012.

Ashton, A. R. (2011). Guanylyl cyclase activity in plants? *Proceedings of the National Academy of Sciences of the United States of America, 108*(19), E96. http://dx.doi.org/10.1073/pnas.1101007108. author reply E97–8.

Aslam, S. N., Erbs, G., Morrissey, K. L., Newman, M.-A., Chinchilla, D., Boller, T., et al. (2009). Microbe-associated molecular pattern (MAMP) signatures, synergy, size and charge: influences on perception or mobility and host defence responses. *Molecular Plant Pathology, 10*(3), 375–387. http://dx.doi.org/10.1111/j.1364-3703.2009.00537.x.

Asselbergh, B., Vleesschauwer, D., & Höfte, M. (2008). Global switches and fine-tuning-ABA modulates plant pathogen defense. *Molecular Plant-Microbe Interactions, 21*(6), 709–719. http://dx.doi.org/10.1094/MPMI-21-6-0709.

Begara-Morales, J. C., Sánchez-Calvo, B., Chaki, M., Valderrama, R., Mata-Pérez, C., López-Jaramillo, J., et al. (2014). Dual regulation of cytosolic ascorbate peroxidase (APX) by tyrosine nitration and S-nitrosylation. *Journal of Experimental Botany, 65*(2), 527–538. http://dx.doi.org/10.1093/jxb/ert396.

Besson-Bard, A., Pugin, A., & Wendehenne, D. (2008). New insights into nitric oxide signaling in plants. *Annual Review of Plant Biology, 59*, 21–39. http://dx.doi.org/10.1146/annurev.arplant.59.032607.092830.

Blatt, M. R., Thiel, G., & Trentham, D. R. (1990). Reversible inactivation of K+ channels of Vicia stomatal guard cells following the photolysis of caged inositol 1,4,5-trisphosphate. *Nature, 346*(6286), 766–769. http://dx.doi.org/10.1038/346766a0.

Boccara, M., Mills, C. E., Zeier, J., Anzi, C., Lamb, C., Poole, R. K., et al. (2005). Flavohaemoglobin HmpX from Erwinia chrysanthemi confers nitrosative stress tolerance and affects the plant hypersensitive reaction by intercepting nitric oxide produced by the host. *The Plant Journal: For Cell and Molecular Biology, 43*(2), 226–237. http://dx.doi.org/10.1111/j.1365-313X.2005.02443.x.

Bojar, D., Martinez, J., Santiago, J., Rybin, V., Bayliss, R., & Hothorn, M. (2014). Crystal structures of the phosphorylated BRI1 kinase domain and implications for brassinosteroid signal initiation. *The Plant Journal: For Cell and Molecular Biology*, (1), 2–5. http://dx.doi.org/10.1111/tpj.12445.

Boudsocq, M., & Sheen, J. (2013). CDPKs in immune and stress signaling. *Trends in Plant Science, 18*(1), 30–40. http://dx.doi.org/10.1016/j.tplants.2012.08.008.

Boudsocq, M., Willmann, M. R., McCormack, M., Lee, H., Shan, L., He, P., et al. (2010). Differential innate immune signalling via Ca(2+) sensor protein kinases. *Nature, 464*(7287), 418–422. http://dx.doi.org/10.1038/nature08794.

Brandt, B., Munemasa, S., Wang, C., Nguyen, D., Yong, T., Yang, P. G., et al. (2015). Calcium specificity signaling mechanisms in abscisic acid signal transduction in Arabidopsis guard cells. *eLife, 4*, 1–25. http://dx.doi.org/10.7554/eLife.03599.

Chen, J., Vandelle, E., Bellin, D., & Delledonne, M. (2014). Detection and function of nitric oxide during the hypersensitive response in *Arabidopsis thaliana*: where there's a will there's a way. *Nitric Oxide: Biology and Chemistry/Official Journal of the Nitric Oxide Society, 43*, 81–88. http://dx.doi.org/10.1016/j.niox.2014.06.008.

Chen, Z.-H., Hills, A., Lim, C. K., & Blatt, M. R. (2010). Dynamic regulation of guard cell anion channels by cytosolic free Ca^{2+} concentration and protein phosphorylation. *The Plant Journal: For Cell and Molecular Biology, 61*(5), 816–825. http://dx.doi.org/10.1111/j.1365-313X.2009.04108.x.

Chin, K., Moeder, W., Abdel-Hamid, H., Shahinas, D., Gupta, D., & Yoshioka, K. (2010). Importance of the alphaC-helix in the cyclic nucleotide binding domain for the stable channel regulation and function of cyclic nucleotide gated ion channels in Arabidopsis. *Journal of Experimental Botany, 61*(9), 2383–2393. http://dx.doi.org/10.1093/jxb/erq072.

Chinchilla, D., Zipfel, C., Robatzek, S., Kemmerling, B., Nürnberger, T., Jones, J. D. G., et al. (2007). A flagellin-induced complex of the receptor FLS2 and BAK1 initiates plant defence. *Nature, 448*(7152), 497–500. http://dx.doi.org/10.1038/nature05999.

Corpas, F. J., Begara-Morales, J. C., Sánchez-Calvo, B., Chaki, M., & Barroso, J. B. (2015). Nitration and S-nitrosylation: two post-translational modifications (PTMs) mediated by reactive nitrogen species (RNS) and their role in signalling processes of plant cells. In K. J. Gupta, & A. U. Igamberdiev (Eds.), *Reactive oxygen and nitrogen species signaling and communication in plants* (Vol. 23). Cham: Springer International Publishing. http://dx.doi.org/10.1007/978-3-319-10079-1.

Craven, K. B., & Zagotta, W. N. (2006). CNG and HCN channels: two peas, one pod. *Annual Review of Physiology, 68*, 375–401. http://dx.doi.org/10.1146/annurev.physiol.68.040104.134728.

Crawford, N., Galli, M., Tischner, R., Heimer, Y., Okamoto, M., & Mack, A. (2006). Response to Zemojtel et al.: plant nitric oxide synthase: back to square one. *Trends in Plant Science, 11*(11), 526–527. http://dx.doi.org/10.1016/j.tplants.2006.09.007.

Dangl, J. (August 1998). Plants just say NO to pathogens. *Nature, 394*, 525–527.

Davies, J. (2014). Annexin-mediated calcium signalling in plants. *Plants, 3*(1), 128–140. http://dx.doi.org/10.3390/plants3010128.

Delledonne, M., Xia, Y., Dixon, R. A., & Lamb, C. (August 1998). Nitric oxide functions as a signal in plant disease resistance. *Nature, 394*, 585–588. http://dx.doi.org/10.1038/29087.

Dubovskaya, L. V., Bakakina, Y. S., Kolesneva, E. V., Sodel, D. L., McAinsh, M. R., Hetherington, A. M., et al. (2011). cGMP-dependent ABA-induced stomatal closure in the ABA-insensitive Arabidopsis mutant abi1-1. *The New Phytologist, 191*(1), 57–69. http://dx.doi.org/10.1111/j.1469-8137.2011.03661.x.

Durner, J., Wendehenne, D., & Klessig, D. F. (1998). Defense gene induction in tobacco by nitric oxide, cyclic GMP, and cyclic ADP-ribose. *Proceedings of the National Academy of Sciences of the United States of America, 95*(17), 10328–10333. Retrieved from http://www.pubmedcentral.nih.gov/articlerender.fcgi?artid=21508&tool=pmcentrez&rendertype=abstract.

Feechan, A., Kwon, E., Yun, B.-W., Wang, Y., Pallas, J. A., & Loake, G. J. (2005). A central role for S-nitrosothiols in plant disease resistance. *Proceedings of the National Academy of Sciences of the United States of America, 102*(22), 8054–8059. http://dx.doi.org/10.1073/pnas.0501456102.

Frederickson Matika, D. E., & Loake, G. J. (2014). Redox regulation in plant immune function. *Antioxidants & Redox Signaling, 21*(9), 1373–1388. http://dx.doi.org/10.1089/ars.2013.5679.

Gao, Q.-F., Fei, C.-F., Dong, J.-Y., Gu, L.-L., & Wang, Y.-F. (2014). Arabidopsis CNGC18 is a Ca^{2+}-permeable channel. *Molecular Plant, 7*(4), 739–743. http://dx.doi.org/10.1093/mp/sst174.

Gao, X., Cox, K., Jr., & He, P. (2014). Functions of calcium-dependent protein kinases in plant innate immunity. *Plants, 3*(1), 160–176. http://dx.doi.org/10.3390/plants3010160.

Garcia-Mata, C., & Lamattina, L. (2007). Abscisic acid (ABA) inhibits light-induced stomatal opening through calcium- and nitric oxide-mediated signaling pathways. *Nitric Oxide: Biology and Chemistry/Official Journal of the Nitric Oxide Society, 17*(3–4), 143–151. http://dx.doi.org/10.1016/j.niox.2007.08.001.

García-Mata, C., & Lamattina, L. (2013). Gasotransmitters are emerging as new guard cell signaling molecules and regulators of leaf gas exchange. *Plant Science, 201-202*, 66–73. http://dx.doi.org/10.1016/j.plantsci.2012.11.007.

Gaupels, F., Kuruthukulangarakoola, G. T., & Durner, J. (2011). Upstream and downstream signals of nitric oxide in pathogen defence. *Current Opinion in Plant Biology, 14*(6), 707–714. http://dx.doi.org/10.1016/j.pbi.2011.07.005.

Gayatri, G., Agurla, S., & Raghavendra, A. S. (2013). Nitric oxide in guard cells as an important secondary messenger during stomatal closure. *Frontiers in Plant Science, 4*, 425. http://dx.doi.org/10.3389/fpls.2013.00425.

Geiger, D., Scherzer, S., Mumm, P., Marten, I., Ache, P., Matschi, S., et al. (2010). Guard cell anion channel SLAC1 is regulated by CDPK protein kinases with distinct Ca^{2+} affinities. *Proceedings of the National Academy of Sciences of the United States of America, 107*(17), 8023–8028. http://dx.doi.org/10.1073/pnas.0912030107.

Geiger, D., Scherzer, S., Mumm, P., Stange, A., Marten, I., Bauer, H., et al. (2009). Activity of guard cell anion channel SLAC1 is controlled by drought-stress signaling kinase-phosphatase pair. *Proceedings of the National Academy of Sciences of the United States of America, 106*(50), 21425–21430. http://dx.doi.org/10.1073/pnas.0912021106.

Gerber, I. B., Zeidler, D., Durner, J., & Dubery, I. A. (2004). Early perception responses of *Nicotiana tabacum* cells in response to lipopolysaccharides from *Burkholderia cepacia. Planta, 218*(4), 647–657. http://dx.doi.org/10.1007/s00425-003-1142-0.

Gilroy, S., Read, N. D., & Trewavas, A. J. (1990). Elevation of cytoplasmic calcium by caged calcium or caged inositol triphosphate initiates stomatal closure. *Nature, 346*(6286), 769–771. http://dx.doi.org/10.1038/346769a0.

Gómez-Gómez, L., & Boller, T. (2000). FLS2. *Molecular Cell, 5*(6), 1003–1011. http://dx.doi.org/10.1016/S1097-2765(00)80265-8.

Gómez-Gómez, L., Felix, G., & Boller, T. (1999). A single locus determines sensitivity to bacterial flagellin in *Arabidopsis thaliana. The Plant Journal: For Cell and Molecular Biology, 18*(3), 277–284. http://dx.doi.org/10.1046/j.1365-313X.1999.00451.x.

Guo, F.-Q., Okamoto, M., & Crawford, N. M. (2003). Identification of a plant nitric oxide synthase gene involved in hormonal signaling. *Science, 302*(5642), 100–103. http://dx.doi.org/10.1126/science.1086770.

Guzel Deger, A., Scherzer, S., Nuhkat, M., Kedzierska, J., Kollist, H., Brosché, M., et al. (2015). Guard cell SLAC1-type anion channels mediate flagellin-induced stomatal closure. *The New Phytologist, 208*(1). http://dx.doi.org/10.1111/nph.13435. n/a–n/a.

Hancock, J. T., Neill, S. J., & Wilson, I. D. (2011). Nitric oxide and ABA in the control of plant function. *Plant Science, 181*(5), 555–559. http://dx.doi.org/10.1016/j.plantsci.2011.03.017.

Hao, F., Zhao, S., Dong, H., Zhang, H., Sun, L., & Miao, C. (2010). Nia1 and Nia2 are involved in exogenous salicylic acid-induced nitric oxide generation and stomatal closure in Arabidopsis. *Journal of Integrative Plant Biology, 52*(3), 298–307. http://dx.doi.org/10.1111/j.1744-7909.2010.00920.x.

Honda, K., Yamada, N., Yoshida, R., Ihara, H., Sawa, T., Akaike, T., et al. (May 2015). 8-Mercapto-cyclic GMP mediates hydrogen sulfide-induced stomatal closure in Arabidopsis. *Plant and Cell Physiology, 56*, 1–9. http://dx.doi.org/10.1093/pcp/pcv069.

Hossain, M. A., Ye, W., Munemasa, S., Nakamura, Y., Mori, I. C., & Murata, Y. (2014). Cyclic adenosine 5′-diphosphoribose (cADPR) cyclic guanosine 3′,5′-monophosphate positively function in Ca^{2+} elevation in methyl jasmonate-induced stomatal closure, cADPR is required for methyl jasmonate-induced ROS accumulation NO production in guard cells. *Plant Biology, 16*(6). http://dx.doi.org/10.1111/plb.12175. n/a–n/a.

Hu, J., Huang, X., Chen, L., Sun, X., Lu, C., Zhang, L., et al. (2015). Site-specific nitrosoproteomic identification of endogenously S-nitrosylated proteins in Arabidopsis. *Plant Physiology, 167*(4), 1731–1746. http://dx.doi.org/10.1104/pp.15.00026.

Huang, J. S., & Knopp, J. A. (1998). Involvement of nitric oxide in Ralstonia solanacearum-induced hypersensitive reaction in tobacco. In P. H. Prior, C. Allen, & J. G. Elphinstone (Eds.), *Bacterial Wilt Disease: Molecular and Ecological Aspects* (pp. 218–224). Berlin: Springer-Verlag.

Hung, C.-Y., Aspesi, P., Hunter, M. R., Lomax, A. W., & Perera, I. Y. (2014). Phosphoinositide-signaling is one component of a robust plant defense response. *Frontiers in Plant Science, 5*, 267. http://dx.doi.org/10.3389/fpls.2014.00267.

Irving, H. R., Kwezi, L., Wheeler, J., & Gehring, C. (2012). Moonlighting kinases with guanylate cyclase activity can tune regulatory signal networks. *Plant Signaling & Behavior, 7*(2), 201–204. http://dx.doi.org/10.4161/psb.18891.

Jeandroz, S., Lamotte, O., Astier, J., Rasul, S., Trapet, P., Besson-Bard, A., et al. (2013). There's more to the picture than meets the eye: nitric oxide cross talk with Ca^{2+} signaling. *Plant Physiology, 163*(2), 459–470. http://dx.doi.org/10.1104/pp.113.220624.

Jeworutzki, E., Roelfsema, M. R. G., Anschütz, U., Krol, E., Elzenga, J. T. M., Felix, G., et al. (2010). Early signaling through the Arabidopsis pattern recognition receptors FLS2 and EFR involves Ca-associated opening of plasma membrane anion channels. *The Plant Journal: For Cell and Molecular Biology, 62*(3), 367–378. http://dx.doi.org/10.1111/j.1365-313X.2010.04155.x.

Joudoi, T., Shichiri, Y., Kamizono, N., Akaike, T., Sawa, T., Yoshitake, J., et al. (2013). Nitrated cyclic GMP modulates guard cell signaling in Arabidopsis. *The Plant Cell, 25*(2), 558–571. http://dx.doi.org/10.1105/tpc.112.105049.

Kadota, Y., Shirasu, K., & Zipfel, C. (2015). Regulation of the NADPH oxidase RBOHD during plant immunity. *Plant and Cell Physiology, 56*(8). http://dx.doi.org/10.1093/pcp/pcv063.

Khokon, M. A. R., Hossain, M. A., Munemasa, S., Uraji, M., Nakamura, Y., Mori, I. C., et al. (2010). Yeast elicitor-induced stomatal closure and peroxidase-mediated ROS production in Arabidopsis. *Plant and Cell Physiology, 51*(11), 1915–1921. http://dx.doi.org/10.1093/pcp/pcq145.

Khokon, M. A. R., Uraji, M., Munemasa, S., Okuma, E., Nakamura, Y., Mori, I. C., et al. (2010). Chitosan-induced stomatal closure accompanied by peroxidase-mediated reactive oxygen species production in Arabidopsis. *Bioscience, Biotechnology and Biochemistry, 74*(11), 2313–2315. http://dx.doi.org/10.1271/bbb.100340.

Krol, E., Mentzel, T., Chinchilla, D., Boller, T., Felix, G., Kemmerling, B., et al. (2010). Perception of the Arabidopsis danger signal peptide 1 involves the pattern recognition receptor AtPEPR1 and its close homologue AtPEPR2. *The Journal of Biological Chemistry, 285*(18), 13471–13479. http://dx.doi.org/10.1074/jbc.M109.097394.

Kwaaitaal, M., Huisman, R., Maintz, J., Reinstädler, A., & Panstruga, R. (2011). Ionotropic glutamate receptor (iGluR)-like channels mediate MAMP-induced calcium influx in

Arabidopsis thaliana. *The Biochemical Journal, 440*(3), 355–365. http://dx.doi.org/10.1042/BJ20111112.

Kwan, Y.-M., Meon, S., Ho, C.-L., & Wong, M.-Y. (2015). Cloning of nitric oxide associated 1 (NOA1) transcript from oil palm (*Elaeis guineensis*) and its expression during Ganoderma infection. *Journal of Plant Physiology, 174*, 131–136. http://dx.doi.org/10.1016/j.jplph.2014.10.003.

Kwezi, L., Meier, S., Mungur, L., Ruzvidzo, O., Irving, H., & Gehring, C. (2007). The *Arabidopsis thaliana* brassinosteroid receptor (AtBRI1) contains a domain that functions as a guanylyl cyclase in vitro. *PLoS One, 2*(5), e449. http://dx.doi.org/10.1371/journal.pone.0000449.

Laha, D., Johnen, P., Azevedo, C., Dynowski, M., Weiß, M., Capolicchio, S., et al. (2015). VIH2 regulates the synthesis of inositol pyrophosphate InsP 8 and jasmonate-dependent defenses in Arabidopsis. *The Plant Cell, 27*(4), 1082–1097. http://dx.doi.org/10.1105/tpc.114.135160.

Lamotte, O., Gould, K., Lecourieux, D., Sequeira-Legrand, A., Lebrun-Garcia, A., Durner, J., et al. (2004). Analysis of nitric oxide signaling functions in tobacco cells challenged by the elicitor cryptogein. *Plant Physiology, 135*(1), 516–529. http://dx.doi.org/10.1104/pp.104.038968.

Lang, Z., & Zuo, J. (2015). Say "NO" to ABA signaling in guard cells by S-nitrosylation of OST1. *Science China Life Sciences, 58*(3), 313–314. http://dx.doi.org/10.1007/s11427-015-4823-6.

Lecourieux, D., Ranjeva, R., & Pugin, A. (2006). Calcium in plant defence-signalling pathways. *The New Phytologist, 171*(2), 249–269. http://dx.doi.org/10.1111/j.1469-8137.2006.01777.x.

Lee, S., Choi, H., Suh, S., Doo, I. S., Oh, K. Y., Choi, E. J., et al. (1999). Oligogalacturonic acid and chitosan reduce stomatal aperture by inducing the evolution of reactive oxygen species from guard cells of tomato and Commelina communis. *Plant Physiology, 121*(1), 147–152. Retrieved from http://www.pubmedcentral.nih.gov/articlerender.fcgi?artid=59362&tool=pmcentrez&rendertype=abstract.

Lemtiri-Chlieh, F., & Berkowitz, G. A. (2004). Cyclic adenosine monophosphate regulates calcium channels in the plasma membrane of Arabidopsis leaf guard and mesophyll cells. *The Journal of Biological Chemistry, 279*(34), 35306–35312. http://dx.doi.org/10.1074/jbc.M400311200.

Lemtiri-Chlieh, F., MacRobbie, E. A. C., Webb, A. A. R., Manison, N. F., Brownlee, C., Skepper, J. N., et al. (2003). Inositol hexakisphosphate mobilizes an endomembrane store of calcium in guard cells. *Proceedings of the National Academy of Sciences of the United States of America, 100*(17), 10091–10095. http://dx.doi.org/10.1073/pnas.1133289100.

Leng, Q., Mercier, R. W., Yao, W., & Berkowitz, G. A. (1999). Cloning and first functional characterization of a plant cyclic nucleotide-gated cation channel. *Plant Physiology, 121*(3), 753–761. Retrieved from http://www.pubmedcentral.nih.gov/articlerender.fcgi?artid=59437&tool=pmcentrez&rendertype=abstract.

Leon, J., Castillo, M. C., Coego, A., Lozano-Juste, J., & Mir, R. (2014). Diverse functional interactions between nitric oxide and abscisic acid in plant development and responses to stress. *Journal of Experimental Botany, 65*(4), 907–921. http://dx.doi.org/10.1093/jxb/ert454.

Li, F., Wang, J., Ma, C., Zhao, Y., Wang, Y., Hasi, A., et al. (2013). Glutamate receptor-like channel 3.3 is involved in mediating glutathione-triggered cytosolic calcium transients, transcriptional changes, and innate immunity responses in Arabidopsis. *Plant Physiology, 162*(3), 1497–1509. http://dx.doi.org/10.1104/pp.113.217208.

Li, L., Li, M., Yu, L., Zhou, Z., Liang, X., Liu, Z., et al. (2014). The FLS2-associated kinase BIK1 directly phosphorylates the NADPH oxidase RbohD to control plant immunity. *Cell Host & Microbe, 15*(3), 329–338. http://dx.doi.org/10.1016/j.chom.2014.02.009.

Lim, C. W., Baek, W., Jung, J., Kim, J.-H., & Lee, S. C. (2015). Function of ABA in stomatal defense against biotic and drought stresses. *International Journal of Molecular Sciences, 16*(7), 15251–15270. http://dx.doi.org/10.3390/ijms160715251.

Lim, C. W., Luan, S., & Lee, S. C. (2014). A prominent role for RCAR3-mediated ABA signaling in response to *Pseudomonas syringae* pv. tomato DC3000 infection in Arabidopsis. *Plant and Cell Physiology, 55*(10), 1691–1703. http://dx.doi.org/10.1093/pcp/pcu100.

Lisjak, M., Srivastava, N., Teklic, T., Civale, L., Lewandowski, K., Wilson, I., et al. (2010). A novel hydrogen sulfide donor causes stomatal opening and reduces nitric oxide accumulation. *Plant Physiology and Biochemistry, 48*(12), 931–935. http://dx.doi.org/10.1016/j.plaphy.2010.09.016.

Liu, J., Elmore, J. M., Fuglsang, A. T., Palmgren, M. G., Staskawicz, B. J., & Coaker, G. (2009). RIN4 functions with plasma membrane H+-ATPases to regulate stomatal apertures during pathogen attack. *PLoS Biology, 7*(6), e1000139. http://dx.doi.org/10.1371/journal.pbio.1000139.

Liu, Y.-K., Liu, Y.-B., Zhang, M.-Y., & Li, D.-Q. (2010). Stomatal development and movement: the roles of MAPK signaling. *Plant Signaling & Behavior, 5*(10), 1176–1180. Retrieved from http://www.pubmedcentral.nih.gov/articlerender.fcgi?artid=3115344&tool=pmcentrez&rendertype=abstract.

Liu, Z., Wu, Y., Yang, F., Zhang, Y., Chen, S., Xie, Q., et al. (2013). BIK1 interacts with PEPRs to mediate ethylene-induced immunity. *Proceedings of the National Academy of Sciences of the United States of America, 110*(15), 6205–6210. http://dx.doi.org/10.1073/pnas.1215543110.

Lozano-Durán, R., & Zipfel, C. (2014). Trade-off between growth and immunity: role of brassinosteroids. *Trends in Plant Science*, 1–8. http://dx.doi.org/10.1016/j.tplants.2014.09.003 (Figure 1).

Lozano-juste, J., Gonzalez-guzman, M., Rodriguez, P. L., Castillo, M., Lozano-juste, J., González-guzmán, M., et al. (September 2015). Inactivation of PYR/PYL/RCAR ABA receptors by tyrosine nitration may enable rapid inhibition of ABA signaling by nitric oxide in plants. *Science Signaling, 8*, 1–9. http://dx.doi.org/10.13140/RG.2.1.1632.6883.

Lozano-Juste, J., & León, J. (2010). Enhanced abscisic acid-mediated responses in nia1nia2noa1-2 triple mutant impaired in NIA/NR- and AtNOA1-dependent nitric oxide biosynthesis in Arabidopsis. *Plant Physiology, 152*(2), 891–903. http://dx.doi.org/10.1104/pp.109.148023.

Lu, D., Wu, S., Gao, X., Zhang, Y., Shan, L., & He, P. (2010). A receptor-like cytoplasmic kinase, BIK1, associates with a flagellin receptor complex to initiate plant innate immunity. *Proceedings of the National Academy of Sciences of the United States of America, 107*(1), 496–501. http://dx.doi.org/10.1073/pnas.0909705107.

Ludidi, N., & Gehring, C. (2003). Identification of a novel protein with guanylyl cyclase activity in *Arabidopsis thaliana. The Journal of Biological Chemistry, 278*(8), 6490–6494. http://dx.doi.org/10.1074/jbc.M210983200.

Ma, W. (2011). Roles of Ca^{2+} and cyclic nucleotide gated channel in plant innate immunity. *Plant Science*. Retrieved from http://www.sciencedirect.com/science/article/pii/S0168945211001658.

Ma, W., Qi, Z., Smigel, A., Walker, R. K., Verma, R., & Berkowitz, G. A. (2009). Ca^{2+}, cAMP, and transduction of non-self perception during plant immune responses. *Proceedings of the National Academy of Sciences of the United States of America, 106*(49), 20995–21000. Retrieved from http://www.ncbi.nlm.nih.gov/pubmed/19933332.

Ma, W., Smigel, A., Tsai, Y.-C., Braam, J., & Berkowitz, G. A. (2008). Innate immunity signaling: cytosolic Ca^{2+} elevation is linked to downstream nitric oxide generation through the action of calmodulin or a calmodulin-like protein. *Plant Physiology, 148*(2), 818–828. http://dx.doi.org/10.1104/pp.108.125104.

Ma, Y., Walker, R. K., Zhao, Y., & Berkowitz, G. A. (2012). Linking ligand perception by PEPR pattern recognition receptors to cytosolic Ca^{2+} elevation and downstream immune signaling in plants. *Proceedings of the National Academy of Sciences of the United States of America, 109*(48), 19852–19857. http://dx.doi.org/10.1073/pnas.1205448109.

Ma, Y., Zhao, Y., Walker, R. K., & Berkowitz, G. A. (2013). Molecular steps in the immune signaling pathway evoked by plant elicitor peptides: Ca^{2+}-dependent protein kinases, nitric oxide, and reactive oxygen species are downstream from the early Ca^{2+} signal. *Plant Physiology, 163*(3), 1459–1471. http://dx.doi.org/10.1104/pp.113.226068.

Mandal, M. K., Chandra-Shekara, A. C., Jeong, R.-D., Yu, K., Zhu, S., Chanda, B., et al. (2012). Oleic acid-dependent modulation of NITRIC OXIDE ASSOCIATED1 protein levels regulates nitric oxide-mediated defense signaling in Arabidopsis. *The Plant Cell, 24*(4), 1654–1674. http://dx.doi.org/10.1105/tpc.112.096768.

Manzoor, H., Kelloniemi, J., Chiltz, A., Wendehenne, D., Pugin, A., Poinssot, B., et al. (2013). Involvement of the glutamate receptor AtGLR3.3 in plant defense signaling and resistance to Hyaloperonospora Arabidopsidis. *The Plant Journal: For Cell and Molecular Biology, 76*(3), 466–480. http://dx.doi.org/10.1111/tpj.12311.

Meier, S., Ruzvidzo, O., Morse, M., Donaldson, L., Kwezi, L., & Gehring, C. (2010). The Arabidopsis wall associated kinase-like 10 gene encodes a functional guanylyl cyclase and is co-expressed with pathogen defense related genes. *PLoS One, 5*(1), e8904. Retrieved from http://www.pubmedcentral.nih.gov/articlerender.fcgi?artid=2811198&tool=pmcentrez&rendertype=abstract.

Melotto, M., Underwood, W., Koczan, J., Nomura, K., & He, S. Y. (2006). Plant stomata function in innate immunity against bacterial invasion. *Cell, 126*(5), 969–980. http://dx.doi.org/10.1016/j.cell.2006.06.054.

Meng, X., & Zhang, S. (2013). MAPK cascades in plant disease resistance signaling. *Annual Review of Phytopathology, 51*, 245–266. http://dx.doi.org/10.1146/annurev-phyto-082712-102314.

Mersmann, S., Bourdais, G., Rietz, S., & Robatzek, S. (2010). Ethylene signaling regulates accumulation of the FLS2 receptor and is required for the oxidative burst contributing to plant immunity. *Plant Physiology, 154*(1), 391–400. http://dx.doi.org/10.1104/pp.110.154567.

Misra, B. B., Acharya, B. R., Granot, D., Assmann, S. M., & Chen, S. (2015). The guard cell metabolome: functions in stomatal movement and global food security. *Frontiers in Plant Science, 6*, 334. http://dx.doi.org/10.3389/fpls.2015.00334.

Miya, A., Albert, P., Shinya, T., Desaki, Y., Ichimura, K., Shirasu, K., et al. (2007). CERK1, a LysM receptor kinase, is essential for chitin elicitor signaling in Arabidopsis. *Proceedings of the National Academy of Sciences of the United States of America, 104*(49), 19613–19618. http://dx.doi.org/10.1073/pnas.0705147104.

Moreau, M., Lee, G. I., Wang, Y., Crane, B. R., & Klessig, D. F. (2008). AtNOS/AtNOA1 is a functional *Arabidopsis thaliana* cGTPase and not a nitric-oxide synthase. *The Journal of Biological Chemistry, 283*(47), 32957–32967. http://dx.doi.org/10.1074/jbc.M804838200.

Moreau, M., Lindermayr, C., Durner, J., & Klessig, D. F. (2010). NO synthesis and signaling in plants—where do we stand? *Physiologia Plantarum, 138*(4), 372–383. http://dx.doi.org/10.1111/j.1399-3054.2009.01308.x.

Mur, L. A. J. (2005). Laser photoacoustic detection allows in planta detection of nitric oxide in tobacco following challenge with avirulent and virulent *Pseudomonas syringae* Pathovars. *Plant Physiology, 138*(3), 1247–1258. http://dx.doi.org/10.1104/pp.104.055772.

Mur, L. A. J., Kenton, P., Lloyd, A. J., Ougham, H., & Prats, E. (2008). The hypersensitive response; the centenary is upon us but how much do we know? *Journal of Experimental Botany, 59*(3), 501–520. http://dx.doi.org/10.1093/jxb/erm239.

Mur, L. A. J., Sivakumaran, A., Mandon, J., Cristescu, S. M., Harren, F. J. M., & Hebelstrup, K. H. (2012). Haemoglobin modulates salicylate and jasmonate/ethylene-mediated resistance mechanisms against pathogens. *Journal of Experimental Botany, 63*(12), 4375–4387. http://dx.doi.org/10.1093/jxb/ers116.

Murata, Y., Pei, Z.-M., Mori, I. C., & Schroeder, J. (2001). Abscisic acid activation of plasma membrane Ca^{2+} channels in guard cells requires cytosolic NAD(P)H and is differentially disrupted upstream and downstream of reactive oxygen species production in abi1-1 and abi2-1 protein phosphatase 2C mutants. *The Plant Cell, 13*(11), 2513–2523. http://dx.doi.org/10.1105/tpc.010210.

Neill, S. J., Desikan, R., Clarke, A., & Hancock, J. T. (2002). Nitric oxide is a novel component of abscisic acid signaling in stomatal guard cells no synthesis. *Plant Physiology, 128*, 13–16. http://dx.doi.org/10.1104/pp.010707.shown.

Palmieri, M. C., Sell, S., Huang, X., Scherf, M., Werner, T., Durner, J., et al. (2008). Nitric oxide-responsive genes and promoters in *Arabidopsis thaliana*: a bioinformatics approach. *Journal of Experimental Botany, 59*(2), 177–186. http://dx.doi.org/10.1093/jxb/erm345.

Parani, M., Rudrabhatla, S., Myers, R., Weirich, H., Smith, B., Leaman, D. W., et al. (2004). Microarray analysis of nitric oxide responsive transcripts in Arabidopsis. *Plant Biotechnology Journal, 2*(4), 359–366. http://dx.doi.org/10.1111/j.1467-7652.2004.00085.x.

Perera, I. Y., Hung, C., Brady, S., Muday, G. K., & Boss, W. F. (2006). A universal role for inositol 1, 4, 5–trisphosphate-mediated signaling in plant gravitropism. *Plant Physiology, 140*, 746–760. http://dx.doi.org/10.1104/pp.105.075119.gravitropism.

Perera, I. Y., Hung, C.-Y., Moore, C. D., Stevenson-Paulik, J., & Boss, W. F. (2008). Transgenic Arabidopsis plants expressing the type 1 inositol 5-phosphatase exhibit increased drought tolerance and altered abscisic acid signaling. *The Plant Cell, 20*(10), 2876–2893. http://dx.doi.org/10.1105/tpc.108.061374.

Polverari, A., Molesini, B., Pezzotti, M., Buonaurio, R., Marte, M., & Delledonne, M. (2003). Nitric oxide-mediated transcriptional changes in *Arabidopsis thaliana*. *Molecular Plant-Microbe Interactions: MPMI, 16*(12), 1094–1105. http://dx.doi.org/10.1094/MPMI.2003.16.12.1094.

Qi, Z., Stephens, N. R., & Spalding, E. P. (2006). Calcium entry mediated by GLR3.3, an Arabidopsis glutamate receptor with a broad agonist profile. *Plant Physiology, 142*(3), 963–971. http://dx.doi.org/10.1104/pp.106.088989.

Qi, Z., Verma, R., Gehring, C., Yamaguchi, Y., Zhao, Y., Ryan, C. A., et al. (2010). Ca^{2+} signaling by plant *Arabidopsis thaliana* Pep peptides depends on AtPepR1, a receptor with guanylyl cyclase activity, and cGMP-activated Ca^{2+} channels. *Proceedings of the National Academy of Sciences of the United States of America, 107*(49), 21193–21198. http://dx.doi.org/10.1073/pnas.1000191107.

Qiao, M., Sun, J., Liu, N., Sun, T., Liu, G., Han, S., & Wang, D. (2015). Changes of nitric oxide and its relationship with H_2O_2 and Ca^{2+} in defense interactions between wheat and *Puccinia triticina*. *PLoS One, 10*(7), e0132265. http://dx.doi.org/10.1371/journal.pone.0132265.

Radi, R. (2013). Peroxynitrite, a stealthy biological oxidant. *The Journal of Biological Chemistry, 288*(37), 26464–26472. http://dx.doi.org/10.1074/jbc.R113.472936.

Ranf, S., Eschen-Lippold, L., Fröhlich, K., Westphal, L., Scheel, D., & Lee, J. (2014). Microbe-associated molecular pattern-induced calcium signaling requires the receptor-like cytoplasmic kinases, PBL1 and BIK1. *BMC Plant Biology, 14*(1), 374. http://dx.doi.org/10.1186/s12870-014-0374-4.

Ranf, S., Eschen-Lippold, L., Pecher, P., Lee, J., & Scheel, D. (2011). Interplay between calcium signalling and early signalling elements during defence responses to microbe- or damage-associated molecular patterns. *The Plant Journal: For Cell and Molecular Biology, 68*(1), 100–113. Retrieved from http://onlinelibrary.wiley.com/store/10.1111/j.1365-313X.2011.04671.x/asset/supinfo/TPJ_4671_sm_FigS1-S12-TableS1-S2.pdf?v=1&s=ebd7c8c2774efb5c894076665f9b717ce1b36f03.

Ranf, S., Gisch, N., Schäffer, M., Illig, T., Westphal, L., Knirel, Y. A., & Scheel, D. (2015). A lectin S-domain receptor kinase mediates lipopolysaccharide sensing in *Arabidopsis thaliana*. *Nature Immunology, Advance on*. http://dx.doi.org/10.1038/ni.3124.

Rasul, S., Dubreuil-Maurizi, C., Lamotte, O., Koen, E., Poinssot, B., Alcaraz, G., et al. (2012). Nitric oxide production mediates oligogalacturonides-triggered immunity and resistance to *Botrytis cinerea* in *Arabidopsis thaliana*. *Plant, Cell & Environment, 35*(8). http://dx.doi.org/10.1111/j.1365-3040.2012.02505.x. no—no.

Romero-Puertas, M. C., Perazzolli, M., Zago, E. D., & Delledonne, M. (2004). Nitric oxide signalling functions in plant-pathogen interactions. *Cellular Microbiology, 6*(9), 795—803. http://dx.doi.org/10.1111/j.1462-5822.2004.00428.x.

Ronzier, E., Corratgé-Faillie, C., Sanchez, F., Prado, K., Brière, C., Leonhardt, N., et al. (2014). CPK13, a noncanonical Ca^{2+} -dependent protein kinase, specifically inhibits KAT2 and KAT1 Shaker K^+ channels and reduces stomatal opening. *Plant Physiology, 166*(1), 314—326. http://dx.doi.org/10.1104/pp.114.240226.

Saito, N., Nakamura, Y., Mori, I. C., & Murata, Y. (2009). Nitric oxide functions in both methyl jasmonate signaling and abscisic acid signaling in Arabidopsis guard cells. *Plant Signaling & Behavior, 4*(2), 119—120. Retrieved from http://www.pubmedcentral.nih.gov/articlerender.fcgi?artid=2637495&tool=pmcentrez&rendertype=abstract.

Sawinski, K., Mersmann, S., Robatzek, S., & Böhmer, M. (2013). Guarding the green: pathways to stomatal immunity. *Molecular Plant-Microbe Interactions, 26*(6), 626—632. http://dx.doi.org/10.1094/MPMI-12-12-0288-CR.

Scherzer, S., Maierhofer, T., Al-Rasheid, K. A. S., Geiger, D., & Hedrich, R. (2012). Multiple calcium-dependent kinases modulate ABA-activated guard cell anion channels. *Molecular Plant, 5*(6), 1409—1412. http://dx.doi.org/10.1093/mp/sss084.

Scuffi, D., Núñez, A., Laspina, N., Gotor, C., Lamattina, L., Garcia-Mata, C., et al. (2014). Hydrogen sulfide generated by l-cysteine desulfhydrase acts upstream of nitric oxide to modulate abscisic acid-dependent stomatal closure. *Plant Physiology, 166*(4), 2065—2076. http://dx.doi.org/10.1104/pp.114.245373.

Shi, C., Qi, C., Ren, H., Huang, A., Hei, S., & She, X. (2015). Ethylene mediates brassinosteroid-induced stomatal closure via Gα protein-activated hydrogen peroxide and nitric oxide production in Arabidopsis. *The Plant Journal: For Cell and Molecular Biology, 82*(2), 280—301. http://dx.doi.org/10.1111/tpj.12815.

Simontacchi, M., García-Mata, C., Bartoli, C. G., Santa-María, G. E., & Lamattina, L. (2013). Nitric oxide as a key component in hormone-regulated processes. *Plant Cell Reports, 32*(6), 853—866. http://dx.doi.org/10.1007/s00299-013-1434-1.

Song, Y., Miao, Y., & Song, C. P. (2014). Behind the scenes: the roles of reactive oxygen species in guard cells. *The New Phytologist, 201*, 1121—1140. http://dx.doi.org/10.1111/nph.12565.

Srivastava, N., Gonugunta, V. K., Puli, M. R., & Raghavendra, A. S. (2009). Nitric oxide production occurs downstream of reactive oxygen species in guard cells during stomatal closure induced by chitosan in abaxial epidermis of *Pisum sativum*. *Planta, 229*(4), 757—765. http://dx.doi.org/10.1007/s00425-008-0855-5.

Stamler, J. S., Lamas, S., & Fang, F. C. (2001). Nitrosylation: the prototypic redox-based signaling mechanism. *Cell, 106*(6), 675—683. http://dx.doi.org/10.1016/S0092-8674(01)00495-0.

Stange, A., Hedrich, R., & Roelfsema, M. R. G. (2010). Ca^{2+}-dependent activation of guard cell anion channels, triggered by hyperpolarization, is promoted by prolonged depolarization. *The Plant Journal: For Cell and Molecular Biology, 62*(2), 265—276. http://dx.doi.org/10.1111/j.1365-313X.2010.04141.x.

Sun, A., & Li, Z. (2013). Regulatory role of nitric oxide in lipopolysaccharides-triggered plant innate immunity. *Plant Signaling & Behavior, 8*(1), e22554. http://dx.doi.org/10.4161/psb.22554.

Sun, A., Nie, S., & Xing, D. (2012). Nitric oxide-mediated maintenance of redox homeostasis contributes to NPR1-dependent plant innate immunity triggered by lipopolysaccharides. *Plant Physiology, 160*(2), 1081–1096. http://dx.doi.org/10.1104/pp.112.201798.

Thor, K., & Peiter, E. (2014). Cytosolic calcium signals elicited by the pathogen-associated molecular pattern flg22 in stomatal guard cells are of an oscillatory nature. *The New Phytologist, 204*(4), 873–881. http://dx.doi.org/10.1111/nph.13064.

Ton, J., Flors, V., & Mauch-Mani, B. (2009). The multifaceted role of ABA in disease resistance. *Trends in Plant Science, 14*(6), 310–317. http://dx.doi.org/10.1016/j.tplants.2009.03.006.

Torres, M. A. (2010). ROS in biotic interactions. *Physiologia Plantarum, 138*(4), 414–429. http://dx.doi.org/10.1111/j.1399-3054.2009.01326.x.

Torres, M. A., & Dangl, J. L. (2005). Functions of the respiratory burst oxidase in biotic interactions, abiotic stress and development. *Current Opinion in Plant Biology, 8*(4), 397–403. http://dx.doi.org/10.1016/j.pbi.2005.05.014.

Trapet, P., Kulik, A., Lamotte, O., Jeandroz, S., Bourque, S., Nicolas-Francès, V., et al. (2014). NO signaling in plant immunity: a tale of messengers. *Phytochemistry*. http://dx.doi.org/10.1016/j.phytochem.2014.03.015.

Vahisalu, T., Kollist, H., Wang, Y.-F., Nishimura, N., Chan, W.-Y., Valerio, G., et al. (2008). SLAC1 is required for plant guard cell S-type anion channel function in stomatal signalling. *Nature, 452*(7186), 487–491. http://dx.doi.org/10.1038/nature06608.

Vatsa, P., Chiltz, A., Bourque, S., Wendehenne, D., Garcia-Brugger, A., & Pugin, A. (2011). Involvement of putative glutamate receptors in plant defence signaling and NO production. *Biochimie, 93*(12), 2095–2101. http://dx.doi.org/10.1016/j.biochi.2011.04.006.

Wang, P., Du, Y., Hou, Y.-J., Zhao, Y., Hsu, C.-C., Yuan, F., et al. (2015). Nitric oxide negatively regulates abscisic acid signaling in guard cells by S-nitrosylation of OST1. *Proceedings of the National Academy of Sciences of the United States of America, 112*(2), 613–618. http://dx.doi.org/10.1073/pnas.1423481112.

Wang, P., Du, Y., Li, Y., Ren, D., & Song, C.-P. (2010). Hydrogen peroxide-mediated activation of MAP kinase 6 modulates nitric oxide biosynthesis and signal transduction in Arabidopsis. *The Plant Cell, 22*(9), 2981–2998. http://dx.doi.org/10.1105/tpc.109.072959.

Wang, P., Zhu, J.-K., & Lang, Z. (2015). Nitric oxide suppresses the inhibitory effect of abscisic acid on seed germination by S-nitrosylation of SnRK2 proteins. *Plant Signaling & Behavior, 10*(6), e1031939. http://dx.doi.org/10.1080/15592324.2015.1031939.

Wang, Y.-F., Munemasa, S., Nishimura, N., Ren, H.-M., Robert, N., Han, M., & et al. (2013). Identification of cyclic GMP-activated non-selective Ca^{2+}-permeable cation channels and associated CNGC5 & CNGC6 genes in Arabidopsis guard cells. *Plant Physiology, 163*(858), 578–590. http://dx.doi.org/10.1104/pp.113.225045.

Wendehenne, D., Pugin, A., Klessig, D. F., & Durner, J. (2001). Nitric oxide: comparative synthesis and signaling in animal and plant cells. *Trends in Plant Science, 6*(4), 177–183. Retrieved from http://www.ncbi.nlm.nih.gov/pubmed/11286923.

Xie, K., Chen, J., Wang, Q., & Yang, Y. (2014). Direct phosphorylation and activation of a mitogen-activated protein kinase by a calcium-dependent protein kinase in rice. *The Plant Cell, 26*(7), 3077–3089. http://dx.doi.org/10.1105/tpc.114.126441.

Yamaguchi, Y., Huffaker, A., Bryan, A. C., Tax, F. E., & Ryan, C. A. (2010). PEPR2 is a second receptor for the Pep1 and Pep2 peptides and contributes to defense responses in Arabidopsis. *The Plant Cell, 22*(2), 508–522. http://dx.doi.org/10.1105/tpc.109.068874.

Yamaguchi, Y., Pearce, G., & Ryan, C. A. (2006). The cell surface leucine-rich repeat receptor for AtPep1, an endogenous peptide elicitor in Arabidopsis, is functional in transgenic tobacco cells. *Proceedings of the National Academy of Sciences of the United States of America, 103*(26), 10104–10109. http://dx.doi.org/10.1073/pnas.0603729103.

Yang, H., Mu, J., Chen, L., Feng, J., Hu, J., Li, L., & et al. (2015). S-nitrosylation positively regulates ascorbate peroxidase activity during plant stress responses. *Plant Physiology, 167*(4), 1604–1615. http://dx.doi.org/10.1104/pp.114.255216.

Ye, W., Muroyama, D., Munemasa, S., Nakamura, Y., Mori, I. C., & Murata, Y. (2013). Calcium-dependent protein kinase CPK6 positively functions in induction by yeast elicitor of stomatal closure and inhibition by yeast elicitor of light-induced stomatal opening in Arabidopsis. *Plant Physiology, 163*(2), 591–599. http://dx.doi.org/10.1104/pp.113.224055.

Yoshioka, H., Mase, K., Yoshioka, M., Kobayashi, M., & Asai, S. (2011). Regulatory mechanisms of nitric oxide and reactive oxygen species generation and their role in plant immunity. *Nitric Oxide: Biology and Chemistry/Official Journal of the Nitric Oxide Society, 25*(2), 216–221. http://dx.doi.org/10.1016/j.niox.2010.12.008.

Yu, M., Yun, B.-W., Spoel, S. H., & Loake, G. J. (2012). A sleigh ride through the SNO: regulation of plant immune function by protein S-nitrosylation. *Current Opinion in Plant Biology, 15*(4), 424–430. http://dx.doi.org/10.1016/j.pbi.2012.03.005.

Yun, B.-W., Feechan, A., Yin, M., Saidi, N. B. B., Le Bihan, T., Yu, M., et al. (2011). S-nitrosylation of NADPH oxidase regulates cell death in plant immunity. *Nature, 3000*(7368), 264–268. http://dx.doi.org/10.1038/nature10427.

Zeier, J., Delledonne, M., Mishina, T., Severi, E., Sonoda, M., & Lamb, C. (2004). Genetic elucidation of nitric oxide signaling in incompatible plant-pathogen interactions. *Plant Physiology, 136*(1), 2875–2886. http://dx.doi.org/10.1104/pp.104.042499.

Zemojtel, T., Fröhlich, A., Palmieri, M. C., Kolanczyk, M., Mikula, I., Wyrwicz, L. S., et al. (2006). Plant nitric oxide synthase: a never-ending story? *Trends in Plant Science, 11*(11), 524–525. http://dx.doi.org/10.1016/j.tplants.2006.09.008. author reply 526–528.

Zeng, W., & He, S. Y. (2010). A Prominent role of the flagellin receptor FLAGELLIN-SENSING2 in mediating stomatal response to *Pseudomonas syringae* pv tomato DC3000 in Arabidopsis. *Plant Physiology, 153*(3), 1188–1198. http://dx.doi.org/10.1104/pp.110.157016.

Zhang, C., Czymmek, K. J., & Shapiro, A. D. (2003). Nitric oxide does not trigger early programmed cell death events but may contribute to cell-to-cell signaling governing progression of the Arabidopsis hypersensitive response. *Molecular Plant-Microbe Interactions, 16*(11), 962–972. http://dx.doi.org/10.1094/MPMI.2003.16.11.962.

Zhang, H., Fang, Q., Zhang, Z., Wang, Y., & Zheng, X. (2009). The role of respiratory burst oxidase homologues in elicitor-induced stomatal closure and hypersensitive response in *Nicotiana benthamiana*. *Journal of Experimental Botany, 60*(11), 3109–3122. http://dx.doi.org/10.1093/jxb/erp146.

Zhang, H., Wang, M., Wang, W., Li, D., Huang, Q., Wang, Y., et al. (2012). Silencing of G proteins uncovers diversified plant responses when challenged by three elicitors in *Nicotiana benthamiana*. *Plant, Cell & Environment, 35*(1), 72–85. http://dx.doi.org/10.1111/j.1365-3040.2011.02417.x.

Zhang, J., Li, W., Xiang, T., Liu, Z., Laluk, K., Ding, X., et al. (2010). Receptor-like cytoplasmic kinases integrate signaling from multiple plant immune receptors and are targeted by a *Pseudomonas syringae* effector. *Cell Host & Microbe*, 7(4), 290–301. http://dx.doi.org/10.1016/j.chom.2010.03.007.

Zhang, W., He, S. Y., & Assmann, S. M. (2008). The plant innate immunity response in stomatal guard cells invokes G-protein-dependent ion channel regulation. *The Plant Journal: For Cell and Molecular Biology, 56*(6), 984–996. http://dx.doi.org/10.1111/j.1365-313X.2008.03657.x.

Zhao, Y., Qi, Z., & Berkowitz, G. A. (2013). Teaching an old hormone new tricks: cytosolic Ca^{2+} elevation involvement in plant brassinosteroid signal transduction cascades. *Plant Physiology, 163*(2), 555–565. http://dx.doi.org/10.1104/pp.112.213371.

Zhou, L., Lan, W., Jiang, Y., Fang, W., & Luan, S. (2014). A calcium-dependent protein kinase interacts with and activates a calcium channel to regulate pollen tube growth. *Molecular Plant,* 7(2), 369–376. http://dx.doi.org/10.1093/mp/sst125.

Zipfel, C., Kunze, G., Chinchilla, D., Caniard, A., Jones, J. D. G., Boller, T., et al. (2006). Perception of the bacterial PAMP EF-Tu by the receptor EFR restricts Agrobacterium-mediated transformation. *Cell, 125*(4), 749–760. http://dx.doi.org/10.1016/j.cell.2006.03.037.

Zottini, M., Costa, A., De Michele, R., & Lo Schiavo, F. (2009). Role of Nitric Oxide in Programmed Cell Death, in Nitric Oxide in Plant Physiology. In S. Hayat, M. Mori, J. Pichtel, & A. Ahmad (Eds.), *Wiley-VCH Verlag GmbH & Co. KGaA*. Weinheim: Germany. http://dx.doi.org/10.1002/9783527629138.ch6.

Zou, J.-J., Li, X.-D., Ratnasekera, D., Wang, C., Liu, W.-X., Song, L.-F., et al. (2015). Arabidopsis CALCIUM-DEPENDENT PROTEIN KINASE8 and CATALASE3 function in abscisic acid-mediated signaling and H_2O_2 homeostasis in stomatal guard cells under drought stress. *The Plant Cell, 27*(5), 1445–1460. http://dx.doi.org/10.1105/tpc.15.00144.

Zuppini, A., Baldan, B., Millioni, R., Favaron, F., Navazio, L., & Mariani, P. (2004). Chitosan induces Ca^{2+} -mediated programmed cell death in soybean cells. *The New Phytologist, 161*(2), 557–568. http://dx.doi.org/10.1046/j.1469-8137.2003.00969.x.

SUBJECT INDEX

'*Note:* Page numbers followed by "f" indicate figures and "t" indicate tables.'

C

D

H

R

U

V

W

X

Y

AUTHOR INDEX

A

C

D

E

F

G

H

I

L

M

N

Q

R

S

U

V

X

Y

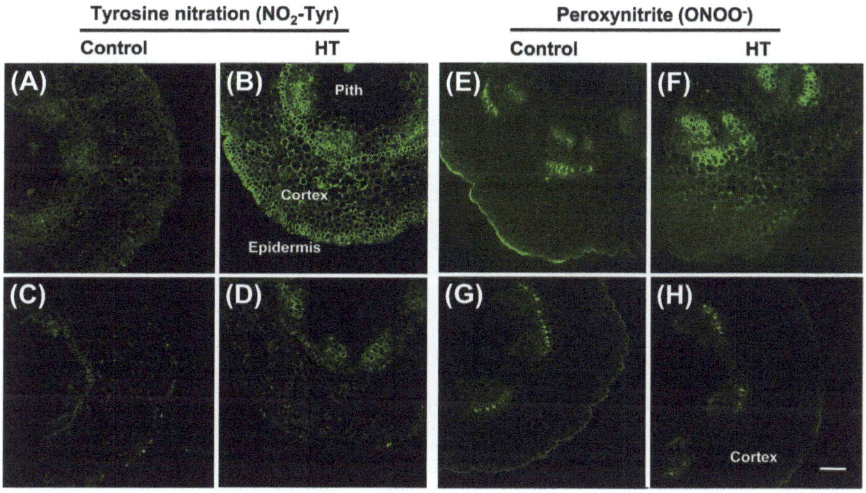

Francisco J. Corpas *et al.*, Figure 3 Confocal laser scanning microscope pictures showing the detection of protein 3-nitrotyrosine (A–D) and peroxynitrite (E–H) in cross sections of hypocotyls from sunflower seedlings subjected to high temperature and control. NO_2-Tyr was detected using a specific antibody and $ONOO^-$ was carried out using 3′-(*p*-aminophenyl) fluorescein as a fluorescent probe. To demonstrate the involvement of SNOs in the generation of protein nitration and peroxynitrite, the hypocotyl samples from sunflower seedlings exposed to high temperature (D and H) and control (C and G) were preincubated in a solution containing 1 mM ascorbate, 10 μM CuCl (molecules which can decompose SNOs) and 200 μM cPTIO (an NO scavenger). Bar = 200 mm. *Reproduced with permission of* Plant Cell and Environment *(2011), 34, 1803–1818.*

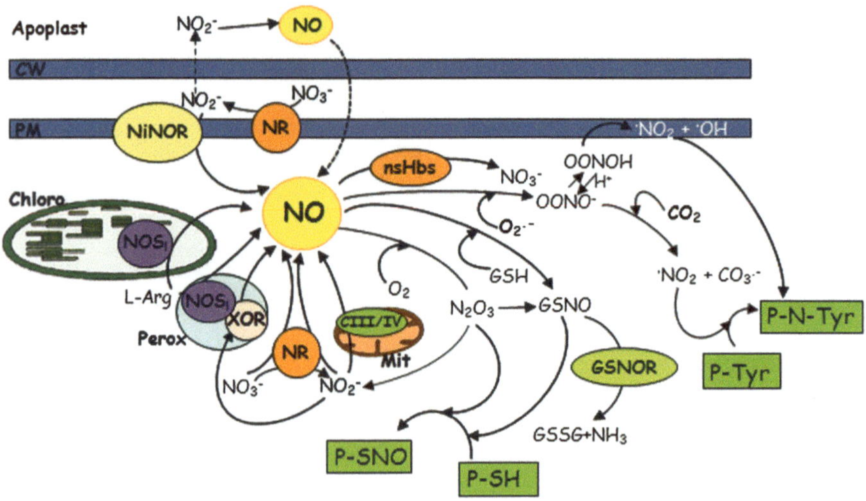

María C. Romero-Puertas and Luisa M. Sandalio, Figure 1 Overview of NO production, metabolism and scavenging in plants. The figure shows a diagram of the main sources described for NO production, including both pathways, oxidative (arginine- or hydroxylamine-dependent) and reductive (nitrate-dependent) and the main scavengers for NO including haemoglobins, oxygen, GSH and superoxide ion. Abbreviations: GSH, glutathione; GSNO, nitrosoglutathione, GSNOR, GSNO reductase; nsHbs, non-symbiotic haemoglobins; NOS_l, activity that resemble NO production as catalyzed by the animal enzyme NOS; NiNOR, plasma membrane-bound NiNOR; NR, nitrate reductase; XOR, xanthine oxidoreductase.

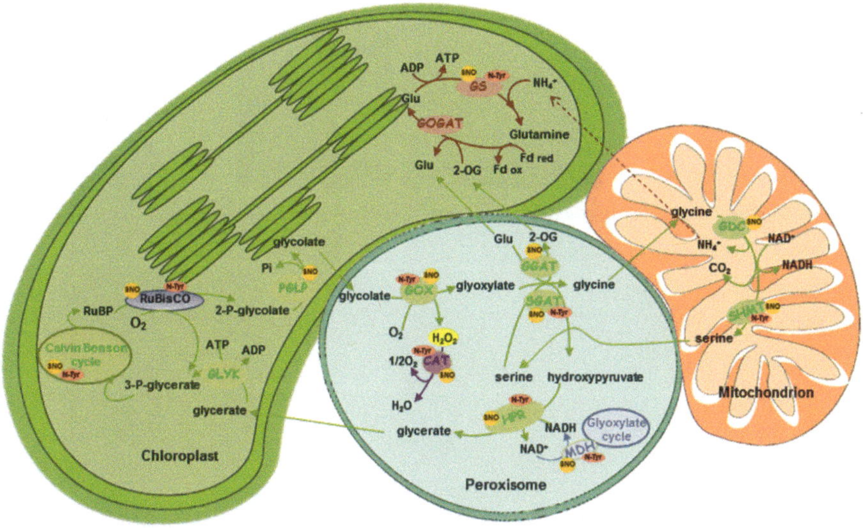

Maria C. Romero-Puertas and Luisa M. Sandalio, Figure 2 Targets of NO-dependent PTMs in the photorespiratory pathway and some related enzymes. Most of the enzymes involved in photorespiration pathway, involving the three organelles, chloroplasts, peroxisomes and mitochondrions, are targets for S-nitrosylation (SNO) and tyrosine nitration (N-Tyr). Abbreviations: CAT, catalase; GDC, glycine decarboxylase complex; GGAT, glutamate glyoxylate aminotransferase; GLYK, glycerate kinase; GOGAT, ferredoxin-dependent glutamate synthase; GOX, glycolate oxidase; GS, glutamine synthetase; HPR, hydroxypyruvate reductase; MDH, malate dehydrogenase; 2OG, 2-oxoglutarate; PGLP, phosphoglycolate phosphatase; RuBP, ribulose-1,5-bisphosphate; RuBisCO, RuBP carboxylase/oxygenase; SGAT, serine glyoxylate aminotransferase; SHMT, serine hydroxymethyltransferase.

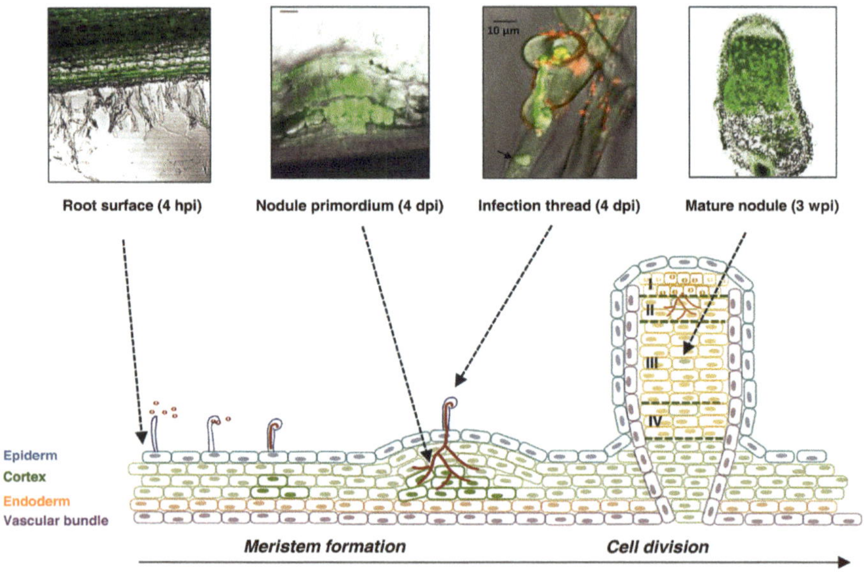

Imène Hichri *et al.*, Figure 1 NO production during the legume–*Rhizobium* symbiotic interaction. NO production sites are detailed for each step of the infection process: on root surface (4 hpi), nodule primordium (4 dpi), infection thread (4 dpi) and mature nodules where NO is present in the infection zone, the N_2-fixing zone, and the interzone III–IV (3 wpi). hpi, hours post-inoculation; dpi, days post-inoculation; wpi, weeks post-inoculation. *The picture of infection thread (4 dpi) is reprinted from Figure 1 in del Giudice et al. (2011) with kind permission from Wiley-Blackwell.*

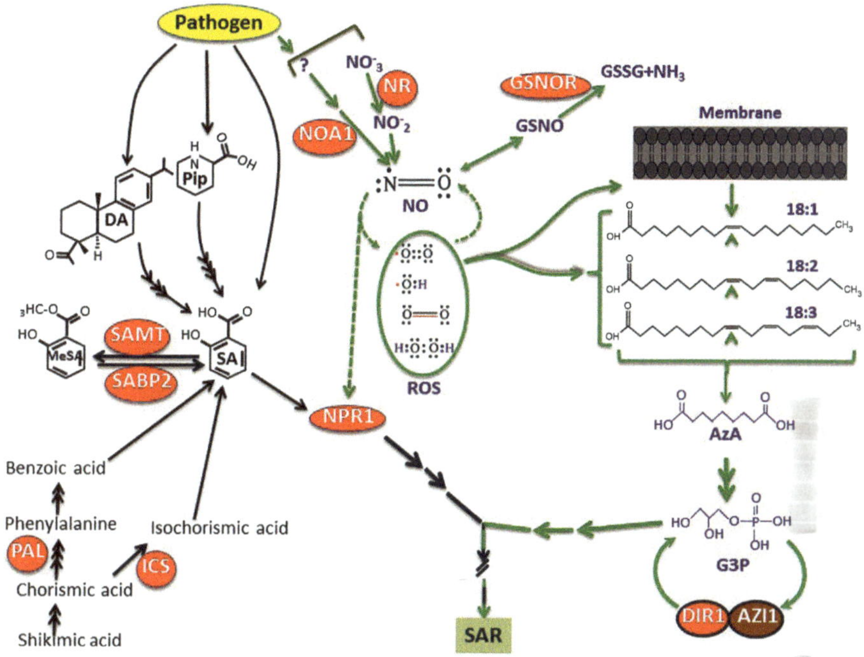

Pradeep Kachroo *et al.*, Figure 1 A simplified scheme showing components of the SAR signaling pathway. Inoculation of avirulent pathogen leads to accumulation of salicylic acid (SA) and nitric oxide (NO). NO acts upstream of reactive oxygen species (ROS), which include the superoxide radical, singlet oxygen, hydroxyl radical and hydrogen peroxide. These acts in an additive manner to catalyse oxidation of C18 unsaturated fatty acids (FA) that contain double bond on carbon 9 (shown by arrowhead). NO and ROS operate in a feedback loop, since mutants defective in ROS biosynthesis do not accumulate NO after pathogen inoculation. Hydrolysis of Δ9 double bond on C18 fatty acids generates AzA, which triggers biosynthesis of G3P via upregulation of genes encoding G3P biosynthetic enzymes, glycerol kinase (GK) and G3P dehydrogenase (G3Pdh). G3P and the lipid transfer proteins DIR1 and AZI1 operate in a feedback loop. DIR1 and AZI1 interact with self and each other. Cellular NO levels are regulated via their storage as GSNO (S-nitrosoglutathione), which can be reduced to glutathione disulfide (GSSG) and NH_3 by GSNOR (S-nitrosoglutathione reductase). The SA and NO/ROS pathways crosstalk at several levels and one of these steps includes nitrosylation of NPR1, a positive regulator of the SA pathway. Critical enzymes and proteins are shown in red. NOA1, nitric oxide associated 1; NR, nitrate reductase; AzA, azelaic acid; G3P, glycerol-3-phosphate; PAL, phenylalanine ammonia lyase; ICS, isochorismate synthase; SABP2, SA binding protein 2; SAMT, SA methyltransferase; DA, dehydroabietinal; Pip, pipecolic acid; NPR1, non-expressor for PR-1.

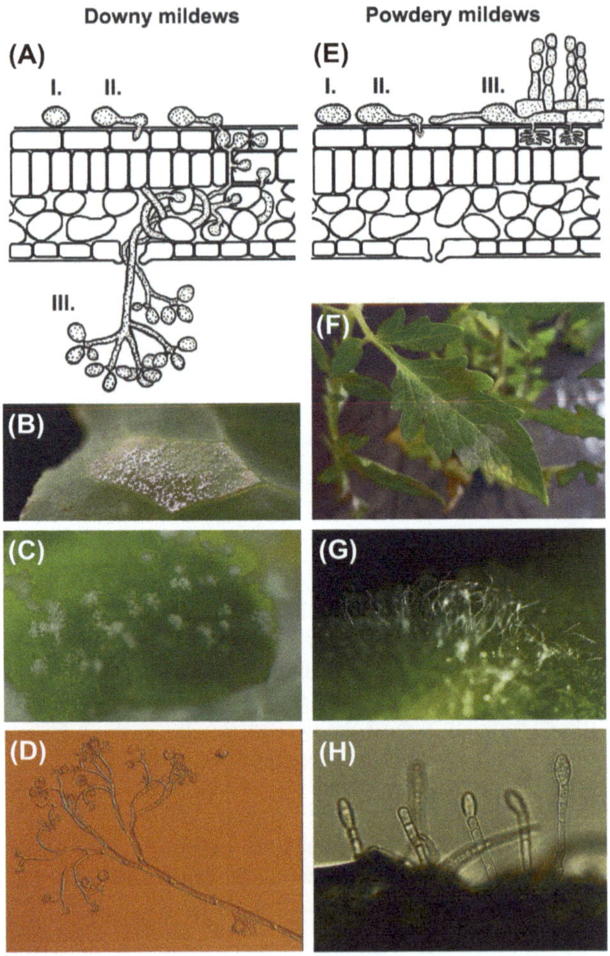

Michaela Sedlářová *et al.*, Figure 1 Comparison of downy (A–D) and powdery mildews (E–H). Developmental phases (A, E) represent: (I) spore deposition, (II) germination followed by penetration, (III) development of mycelia (A – endotrophic, E – ectotrophic) and formation of asexual spores. Typical symptoms on leaves include sporulation on abaxial side, often limited by veins, for downy mildews (B, C) and sporulation on adaxial side by powdery mildews (F, G). Details of conidiophore morphology are demonstrated for *Bremia lactucae* (D) and *Oidium neolycopersici* (H). *Drawing and photos: M. Sedlářová.*

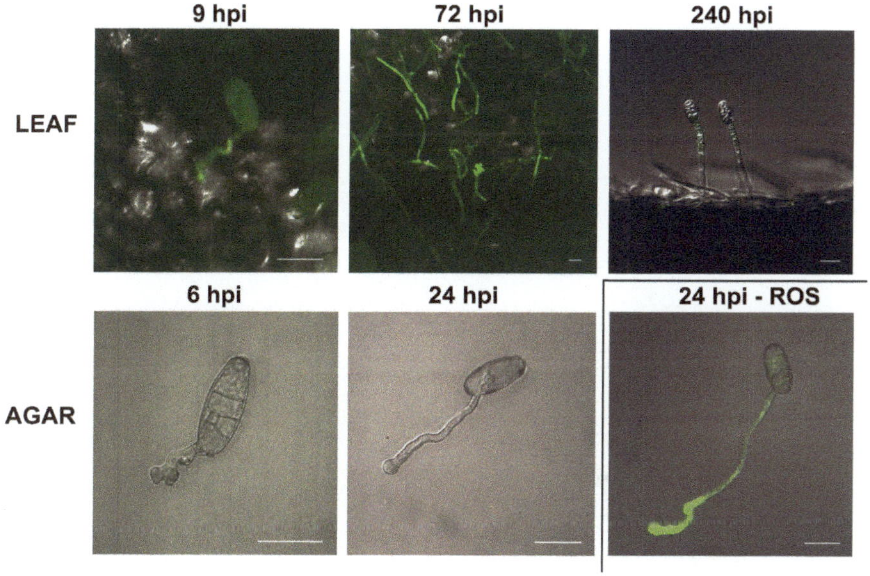

Michaela Sedlářová *et al.*, Figure 2 Nitric oxide (NO) and powdery mildews development. The presence of NO is observed during the germination (9 hpi) and formation of haustoria and mycelium (72 hpi) when grown on host plants or detached leaves. A weak signal for NO was also detected during the conidiogenesis (240 hpi). On the contrary, NO was not observed in *Oidium neolycopersici* conidia germinated on agar medium enriched with tomato leaf extract during 24 hpi, while a high ROS signal was localized in the tips of germ tubes. Bar represents 20 μm. *Photos: M. Sedlářová.*

FSC
www.fsc.org
MIX
Papier | Fördert gute Waldnutzung
FSC® C083411

Kolibri 360 GmbH
Ferdinand-Jühlke-Straße 7
99095 Erfurt, Deutschland
produktsicherheit@kolibri360.de